The Neurocognitive Theory of Dreaming

The Neurocognitive Theory of Dreaming

The Where, How, When, What, and Why of Dreams

G. William Domhoff

The MIT Press
Cambridge, Massachusetts
London, England

The MIT Press would like to thank the anonymous peer reviewers who provided comments on drafts of this book. The generous work of academic experts is essential for establishing the authority and quality of our publications. We acknowledge with gratitude the contributions of these otherwise uncredited readers.

This book was set in Stone by Westchester Publishing Services, Danbury, CT. Printed and bound in the United States of America.

Library of Congress Cataloging-in-Publication Data

Names: Domhoff, G. William, author.
Title: The neurocognitive theory of dreaming : the where, how, when, what, and why of dreams / G. William Domhoff.
Description: Cambridge, Massachusetts : The MIT Press, 2022. |
   Includes bibliographical references and index.
Identifiers: LCCN 2021051383 | ISBN 9780262544214 (paperback)
Subjects: LCSH: Dreams. | Cognitive neuroscience. | Neural networks
   (Neurobiology)
Classification: LCC BF1078 .D583 2022 | DDC 154.6/3—dc23/eng/20211122
LC record available at https://lccn.loc.gov/2021051383

10 9 8 7 6 5 4 3 2 1

To Lizzy,
my loving and caring wife, for so many reasons,
which include the help she gave me in completing this book.

Contents

Introduction: Establishing a Context, Providing an Overview

This book presents a detailed statement of the neurocognitive theory of dreaming, along with the empirical findings that support it. The theory has three levels. It begins with the neural substrate that subserves dreaming. Next, the specific cognitive processes supported by this neural substrate are outlined. Finally, the dream reports that research participants provide, either verbally or in writing, are used to carry out quantitative studies of dream content.

The neurocognitive theory of dreaming is therefore based on "three distinct, but fundamentally interrelated, levels of analysis," which are generally considered to be the hallmarks of much current theorizing in cognitive neuroscience (Ochsner & Kosslyn, 2014b, p. 2). There is an underlying neural substrate, along with the cognitive processes that the underlying substrate supports. Finally, a neurocognitive theory has a behavioral level, which in the case of dreaming can only mean verbal and written dream reports. There are numerous links between levels, and more links are likely to be discovered (Ochsner & Kosslyn, 2014a, pp. 478–479, 481).

The theory also has a developmental dimension. The neural substrate that supports dreaming matures only gradually during childhood and early adolescence. The cognitive skills necessary for dreaming also develop very gradually. As a result, the dreams reported by children and adolescents in controlled laboratory settings change in frequency, complexity, and content until the participants reach middle adolescence.

Different groups of researchers in different disciplines carried out the studies that are the basis for the neurocognitive theory of dreaming. Since the theory draws upon a multiplicity of sources, it is first and foremost a synthesis. It is therefore useful to proceed in a step-by-step fashion.

Although each of these research literatures is considered to be solid and well-grounded by the specialists in them, there may be differences among both cognitive neuroscientists and psychological scientists as to the degree of their relevance and importance in understanding dreaming and dream content. In addition, some of the findings that emerge from these research fields lead to counterintuitive conclusions. For these reasons, the theory may be perceived as controversial to varying degrees on some issues.

Due to the focus on explaining a theory and presenting the evidence for it, the book does not draw upon all aspects of the research literature on dreaming. It therefore does not provide a comprehensive account of everything that has been written about dreams, nor a historical overview of the development of this large literature. The primary focus is on replicated findings that are based on large sample sizes and make use of reliable and validated methodologies. In addition, there is an emphasis on the use of statistical analyses that are appropriate to the level of measurement used in the various studies.

Within the context provided by the previous five paragraphs, the remainder of this brief introduction presents an overview of the where, how, when, what, and why of dreams. The "where" of dreaming is located in a relatively small portion of the human brain. During dreaming, the neural substrates that support waking sensory input, task-oriented thinking, and movement are relatively deactivated. This discovery in the mid-1990s, based on new neuroimaging technologies, was not anticipated on the basis of the many earlier electroencephalogram (EEG) studies of the brain during waking and sleeping. Any hesitancy about accepting the accuracy of these unexpected results with regard to dreaming was soon abandoned on the basis of the findings in systematic studies of the effects of brain lesions on dreaming.

These lesion studies, which were carried out by a neuropsychologist shortly before the neuroimaging studies were undertaken by other researchers but published at the same time as the neuroimaging studies were first being reported, pointed to the same neural substrates as necessary or not necessary for dreaming (Solms, 1997). The convergent findings from neuroimaging and lesion studies took on even greater significance on the basis of a further discovery by still other researchers, which occurred shortly thereafter. The neural network that subserves dreaming is part of a larger neural substrate, called the "default network." During waking, this network enables the cognitive processes that have a large role in supporting memory, imagination, and other forms of internally generated thought, including mind-wandering and daydreaming (Gusnard, Akbudak, Shulman, & Raichle, 2001; Gusnard & Raichle, 2001; Raichle et al., 2001).

As to the "how" of dreaming, the primary cognitive process that produces dreaming is "simulation." Simulation is a type of thinking that places a person in imagined hypothetical situations. These imagined situations are often used to compare various possible alternative outcomes. Still other studies suggest that the simulation process is sometimes enhanced to the point that it can be characterized as "embodied simulation." Embodied simulation not only involves imagination and narrative flow; it also includes a greater activation of secondary sensory and sensorimotor areas, which support both vivid mental imagery, such as visual and auditory imagery, as well as imagined movements. During dreaming, the mental imagery involved in embodied simulation is usually so vivid that dreaming is subjectively experienced as the person in action within a real environment.

Turning to the "when" of dreaming, dreaming occurs spontaneously whenever six specific conditions are met. These six conditions, which are stated very generally here, primarily concern the maturity of the neural substrate that supports dreaming, the absence of external demands on the brain, and an adequate level of cortical activation. As a final condition, there has to be a loss of conscious self-control, which involves the relative deactivation of the neural substrates that are involved in supporting focused attention. The six necessary conditions for dreaming are explained in detail as the analysis unfolds.

The "what" of dreaming is called "dream content." It is analyzed quantitatively by placing all aspects of dream reports into a wide range of carefully defined categories. The ensuing statistical analyses reveal that dreams in large part enact personal concerns. These personal concerns usually relate to important people and avocations in the dreamers' lives. Dreams thereby dramatize the dreamers' conceptions of themselves and their relationships with other people. There are large individual differences in people's conceptions and concerns. For example, an individual dreamer's social interactions with the people she or he dreams about may portray the dreamer as generally friendly with people or as being hostile and aggressive. Then, too, a dreamer may frequently initiate friendly interactions with one dream "character," as the people and animals in dreams are called, but not with another. A person's dreams may show her or him engaging in supportive interactions with one character but ignoring attempts by another character to initiate friendly interactions. When the frequency and nature of a dreamer's interactions with each specific character are precisely quantified and then turned into percentages or ratios to correct for the varying lengths of dream reports, the dream content in 125 or more dream reports from a group or individual can provide a reasonably precise portrait of the dreamers' conceptions and personal concerns.

The "why" of dreaming may be the most counterintuitive outcome of empirical dream research. Numerous systematic studies inadvertently cast doubt on all adaptive theories of dreaming. Nor is there any positive evidence for any adaptive theory of dreaming. These various studies lead to the hypothesis that dreaming may be a by-product of the selection for waking imaginative cognitive capacities. Waking imagination has reproductive and survival value because it allows humans to rethink the past, bring together the past and the future in their present thinking, and to plan and prepare for the future (Spreng, Madore, & Schacter, 2018).

Although dreaming may not have any adaptive functions, research by anthropologists, historians, and comparative religion scholars demonstrates that dreaming has psychological and cultural uses, which were invented in the course of human history. The most important of these uses are found in religious ceremonies and in healing practices. In order to develop a complete theory of dreaming, it is therefore essential to distinguish five separate issues from each other: neural substrates, cognitive processes, the psychological meaning contained in dream content, evolutionarily adaptive functions, and historically invented cultural uses. It is then possible to explore how these different issues are intertwined.

In addition to explaining the where, how, when, what, and why of dreaming, the book includes chapters on the degree to which there is "symbolism" in dreams, the development of dreaming in children, and the relative frequency of emotions in the dreams of both children and adults. Following those chapters, there is a chapter that discusses the two main traditional and comprehensive theories of dreaming, as well as a chapter on the several theories concerning the possible adaptive functions of dreaming. A brief concluding chapter then highlights the differences between the neurocognitive theory of dreaming and other theories of dreaming.

Chapter 1, to which the book now turns, discusses the definitions, distinctions, and limitations that should be kept in mind while reading through the step-by-step theoretical chapters that follow it.

1 Definitions, Distinctions, and Limitations

Introduction

This chapter lays the groundwork for the seven detailed chapters that follow. Those seven chapters focus on the several dimensions of the theory and on the many empirical studies that support each of those dimensions. Then, the final three chapters demonstrate that many of these studies raise doubts about the assumptions underlying alternative theories. This order of presentation makes it possible to state the theory in a more direct and hopefully a more clear fashion, and it postpones any back-and-forth arguments about other theories. This approach also may provide readers new to the topic with a basis for drawing their own conclusions concerning the assumptions and research studies that provide the starting points for the other theories of dreaming.

This chapter begins with an overview of the several different types of thinking that occur during waking. It then discusses the many different types of mental activity that occur throughout the night and how they do or do not relate to dreaming. These definitions and distinctions serve as a starting point for more precise statements in later chapters, which factor in the neural substrates and cognitive processes that support different types of mental activity. The comparisons of the various forms of thought during waking and sleep lead to a definition of dreaming within a neurocognitive theoretical framework. The contrast between dreaming and all forms of waking thought provides a context for concluding it is necessary to guard against a "wake-state bias" before drawing any conclusions about dreaming (Windt, 2015).

These early sections of the chapter set the stage for explaining why many cognitive neuroscientists believe it is necessary to create a "foundational" theory, based on representative samples of nonclinical populations. In the case of dream research, this approach bypasses the use of any psychiatric conditions as a way to understand dreaming. The chapter concludes with a discussion of the unusual combination of limitations researchers face in studying a unique mental state. To date, there is no known way to induce dreaming, as is possible with most waking cognitive processes. Nor can dreaming be observed by researchers, or reported upon by dreamers, while it is happening. As of now, and for the foreseeable future, the fact that dreaming has occurred can only be known to a research scientist with a very high degree of certainty if a research participant recalls a dream and is willing to report it as fully and accurately as possible.

Mental Activity during Waking and during the Night

There are two general types of thinking during waking, both of which have several slightly different subtypes. Directed thinking is task-oriented and goal-oriented. It usually concerns issues presented by the external world. Goal-directed thinking is also hierarchical "in the sense that it aims at the achievement of a number of goals and subgoals" (Christoff, 2014, p. 319). Internally generated thought, on the other hand, is often internally focused and spontaneous. However, it sometimes involves thinking through complex issues related to the waking world as well. Internally generated thought includes mind-wandering, which is characterized by frequent topic changes and underdeveloped thoughts. Daydreaming is a more specific and sustained form of mind-wandering, which is more likely to focus on personal concerns and to simulate those personal concerns in scenarios that have a narrative flow (Klinger, 2009; Singer, 1975). Creativity is a hybrid form of thought, which may begin spontaneously and remain internally focused for varying degrees of time. Eventually, creativity involves aspects of directed thought, which gives this hybrid state a back-and-forth quality in terms of the neural substrates that support it (Abraham, 2018; Beaty, Chen, Qiu, Silvia, & Schacter, 2018; Christoff, 2014; M. L. Meyer, Hershfield, Waytz, Mildner, & Tamir, 2019).

Just as there are several forms of waking thought, there are several different forms of mental activity during sleep. More exactly, there are several forms of mental activity during the night, which begin with the fact that some thinking occurs during one or more of the several brief arousals throughout the night. Brief arousals can last from a few seconds to 30 seconds or more, and they occur from childhood through adulthood (Bli-wise & Scullin, 2017, pp. 26–27; Mathur & Douglas, 1995). The thoughts during brief arousals are often benign and are soon forgotten, but they can include frightening thoughts, expressions of concern over the next day's tasks, or general worries about the future. On rare occasions they include hallucinations, and the person is able to describe the contents (Iranzo, 2017, p. 1011). Brief arousals sometimes include talking that is not part of the dreaming process, as first verified through immediate questioning of the participants in studies in a sleep-dream laboratory (Arkin, 1981).

Nor do "sleep terrors," which are experienced by a significant minority of children in the first two or three years of life, and nearly 1% of adults, occur during sleep. They usually occur early in the night. They are "accompanied by a piercing scream," or crying, along with a "behavioral manifestation of fear." However, those who go through this experience usually have "little or no memory of the event" once they are calmed or when they wake up in the morning (Vaughn & D'Cruz, 2017, p. 583). Once thought to be nightmares, these frightening events are instead "disorders of arousal" (Avidan, 2017; Roger Broughton, 1968; C. Fisher, Kahn, Edwards, & Davis, 1973).

The vivid and often upsetting mental imagery and thoughts that accompany sleep paralysis are not a form of dreaming either. Sleep paralysis occurs when the sleepers have "complete awareness of their surroundings," or while "feeling partially asleep with awareness." And yet, they are "unable to move even the fingers or to vocalize" (Vaughn & D'Cruz, 2017, p. 582). Episodes of sleep paralysis often happen during the morning awakening. Those who experience sleep paralysis cannot move because of an inhibition of muscle movements (atonia). Atonia occurs during one particular stage of sleep, which is called "REM (rapid eye movement) sleep" because of the fast and erratic eye movements that are one aspect of it, as discussed in the next chapter. However, REM-based atonia usually subsides during an awakening. In other words, on some occasions it is possible to be mentally awake and yet retain the atonia that accompanies REM sleep. Although sleep paralysis is often very frightening, it can include feelings of warmth, with a sensual tinge, along with the sense that there is some human, animal, or alien in the room.

In addition to these three types of wake-like mental activity during the course of a night's sleep, there are several different types of mental activity during sleep, as discovered through planned awakenings in sleep-dream laboratories. Sometimes participants may report isolated, static mental imagery, which is usually visual in nature. At other times they may report a single thought or say they were thinking (Kamiya, 1961). There is also sleep talking during sleep. Sometimes it is related to ongoing dream content, as revealed by the dream reports that are collected after awakenings during the sleep-talking episode. At other times, sleep talking is not related to dreaming, as also documented by immediate awakenings in a sleep-dream lab (Arkin, 1981).

Along with the static imagery, brief thoughts, and sleep talking that occur during sleep, participants sometimes report mind-wandering after planned awakenings. They usually do not experience this mind-wandering as dreaming, even though they may report a degree of narrative flow in this form of thinking during sleep (Rechtschaffen, Verdone, & Wheaton, 1963). The fact that there can be mind-wandering during both waking and sleep is a useful discovery with regard to the construction of a neurocognitive theory of dreaming.

What Is Dreaming? A Neurocognitive Definition

There are over a dozen different definitions of dreaming. They range from very general definitions that encompass any form of mental activity that is reported after an awakening to more circumscribed definitions that emphasize a series of sensory mental images that have a narrative-like structure (Pagel et al., 2001). Based on the above discussion of the various kinds of mental activity that occur during both waking and sleep, dreaming is most generally defined within the context of the neurocognitive theory of dreaming as a unique form of spontaneous, internally generated thought. It shares features in common with mind-wandering and even more with daydreaming, although dreaming more frequently involves ongoing personal concerns (Fox, Nijeboer, Solomonova, Domhoff, & Christoff, 2013). More specifically, dreaming is defined in the neurocognitive theory of dreaming as an intensive and enhanced form of mind-wandering and daydreaming, in which dreamers experience themselves as being in hypothetical scenarios that almost always include other human beings and/or animals. In addition, the other human beings and animals are usually interacting with the dreamer in the context of vivid sensory environments.

The evidence for the usefulness and accuracy of this definition of dreaming is presented in chapters 2 and 3. Dreaming can occur during the sleep-onset process, which shares aspects in common with waking, and is not yet sleep. Dreaming also can occur for at least several seconds during periods of drifting waking thought, when the participant is alone in a controlled laboratory setting. These findings suggest that sleeping and dreaming are not inherently connected, although the conditions leading to dreaming occur most frequently during sleep.

Guarding against a Wake-State Bias

Due to the unique nature of dreaming, it is essential to avoid sliding into a "wake-state bias" in theorizing about the content, meaning, and possible adaptive functions of dreaming. The wake-state bias involves "The practice of projecting beliefs about waking experience onto dreaming without critical reflection or solid empirical evidence" (Windt, 2015, p. 200). This bias seems to make sense to most people and is compelling because dreams appear to be based on perceptual information, just as thinking often is during waking. As a result, the experience often feels as if it is "real." In addition, dreams often contain settings, objects, and activities that occur during waking, such as home settings, familiar furniture, and social interactions with family members and friends. Interpersonal interactions give dreaming a sense of familiar narrative content, which also has parallels with waking thought. More generally, dreams create a "subjective impression of verisimilitude" (Windt, 2015, pp. 309–310). Finally, and just as important as any of the other factors, the "linguistic form of dream reports and their formulation in the first-person, past-tense," leads people to "inadvertently import wake-state bias into theoretical accounts of dreaming." People thereby lay "the groundwork for regarding dreams as meaningful and personally significant experiences alongside waking experiences" (Windt, 2015, p. 500).

Based on the sense of verisimilitude within the dreaming experience, the concept of wake-state bias can be extended to include issues such as the appearance of metaphor and emotions in dreams, which are discussed in chapters 5 and 8. It also may be relevant in the case of one or more of the adaptive theories of dreaming discussed in chapter 10.

Foundational Models and Transitional Models

This book makes use of a distinction between "foundational" and "translational" research, which is often employed in cognitive neuroscience. Foundational research is focused on building "a model of normal behaviors, typically in normal adults"; translational research concerns "translating that model to a population of interest" (Ochsner & Kosslyn, 2014a, p. 481). This two-step strategy "allows initial research to focus on understanding core processes—considered in the context of different levels of analysis" (Ochsner & Kosslyn, 2014a, p. 482). Within this framework, the neuroimaging studies used in building the neurocognitive theory of dreaming are based on large samples from normally functioning adults, adolescents, and children. The studies of dream content are based on large samples obtained from similar types of participants. Virtually all of the studies have been replicated at least once. The few exceptions are clearly indicated when they are discussed.

Based on this distinction, psychiatric patients and their various symptoms, such as hallucinations, delusions, and confabulations, are not useful for comparisons with dreams. These abnormalities may be due to many different factors, not all of which are likely known as yet. Psychiatric conditions are also problematic because they involve a comparison of an unusual waking state with a normal cognitive process, which occurs spontaneously when several conditions are met. Although such comparisons sound plausible at first glance and often have been made in Western philosophy and medicine for at least the past several centuries, they have turned out to be superficial and wrong, as shown throughout this book.

The two exceptions to the exclusion of those diagnosed with one or another psychiatric illness are noted at the places where the studies of them are cited. The first and most important exception occurs in a lengthy discussion in chapter 2, which focuses on the impact of brain lesions on dreaming. Most of the patients discussed are people who experienced lesions due to accidents or strictly neurological pathologies. However, one subset of these patients was subjected to a form of neurosurgery. This surgical procedure was used with psychiatric patients from the late 1930s to the early 1960s in the United States, Canada, and parts of Western Europe. Secondly, there are also references to neuroimaging and electroencephalogram (EEG) studies of patients suffering from posttraumatic stress disorder (PTSD).

The Limitations on Scientific Studies of Dreaming and Dream Content

Every scientific field faces one or more unique barriers, which often can be dealt with through new and constantly improving instrumentation. However, there is a unique combination of obstacles inherent in dream research. This is the case even though new technologies first made the scientific study of dreaming and dream content possible in a systematic way. Even with a focus on foundational studies, the confluence of issues facing dream researchers place limits on how much technological advances can do to overcome the difficult challenges dream research faces. These obstacles keep dream research on the margins in the neurocognitive and psychological sciences.

Foremost among the limits on dream research, there are very few instances in which dream researchers have been able to do experiments. They cannot manipulate independent variables because it is not possible to make dreams happen. Nor are there any physiological or behavioral measures that can serve as reliable and valid dependent variables. In the face of these problems, experimental studies are sometimes attempted but they often cannot provide adequate controls for potential confounding factors. (The occasional claim that topics or images suggested to participants can be detected in brain-wave patterns are based on small pilot studies (e.g., Horikawa & Kamitani, 2017; Horikawa, Tamaki, Miyawaki, & Kamitani, 2013). Any possibility that dream content might be traced through brain patterns in any useful detail remains a distant future possibility.)

Therefore, the only way to know the effects of any manipulations of variables is through verbal reports by participants. However, participants do not invariably recall a dream or may in some instances choose to alter or omit parts of their dream report. There are also problems in collecting large numbers of dream reports using either questionnaires or two-week dream diaries. These problems are discussed in the relevant contexts in later chapters. Moreover, there may be limits on the human ability to make accurate judgments about personal memories in general (Alexandra & Chua, 2018), which makes it problematic to rely on dreamers' judgments about their dream reports. Such judgments are based on metamemory and involve some of the most recently evolved prefrontal areas, including the frontal pole (Mansouri, Koechlin, Rosa, & Buckley, 2017).

Some scientific fields depend very heavily on direct observations by researchers, aided in more recent decades by sound-activated audio technologies and/or movement-activated visual equipment. Obviously, however, dreams cannot be observed while they are happening. Nor can participants make reports of ongoing subjective experience as it is unfolding, as is possible in many waking studies in other fields within psychological science.

Still, as noted at the outset of this section, the study of dreaming is similar to other scientific fields of research in that it has been aided by advances in technology. By the late 1950s, the widespread use of the polysomnograph (which records electrical wave patterns from the brain, the heart, muscles, and eyes) made it possible to map the stages of sleep throughout the night and to do detailed studies of the sleep-onset process. The polysomnograph also made it possible to collect dream reports in a more controlled way in sleep-dream laboratories and to detect the specific types of brain waves (discussed in chapter 2) that are present during drifting waking thought, the sleep-onset process, and dreaming. Neuroimaging studies in the 1990s made it possible to move beyond, and even supersede, the general electrical patterns provided by the EEG. Neuroimaging studies therefore are given a primary role in this book, although past EEG findings are incorporated into the context provided by neuroimaging findings whenever it is possible. At about the same time as neuroimaging technologies became more generally available in the 1990s, the growing availability of desktop computers made it possible to do faster and more accurate quantitative analyses of very large samples of dream content almost instantaneously, as well as to use more sophisticated ways to determine p values and effect sizes (Noreen, 1989).

With the necessary definitions and distinctions clearly in mind, along with the limitations facing dream research, the next seven chapters provide a gradually unfolding account of the neurocognitive theory of dreaming.

2 Neural Substrates, Embodied Simulation, and Cognitive Insufficiencies

In order to provide the full context for demonstrating the considerable accuracy with which the neural substrate that subserves dreaming can be demarcated, this chapter begins with the wide range of replicated neuroimaging studies of the waking state. Then, neuroimaging studies of sleep are used to make comparisons with the waking studies. Neuroimaging studies in both the wake and sleep states start at the level of neural substrates and move to the cognitive networks they enable. These two levels provide the foundation for the neurocognitive theory of dreams, which then incorporates the replicated findings from large-scale quantitative studies of dream content and developmental studies of dreaming in children and teenagers.

To start with, it has been established that various combinations of neural substrates subserve seven main brain networks during waking. These findings led researchers to focus on the four major interacting "association" networks that are the basis for the cognitive processes of attention, directed thinking, and imagination. Based on this large literature, the chapter then focuses on the particular association network, the "default network," that is the primary basis for dreaming. The fact that portions of the default network are the primary support basis for dreaming was first demonstrated empirically in a meta-analysis of six earlier neuroimaging studies of the highly activated stage of sleep called REM sleep, and the studies on which this meta-analysis was based have been replicated. (As mentioned in the previous chapter, REM sleep is named on the basis of the rapid, irregular, and unpredictable eye movements that are one striking feature of it; however, there are also changes in heart rate, breathing patterns, and much else during REM sleep [J. Siegel, 2017a].)

In addition, the chapter shows that portions of the default network support the cognitive process of dreaming during the sleep-onset process and in one of the non-REM stages of sleep, named "NREM 2 sleep." It also shows that dreaming happens far less often, if at all, during slow-wave sleep, which is also sometimes called "N3" in the research literature. Moreover, portions of the default network can become ascendant during brief episodes within periods of drifting waking thought in a laboratory setting. These brief episodes occur when a participant is left alone in a room, with the waking state monitored by brain-wave patterns, using the channels of the electroencephalogram (EEG) (Foulkes, 1985; Foulkes & Fleisher, 1975; Foulkes & Scott, 1973).

Although the findings and concepts presented in this chapter on the neural substrates that are organized into association networks represent a major advance in the understanding of the functioning human brain compared to 30 years ago, these findings also lead to the realization of just how much needs to be learned before any solid conclusions are fully warranted. There are also continuing discussions and studies relating to what should be included, or not included, in one or more of the association networks, as well as ongoing discussions about if and how nomenclature should be updated as new discoveries lead to new understandings (e.g., Uddin, Yeo, & Spreng, 2019). Within this context, it should be clear that the neurocognitive theory of dreaming is not presented as a series of established theorems but as a set of well-grounded hypotheses that can be tested, modified, or abandoned.

In addition to the baseline studies of neural substrates, which are based on 1,000 nonpatient participants, the neural level of the neurocognitive theory of dreaming also relies on studies of hundreds of people with circumscribed (focal) brain lesions in varying locations in every part of the brain. Their lesions are primarily due to injuries and illnesses, although one sample of fewer than 25 psychiatric patients, from over five decades ago, all had lesions in the same area of their brains due to what turned out to be misguided attempts by neurosurgeons to relieve their symptoms. The lesion studies lead to the same conclusions as do the neuroimaging studies concerning the neural substrates that support dreaming. Lesion studies also add useful first-person testimony regarding the impact of lesions in different neural substrates on dreaming. Most importantly, these impacts include the complete loss of dreaming due to lesions in either of two regions in the neural substrates that comprise the default network. In addition, lesions in still other parts of the default network lead to alterations in dreaming, such as the loss of different visual aspects of dreaming.

The findings on the boundaries of the neural substrate that subserves dreaming make it possible to compare this substrate with the neural substrates underlying several types of atypical states of waking consciousness. These comparisons show that the neural substrates underlying atypical wake states are each somewhat different from each other and that all of them are different from the neural substrate that supports dreaming. These comparisons are necessary because such states are frequently compared to dreaming or even used as a basis for studying dreaming and dream content. In a similar fashion, the neural substrates that enable two different kinds of atypical dreamlike states during sleep are shown to be somewhat different from the neural substrate that supports dreaming. These findings suggest that atypical states during either sleep or waking should not be utilized in developing a foundational model of dreaming.

The chapter also discusses the cognitive process of simulation, which is the cognitive basis of dreaming. Simulation is defined for the purposes of this book as "a particular type or subset of thinking that involves imaginatively placing oneself in a hypothetical scenario and exploring possible outcomes" (Schacter, Addis, & Buckner, 2008, p. 42). Simulation thereby includes mental imagery and an unfolding narrative flow. Moreover, other studies suggest that the simulation process is sometimes enhanced by the increased activation of secondary sensory regions in the brain. At that point, simulation can be characterized as "embodied simulation," in the psychological sense of the term "embodied" (Gibbs, 2006).

Building on comparisons of the activation patterns during waking and dreaming, the concept of "cognitive insufficiencies" is introduced. This concept is used to suggest that one or more types of cognitive insufficiency may explain the differences between waking thought and certain aspects of dreaming. The varying bases for cognitive insufficiencies begin with the truncated nature of the neural substrate that supports dreaming. The concept also includes focal brain lesions. Cognitive insufficiencies also may be due to the gradual maturation of the brain during childhood or the gradual cognitive development of the capacities that support the process of simulation. Examples of the effects of various cognitive insufficiencies are then highlighted. Other, more important instances of possible cognitive insufficiencies are discussed in ensuing chapters.

The Network Organization of the Human Brain

Based on a large-scale neuroimaging study that included 1,000 participants, 17 neurocognitive networks were identified. A majority of these networks are subnetworks within seven general networks (Schaefer, Kong, Eickhoff, & Yeo, 2018; Uddin et al., 2019; Yeo et al., 2011, p. 1135, fig. 11). The seven general networks can be divided into two basic types. First, there are two sensory networks, the visual and sensorimotor networks. These two sensory networks, which include only 35% of the cerebral cortex, are localized in specific brain regions and function in a linear and hierarchical fashion (Yeo et al., 2011, p. 1135). There are also "four major association networks," which are very different from the sensory and sensorimotor networks in that they are neither linear nor hierarchical (Yeo et al., 2011, p. 1149). They are instead characterized by highly complex internal relations, which are multiple and interdigitated (i.e., interweaved, interlocked) (Yeo et al., 2011, pp. 1136–1137, 1150). The fifth of the five association networks is marginal to the issues discussed in this chapter, so it is discussed more fully in chapter 8.

Still other research reveals that the association networks are small-world networks, which differ from random networks in that they have short paths (links) between two nodes (points, brain regions), along with a tendency for two nodes that have links with a third node to have links to each other. There is also a tendency for the nodes that have many links to be linked to each other (clustering), which often leads to a large general component in the network (Watts & Strogatz, 1998). The fact that the association networks are small-world networks makes it possible to analyze them with the techniques developed by graph theorists (Bullmore & Sporns, 2009; Sporns, 2011). As noted in a replication and extension study of these networks, "the human brain network is modular," in that it is "comprised of tightly interconnected nodes." These nodes have "many connections within their own communities," but also central hubs that lead to connections "diversely distributed across communities" (Bertolero, Yeo, Bassett, & D'Esposito, 2018, pp. 1–2). These results were confirmed in an analysis based on a new "gradient-weighted" methodology (Schaefer et al., 2018). In addition, these networks have subcortical and cerebellar dimensions to them. The most important of these subcortical areas for purposes of this book include the hippocampus and parahippocampus, the thalamus and hypothalamus, and the amygdala, the basal ganglia, and the pons, the latter of which is located in the brainstem. More exactly, there are 358 closely interacting subcortical regions, some of which have not been fully characterized as yet (Ji, Spronk, Kulkarni, Anticevic, & Cole, 2019; Schaefer et al., 2018; Uddin et al., 2019). As discussed below, most of the subcortical regions appear to be relatively deactivated during sleep.

The first of the five association networks, the frontoparietal control network, has control and integration functions in relation to the sensory and sensorimotor networks, as well as in relation to the other association networks. It provides support for the cognitive "executive network." It works closely with three "attention" networks involved in facilitating attention, beginning with the dorsal attention network. Although it originally appeared that there were two other attention networks, the ventral attention network and the salience network, it was subsequently discovered that these two attention networks have a "high degree of spatial overlap" and are "highly functionally interrelated," so they "perhaps can be considered as a single bilateral network" (Kucyi, Hodaie, & Davis, 2012, pp. 3388–3389). Another study concludes there is a "salience/ventral network," which is the phrase that will be used in this book (Schaefer et al., 2018, p. 3103). The ventral portion of the salience/ventral network functions "mainly during exogenous salience detection," whereas the salience portion "plays a broader role, engaging across domains during processing of personally relevant inputs" (Uddin et al., 2019, p. 12). The salience portion of the salience/ventral network and its crucial alerting function have been fully documented in a wide range of studies. It can change the functioning of all brain networks and basic bodily functions in milliseconds by sending warning signals to the frontoparietal control network and the dorsal attention network (Menon & Uddin, 2010; Seeley et al., 2007; Touroutoglou, Hollenbeck, Dickerson, & Barrett, 2012; Uddin, 2015).

In contrast to the other three main association networks, the default network is internally focused. Its contours have been outlined in several studies, including in a meta-analysis. These studies demonstrate that the default network includes two distinct but functionally connected subsystems, the dorsal medial prefrontal cortex subsystem and the medial temporal cortex subsystem (Andrews-Hanna, Irving, Fox, Spreng, & Christoff, 2018; Andrews-Hanna, Reidler, Sepulcre, Poulin, & Buckner, 2010; Andrews-Hanna, Smallwood, & Spreng, 2014; M. L. Meyer, Hershfield, Waytz,

Mildner, & Tamir, 2019). The dorsal medial subsystem within the default network is differentially activated by instructions to think about the person's present situation or present mental state ("present self") and also may be supportive of verbal, abstract, and reflective aspects of internally generated thought. It includes the dorsal medial prefrontal cortex and three other brain regions. The medial temporal cortex subsystem of the default network, which is differentially activated by thinking about personal situations and decisions in the future ("future self"), includes the ventral medial prefrontal cortex and four other brain regions, and it may be more involved in supporting image-based, concrete, specific forms of thought (Andrews-Hanna & Grilli, 2021; Andrews-Hanna, Reidler, Sepulcre, et al., 2010, pp. 554, 559; Andrews-Hanna et al., 2014; Fox, Spreng, Ellamil, Andrews-Hanna, & Christoff, 2015). However, most of the time the two subsystems are working together. Finally, the default network also includes a set of regions that serve as "zones of integration," which are connected to the other four association networks (Andrews-Hanna et al., 2014, p. 35). The primary brain regions within the two subsystems and the zones of integration are listed in table 2.1.

In addition, the default network has close links to the nearby fifth association network, the limbic network, which includes the orbitofrontal cortex, insula, medial prefrontal cortex, anterior cingulate cortex, and the temporal pole at the cortical level, and a prominent subcortical neural substrate, the amygdala. Different parts of the limbic network are involved with other association networks in supporting a variety of cognitive processes, such as memory, reward, and empathy (Bickart, Dickerson, & Barrett, 2014, p. 238;

Table 2.1
The primary brain regions in the two subsystems and zones of integration of the default network

Dorsal medial prefrontal cortex subsystem	Medial temporal cortex subsystem	Zones of integration
dorsal medial prefrontal cortex	parahippocampal cortex	anterior medial prefrontal cortex
inferior temporal cortex	hippocampal formation	angular gyrus
temporal pole	retrosplenial cortex	anterior temporal lobes
temporoparietal junction	posterior inferior parietal lobule	superior frontal gyrus
		posterior cingulate cortex

Dixon, Thiruchselvam, Todd, & Christoff, 2017; Inman et al., 2018). Since the limbic network also has a role in supporting waking emotions, it is discussed more fully in chapter 8.

The default network provides significant support for imagination, memory, and mental imagery (Andrews-Hanna, Reidler, Huang, & Buckner, 2010; Andrews-Hanna, Reidler, Sepulcre, et al., 2010; Andrews-Hanna et al., 2014). The default network also includes brain areas that provide support for the self-system (Abraham, 2013; D'Argembeau, 2020; M. L. Meyer & Lieberman, 2018). The default network is also unique compared to the other three major association networks in that it is more "internally integrated," "self-regulated," and "self-contained" (Power, Cohen, Nelson, & Petersen, 2011, pp. 665, 671). Similarly, the default network has "the highest structure-function agreement within the whole brain" (Horn, Ostwalda, Reisert, & Blankenburg, 2013, p. 6). It is at "the top of a representational hierarchy" of cerebral cortical organization, based on a principal gradient analysis, which also placed the sensory networks at the base of the pyramid (Margulies et al., 2016, p. 12574).

Finally, the default network has more frequent changes in its relationships with the other association networks than do the other association networks with each other (Denkova, Nomi, Uddin, & Jha, 2019). It contains more individual differences within it than do the other association networks, (Cai et al., 2019, p. 4843). Perhaps as a result of the individual differences within it, the default network played a "primary role," along with the frontoparietal control network, in identifying individual differences when 140 participants, ages 12–30, were scanned for a second time 12 to 18 months after the initial scan (Jalbrzikowski, Liu, Foran, Roeder, & Luna, 2020, p. 1). Figure 2.1 presents a very general overview of the location of the four major association networks. The primary purpose of figure 2.1 is to provide readers with a general sense of the degree to which these four association networks have different spatial locations.

The Control of the Default Network by the Other Three Major Association Networks

The frontoparietal control network is at the center of the system of brain networks during alert waking thought. It works closely with the dorsal attention network and the salience/ventral network in maintaining a focus on the external world, and while making its executive decisions. For all the connections among these three networks, however, the frontoparietal

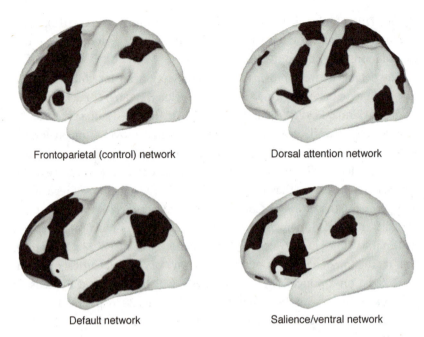

Frontoparietal (control) network   Dorsal attention network

Default network   Salience/ventral network

Figure 2.1
The general locations of the four major association networks within the brain.
Adapted from Buckner, Krienen, & Yeo, 2013.

control network is most frequently interactive with the default network. The
first piece of evidence for this point is a connectivity correlation of .70 when
the four main association networks are not involved in one or another task,
even though the mind is drifting, or perhaps daydreaming, or even momentarily dreaming, during this "resting state" (Lee et al., 2012, p. 2).

By way of contrast, the default network is negatively correlated with
the other association networks during non-task conditions, starting with the
sensorimotor network (–.82) and the visual network (–.60). Finally, there is
a –.20 correlation with the ventral portion of the salience/ventral network
(Lee et al., 2012, p. 2). Taken together, these results suggest that the default
network is more loosely controlled when an individual is not involved in
a demanding external task. (The salience portion of the salience/ventral
network was not included in this study.)

The constantly changing relationship between the frontoparietal control network and the default network during waking is further illuminated
in an experimental study that used transcranial magnetic stimulation to

alter the relationship between them (A. Chen, Oathes, Chang, Bradley, & Zhou, 2013). As demonstrated by the changes in brain activity during this experimental manipulation, the stimulation of the frontoparietal control network decreased its connectivity to the default network, which suggests a greater emphasis on focused attention. On the other hand, inhibitory stimulation of the frontoparietal control network increased the activation level of the default network, which suggests more internally oriented and imaginative thinking may be occurring when the frontoparietal control network is less activated (A. Chen et al., 2013).

The relationship between the frontoparietal control network and the default network has been refined and importantly extended in a study based on four independent datasets. The datasets contain the network patterns that resulted from eight different task conditions and one non-task condition. Comparing the various task conditions with the non-task condition, the researchers discovered that one of the subsystems of the frontoparietal control network regulates the dorsal attention network during attention and action and that another subsystem interacts with the default network when the focus is on personal thoughts (Dixon, Vega, Andrews-Hanna, Spreng, & Christoff, 2018). However, the two subsystems work together closely to modulate the default network in the context of increased sensory input. This finding on the role of sensory systems in sending signals to the frontoparietal control network, with regard to attenuating the activation of the default network, fits with an earlier study by the same research group. It showed that the relationship between the dorsal attention network and the default network varies widely across different cognitive states but that their relationship is coordinated with interactions involving the frontoparietal control network (Dixon, Andrews-Hanna, Spreng, Christoff, & Irving, 2017, p. 632).

In keeping with these findings, still another study of the dorsal attention network examined its role in regulating the default network, through its interactions with the default network's posterior areas. This study turned out to be all the more useful for dream researchers because the participants were instructed to keep their eyes closed, which is a basic starting point in the sleep-onset process. When gradual decoupling begins to occur between the dorsal attention network and the posterior areas of the default network, the anterior regions of the default network become more connected with each other and more activated, which leads to a strong inverse relationship between the anterior portions of the default network and the dorsal attention network (Chang, Liu, Chen, Liu, & Duyn, 2013, p. 230). To complete the picture, the decoupling of the posterior portions of the default network from the dorsal attention network is accompanied by a simultaneous disconnection from the posterior thalamic region. This subcortical region is important in supporting the salience/ventral network, which suggests the salience portion of that network is gradually reducing its ability to prepare the body for immediate action (Chang et al., 2013, p. 230; Menon & Uddin, 2010; Uddin, 2015; Zabelina & Andrews-Hanna, 2016, pp. 88, 92n30). As discussed in the next subsection, this process has similarities with the sleep-onset process. It is also noteworthy in considering the default network during sleep that one key part of the salience/ventral network, the right anterior insula, is known to be "a critical node for suppressing default network activity and reallocating attentional resources to salient events" (Andrews-Hanna et al., 2014, p. 42, box 3). This finding very likely means that the salience/ventral network would have to be deactivated before dreaming could occur.

As a final note on the Chang et al. (2013) study, it is important to add that the decoupling of the anterior default network from the dorsal attention network with eyes closed is also accompanied by the increasing activity of two brain waves, called "alpha" and "theta," which are defined and displayed in table 2.3 in a later section of this chapter. Alpha and theta are useful because one or both of them can be linked to functional magnetic resonance imaging (fMRI) findings on the sleep-onset process, REM sleep, and NREM 2 sleep, all of which are associated with dreaming. They create bridges between neuroimaging findings and EEG studies, and therefore can be employed as indicators in some instances for studying both drifting waking thought and dreaming (see Domhoff & Fox, 2015, p. 348, for a summary of several studies that demonstrate this point).

The Association Networks during Sleep Onset, Sleeping, and Dreaming

In alert waking states, the four main association networks constantly interact in rapidly changing and complex ways. These complex interactions often involve connections among specific subsections in different networks, but always with the other three main association networks controlling the default network. During the sleep-onset process and during brief episodes of drifting waking thought, the association networks that concern control, integration, attention, and alertness become increasingly deactivated. However, portions of the default network remain activated during the sleep-onset process. They remain activated into NREM 2 sleep. These neural substrates are relatively deactivated, and more disconnected, in the ensuing stage of slow-wave sleep. Then they activate about four to six times each night during the periodic REM periods, which occur about every 90 minutes throughout sleep. Portions of the default network are also activated in NREM 2 in the two or three hours before morning awakening. In terms of the neurocognitive theory of dreaming, it is the default network that matters, because portions of it provide the primary support for the cognitive processes that are involved in dreaming. (The overview provided in this paragraph is documented in detail in later sections in this chapter.)

Before turning to the issue of dreaming, however, it is first necessary to stress that "activation" and "deactivation" can be measured only at the ordinal level of measurement, in which one brain region or network is more or less active than another one. It is not possible on the basis of neuroimaging studies, at least as yet, to determine the degree to which one region or network is more activated or deactivated than the other. This point is important to keep in mind because it may turn out that the degree to which one or another brain region or network is deactivated in relation to the default network is essential for a full understanding of the process of dreaming.

The Neural Substrate That Supports Dreaming

As already mentioned in the previous section, the control of the default network by other association networks during waking gradually declines when individuals become drowsy and begin to fall asleep. From that point on, the other association networks become relatively deactivated. As a result, portions of the default network are ascendant during sleep onset and into NREM 2 sleep, and people spontaneously begin to dream. As noted earlier, the default network is highly activated during the four to six REM periods that occur during a typical night of sleep and in the increasing amount of NREM 2 sleep toward morning. It is important to note that REM periods become longer throughout the night, which parallels the increasing amount of NREM 2 sleep as the morning awakening approaches.

Just as neuroimaging studies have revolutionized the understanding of the waking brain and its many various mental states, they have also done so in the case of sleep and dreaming. Neuroimaging studies make it possible to study whole-brain networks with time frames of a few seconds. They make it possible to study spatiotemporal complexity in the transition to sleep and in transitions in sleep stages (Stevner et al., 2019, pp. 1–2). For purposes of the findings that relate to a neurocognitive theory of dreaming, this complexity is perhaps most usefully indexed by the degree of "functional connectivity" within and between neural substrates—that is, the degree to which the usual links are activated within a neural network. (The aforementioned Chang et al. [2013] study of the decoupling of the dorsal attention network from the anterior default network is a good demonstration of a decline in functional connectivity.) More generally, and most importantly in terms of a neurocognitive theory of dreaming, neuroimaging studies make it possible to study the relative activation levels and the degree of functional connectivity of brain networks during the sleep-onset process and the different stages of sleep.

The EEG, on the other hand, "involves considerable averaging of brain activity in terms of both time and space—arguably leading to an incomplete representation of brain activity" (Stevner et al., 2019, p. 2). Although the EEG is usually thought to provide a faster temporal resolution than the fMRI, in fact scalp electrodes are "generated several centimeters below the recording electrodes," and as a result "cortical current must go through different resistive layers" (Burle et al., 2015, p. 211). Nevertheless, the polysomnograph, and in particular its capacities for detecting and recording electrical currents in the brain, remains a very good, fast, and relatively inexpensive way to determine sleep stages. The polysomnograph thereby makes it possible to carry out planned awakenings to obtain reports of mental activity, including dreams. It is far less intrusive than fMRI, and there are portable polysomnographs as well. In addition, as noted earlier in the chapter, there are bridges between fMRI network findings and EEG findings on the appearance of alpha and theta waves. These two brain waves are useful for the development of the neurocognitive theory of dreaming by providing additional links between drifting waking thought, the sleep-onset process, NREM 2 sleep, and REM sleep.

The neural foundations for the neurocognitive theory of dreaming emerged within the context of neuroimaging studies of REM sleep in the 1990s (Braun et al., 1997, 1998; Maquet et al., 1996; Nofzinger, Mintun, Wiseman, Kupfer, & Moore, 1997). The findings were supplemented by lesion studies of alterations in dreaming (Solms, 1997). The discovery, very shortly thereafter, of the default network and its relationship to imagination and mind-wandering, further expanded the horizons (Gusnard, Akbudak, Shulman, & Raichle, 2001; Raichle et al., 2001). These studies were bolstered by later studies of the connection between mind-wandering and the default network (Andrews-Hanna, Reidler, Huang, et al., 2010; Andrews-Hanna, Reidler, Sepulcre, et al., 2010; Buckner, Andrews-Hanna, & Schacter, 2008; Christoff, Gordon, Smallwood, Smith, & Schooler, 2009; M. Mason et al., 2007). These studies led to the earlier versions of the neurocognitive theory of dreaming (Domhoff, 2000; 2001; 2003, chap. 1; 2011).

The hypothesis that portions of the default network are the key part of the neural substrate that subserves dreaming was first supported in a meta-analysis that compared brain patterns, based on six neuroimaging studies of REM sleep, with findings from mind-wandering studies (Fox et al., 2013, p. 8). It was further supported by studies of people who define themselves as frequent and infrequent dream recallers (Eichenlaub, Bertrand, & Ruby, 2014; Vallat, Eichenlaub, Nicolas, & Ruby, 2018). The neuroimaging findings in several of the waking studies, which served as a control group, support the supposition that mind-wandering and dreaming share some commonalities. They draw on semantic memory, have a loosely narrative structure, simulate social interactions, and frequently express personal concerns (Fox et al., 2013, p. 1). Finally, the findings concerning the activation patterns during REM sleep were confirmed and extended in a later replication study that used different analytical techniques to examine several of the extant REM databases. It concluded that REM sleep creates "a network that includes retrosplenial cingulate cortex, parahippocampal gyrus, and extrastriate visual cortices." This network corresponds "to components of the default mode network and [secondary] visual networks" (Uitermarkt, Bruss, Hwang, & Boes, 2020, p. 1). At the same time, the study confirmed that the frontoparietal and the salience/ventral network are relatively deactivated during REM sleep (Uitermarkt et al., 2020, p. 5).

The neural network that subserves dreaming is located primarily within the two functional subsystems within the default network—the dorsal medial prefrontal cortex system and the medial temporal cortex system, as discussed above (Andrews-Hanna, Reidler, Sepulcre, et al., 2010). These two functional subsystems provide support for the cognitive processes of mentalizing, social cognition, and imagination. They thereby subserve the ability to enact thoughts related to the past, present, and future, and to infer other people's thoughts and intensions (Andrews-Hanna et al., 2014; Lieberman, 2013; Moulton & Kosslyn, 2011; Spreng et al., 2018). In addition, the inclusion of language areas, located in regions in the temporal lobes, is consistent with the frequency, correctness, and specificity of language use in dreams (Foulkes et al., 1993; B. Meier, 1993). There is also a greater relative activation of the medial prefrontal cortex, which serves as a central hub in the widely distributed waking self-system. This finding may help explain why dreamers are usually at the center of their dream scenarios (Abraham, 2013; D'Argembeau, 2020; Gusnard et al., 2001; Jenkins & Mitchell, 2011). The two subsystems are supplemented by the caudate nucleus, located in the basal ganglia, which enables the initiation of movement, among other functions. They also recruit the secondary visual cortices, including the lingual gyrus. The lingual gyrus, which is located in the medial occipital lobe, is involved in complex visual imagery.

At the same time as the two systems of the default network become more activated and integrated, several of the core regions in the default network, the ones that serve as zones of integration with the other association networks, become increasingly deactivated. This finding is especially critical in the case of the posterior cingulate cortex. Some regions within it, as also noted previously, are involved in awareness and alert attention, in conjunction with brain systems outside the default network (Chang et al., 2013; Leech & Sharp, 2014, pp. 5, 11–12). The relative deactivation of the posterior cingulate cortex, and its subsequent decoupling from the dorsal attention network, along with the relative deactivation of regions in anterior areas of the frontoparietal control network, which are also involved in sustaining awareness, may be the final steps in the complex neurocognitive process of losing conscious self-control. In addition, as the frontoparietal control network and dorsal attention networks deactivate, there is an attenuation of the salience/ventral network as well. Based on these findings, dreaming can be considered an intensified form of mind-wandering and daydreaming precisely because it is not constrained by sensory and sensorimotor input or by the frontoparietal, dorsal attention, and salience/ventral networks.

Within the context provided by the two previous paragraphs, table 2.2 presents a comparison between the neural substrates that support waking and the neural substrates that support dreaming. As shown in the table, the neural substrates that support dreaming also include the caudate nucleus,

Table 2.2
A comparison of the brain regions that subserve waking and dreaming

Brain area	Default network	Neural substrate that supports dreaming
Frontal lobe		
rostrolateral cortex (anterior, frontopolar)	activated	relatively deactivated
dorsolateral frontal cortex	activated	relatively deactivated
orbitofrontal cortex	activated	relatively deactivated
right anterior insula	activated	relatively deactivated
rostral anterior cingulate cortex	activated	relatively deactivated
dorsal cingulate cortex	activated	relatively deactivated
inferior frontal gyrus	activated	relatively deactivated
dorsomedial prefrontal cortex	activated	activated
ventral medial prefrontal cortex	activated	activated
Parietal lobe		
posterior cingulate cortex (inferior parietal)	activated	relatively deactivated
retrosplenial cingulate cortex	activated	activated
lateral parietal cortex (supramarginal gyrus and temporoparietal junction)	activated	activated
Temporal lobe		
lateral temporal cortex/temporal pole	activated	relatively deactivated
hippocampus (medial temporal cortex)	activated	activated
parahippocampal cortex (medial temporal cortex)	activated	activated
entorhinal cortex (medial temporal cortex)	activated	activated
*Occipital lobe**		
primary visual area (V1)	activated	relatively deactivated
extrastriate (V2–5, secondary visual areas)	activated	activated
lingual gyrus (visual processing)	activated	activated
Subcortical regions		
caudate nucleus	activated	activated
pons/midbrain	activated	activated

* Not a part of the default network, but secondary areas are important in dreaming.

which is located in a subcortical region, and parts of the midbrain and the pons, which are part of the subcortical brain arousal systems that are necessary for cortical activation in both waking and in some sleep stages (Edlow et al., 2012; B. E. Jones, 2020; McGinty & Szymusiak, 2017; J. Siegel, 2017a). Also, secondary visual areas remain activated during dreaming, although they are not part of the default network.

Lesion and Electrical Brain Stimulation Studies Support the Neuroimaging Findings

This section provides further evidence concerning the nature of the neural substrate that supports dreaming, based on two very different methods. Lesion studies in effect replicate and extend the findings based on neuroimaging studies. Electrical brain stimulation studies of waking epilepsy patients contribute new information, which relates to the spontaneous nature of dreaming.

Lesion Studies of Dreaming

The lesion method has been called "a gold standard for assessing causality" when the studies are done correctly. Such studies require the use of patients with focal lesions, and an examination of the patient's cognitive capacities before neural plasticity has led to compensations for the lesions (Lieberman, Straccia, Meyer, Du, & Tan, 2019, p. 314). In the case of lesion studies of the neural substrate that enables dreaming, the focal lesion studies over the previous 75–80 years reach close to that high standard. Because patients can provide accounts of how their lesions do or do not affect their dreaming, these studies also provide an important addition to understanding the effects of lesions in different locations in the neural substrate that subserves dreaming.

The most important large-scale systematic study of lesions and dreams provides a context for discussing earlier and later clinical studies of individuals and groups. It was based on complete clinical evaluations of 361 new neurological patients in a hospital in Cape Town, South Africa, in the late 1980s. All of the new patients were examined and assessed by the researcher himself (Solms, 1997). They were also asked to fill out a questionnaire regarding possible changes in their dreaming, and they were asked questions about dreaming during interviews as well. Most of the patients continued dreaming in their normal fashion, but some did not. The final report included all past case studies as well as the results of the new study. Drawing together all of the results, it can be concluded that lesions *outside* the neural network that subserves dreaming, in regions such as the dorsolateral prefrontal cortex, the primary visual cortex, the primary sensorimotor cortices and some regions in the brainstem, have no impact upon dreaming (Solms, 1997, pp. 82, 153, 219–223, 237). Nor do localized injuries in the hypothalamus or cerebellum have any effect (Solms, 1997, p. 82, table 9.3, and pp. 153–154). These results circumscribe the possible neural substrates that are necessary for dreaming. They also provide, in effect, the experimentally based control group in terms of eliminating most neural substrates as possible supports for dreaming. They thereby make it possible to formulate reasonably confident assertions about the neural substrates that do support dreaming.

In a subsequent study, the research group at the University of Cape Town discovered that focal damage to the basolateral amygdala in eight patients showed this portion of the amygdala is *not* essential to dreaming (Blake, Terburg, Balchin, van Honk, & Solms, 2019). This University of Cape Town finding is consistent with the fact that those who suffer from the rare disorder of Urbach Wiethe disease, which causes bilateral damage in the amygdala, also continue to dream. This point is seen in the case of the best-studied Urbach Wiethe patient in the United States, who continued to dream: "Over the course of many years' worth of interactions with Patient SM, I can say with certainty that she does report dreaming" (Feinstein, 2015). She also had a normal sleep cycle, according to a hospital-administrated sleep study. But she was not awakened for dreams during this clinical assessment. Since Urbach Wiethe patients are most notable for their lack of fear in waking life, it is of interest to know that an important neural substrate underlying reactions to many types of fearful situations in waking life is not necessary for dreaming.

Conversely, injuries *inside* the neural network that subserves dreaming lead to changes in dreaming. These changes include a global loss of dreaming due to injuries in either the ventral medial prefrontal cortex or in the supramarginal gyrus and temporoparietal junction, which are located in the lateral parietal cortices (Solms, 1997, chaps. 4 and 16). In addition, injuries in regions of the secondary visual cortex lead to the loss of visual imagery in both dreaming and waking. This finding is best established by a study that included both waking cognitive testing and awakenings from REM in a laboratory setting (Kerr, Foulkes, & Jurkovic, 1978). The experiential reports of continued dreaming, loss of dreaming, or the loss of visual imagery in dreaming provide crucial subjective evidence that the neural network detected in neuroimaging studies is indeed related to the cognitive process of dreaming (Solms, 1997, chaps. 3 and 11).

The loss of dreaming is an important issue in terms of the neurocognitive theory of dreaming. This is because the ability to abolish a phenomenon is one of the criteria in biology for establishing that a bodily area or brain region has the function, or functions, that are claimed for it. In fact, as already noted, dreaming can be abolished due to focal injuries in either of two separate regions, the ventral medial prefrontal cortex or in the areas in and around the temporoparietal junction. It is first essential to repeat that both areas fall within the default network. Moreover, some of the cases involving the loss of dreaming were caused by a surgical operation carried out by neurosurgeons from the late 1930s through in the early 1960s in the United States, Canada, and parts of Western Europe. These operations were tragically thought to help mental patients. The operation severed the ventral medial prefrontal cortex from a region posterior to it. It led to the unintended loss of dreaming in 70–90% of hundreds of such patients, along with a large decline in fantasy during waking (see Frank, 1950; Solms, 1997, chap. 5; and Walsh, 1994, pp. 158–168, for a full discussion of the exact nature and location of the operation).

The findings concerning the loss of dreaming due to neurosurgery in the ventral medial prefrontal cortex are supported in a two-night study of reports after REM awakenings, which involved 13 schizophrenic patients (1 woman, 12 men) in a hospital in Montreal, Canada. These patients had undergone the neurosurgery that severed the ventral medial prefrontal cortex from nearby areas posterior to it 10 to 29 years earlier. Even though there had been ample time for neural plasticity to restore dreaming, nine did not report any dreams, three reported four instances in which they thought they might have been dreaming, and only one reported a dream after a group total of 97 REM awakenings (Jus, Jus, Villeneuve, et al., 1973). On the other hand, 10 of the 13 nonsurgical schizophrenic patients in a control group, who were matched with the surgery patients for age, gender, duration of illness, medications, scores on memory tests, and a score on an intelligence scale, reported the contents for at least one dream. One patient in the control reported dreams after six of nine awakenings (Jus, Jus,

Villeneuve, et al., 1973, pp. 278–279, 286–287, tables IV and V). The same research team then replicated the study with 10 male surgical patients, who were given reserpine, a medication believed at the time to improve dream recall (Jus, Jus, Gautier, et al., 1973). The researchers again found few or no instances of dream recall.

These findings were also corroborated in eight cases of nondreaming patients in the large-scale study at the University of Cape Town, who suffered similar prefrontal neurological damage through injuries (Solms, 1997, p. 145). Importantly, none of these eight patients was any more likely to display waking memory impairments than patients who continued to report dreaming. It is therefore unlikely that they were simply forgetting their dreams (Solms, 1997, pp. 34–35, 160–161). Two other cases with injuries in the same region reported a very large reduction in the "frequency and duration of dreams" (Solms, 1997, p. 149).

The study at the University of Cape Town also included 47 patients with lesions in either parietal lobe. They reported a loss of dreaming, which is consistent with several cases in the historical literature (Solms, 1997, pp. 141–145). A follow-up study of 10 of these patients a year later found that all of them had recovered their capacity to dream. This finding provides a striking contrast to the continuing absence of dreaming for those who had lesions in the ventral medial prefrontal cortex (Solms, 1997, p. 145). The study found that most of these parietal-lobe cases had lesions localized to the supramarginal gyrus or the temporoparietal junction, which suggested that these more circumscribed areas might be sufficient to lead to the loss of dreaming (Solms, 1997, p. 143). This possibility was supported in the case of the temporoparietal junction by two later neurological case studies of individual patients. They lost dreaming with no other parietal involvement beyond the temporoparietal junction (Bischof & Basset, 2004; Poza & Marti-Masso, 2006).

In keeping with the later emphasis on association networks, which were discovered through neuroimaging studies, Solms (1997, pp. 47–48) concluded as follows on the basis of his findings on the loss of dreaming due to lesions in two different areas in what is now known as the default network:

> Modern neuropsychological research demonstrated conclusively that mental faculties are subserved by complex functional systems or networks, which consist of constellations of cortical and subcortical structures working together in a concerted fashion. Such systems or networks can be disrupted by damage to any one of their component parts.

These lesion results fully support the neuroimaging findings with what are, in effect, accidental experimental studies of dreaming. They also demonstrate causality. Due to the large number of lesion studies, it is as if researchers—not the cruel accidents of nature or the misguided efforts of previous generations of neurosurgeons—had carried out a very large number of carefully designed studies. These accidents and operations coincidentally test specific hypotheses regarding the impact of lesions in different parts of the neural substrate that subserves dreaming. They reveal the areas that cause dreaming to be abolished. They also demonstrate that dreaming can be altered in some specific ways, such as the loss of one or another visual dimension. However, these more secondary results are not essential for the theoretical purposes of this book (see Domhoff, 2018a, chap. 5; and Solms, 1997, for information on the other ways in which lesions can alter dreaming). Moreover, as noted above, lesion studies also include the necessary control studies that rule out other regions (such as the dorsolateral prefrontal cortex, the primary visual cortex, and the hypothalamus) as the neural substrates that enable dreaming.

Electrical Brain Stimulation Studies: The Spontaneity of Dreaming
Electrical brain stimulation studies make use of implanted electrodes to help locate the small areas in the brain that may be the basis for an individual's epileptic seizures. This medically necessary procedure has inadvertently contributed to an understanding of the neural substrate that supports dreaming. In a study of a large database that resulted from such probes, the experiential reactions patients report as dreamlike were all located in an important area in the default network, the temporal lobe. More specifically, most of the 42 instances occurred in the medial temporal cortex, which is one of the areas activated during dreaming (Curot et al., 2018, pp. 9–10). This result is consistent with the neuroimaging and lesion findings. Even more important in this context, the results of electrical brain simulation studies in another medical laboratory discovered the medial temporal lobe is the primary area for the generation of spontaneous waking thought as well (Fox, 2018, pp. 170, 173). Taken together, these two studies of otherwise normal patients very likely provide a window into the spontaneity of dreaming.

Based on the findings with the two methods discussed in this section, lesion studies and electrical brain stimulation studies, figure 2.2 presents an overview of the areas of the brain that are needed and not needed for dreaming.

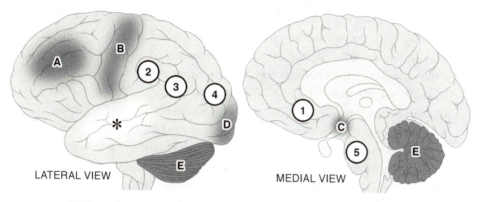

NOT needed for dreaming:
A. dorsolateral prefrontal cortex
B. primary motor cortex
C. amygdala
D. primary visual cortex
E. cerebellum

Needed for normal dreaming:
1. ventral medial prefrontal cortex
2. supramarginal gyrus
3. temporoparietal junction
4. secondary visual cortex
5. midbrain/pons

* waking dream imagery in epilepsy patents in various regions in the temporal cortex, including the hippocampus and the entorhinal cortex.

Figure 2.2
Areas of the brain that are needed and not needed for normal dreaming. Brain hemisphere illustrations by Patrick J. Lynch, medical illustrator, and C. Carl Jaffe, MD, cardiologist, Yale University Center for Advanced Instructional Media.

Dreaming during Sleep Onset, NREM 2 Sleep, and Drifting Waking Thought

Based on the neuroimaging findings overviewed in table 2.2 and the lesion and electrical brain stimulation studies overviewed in figure 2.2, the next three subsections discuss the occasions other than REM sleep when dreaming occurs. The fact that dreams are reported after 80–90% of lab awakenings during the four to six REM periods serves as a baseline starting point for these discussions (e.g., Dement & Kleitman, 1957a, 1957b; Kamiya, 1961; Strauch & Meier, 1996).

The Sleep-Onset Process, the Default Network, and Dreaming
A detailed neuroimaging study of the sleep-onset process in 57 participants (39 women, 18 men) found there is an increasing ascendancy of the default network while other networks become increasingly deactivated. This finding led the researchers to suggest that the default network might

"serve a gate-function for the entry into NREM sleep" (Stevner et al., 2019, p. 12). This study thereby replicated and extended an earlier fMRI/EEG study, which included 25 participants (Sämann et al., 2011). In addition, the results in both of these studies are consistent with several studies with smaller sample sizes (Horovitz, 2008; Horovitz et al., 2009; Larson-Prior et al., 2009; Vecchio, Miraglia, Gorgoni, Ferrara, & De Gennaro, 2017).

The results from these studies provide a context for the findings in several studies of both EEG changes and subjective reports during sleep onset, which used 5-second epochs to provide a detailed mapping of the sleep onset process in several studies involving dozens of participants. (This work originally reported that there are nine stages in the sleep-onset process but later determined that one of them was due to a more vigilant state on the first night in the sleep-dream lab, so the number of stages was reduced to eight [Hayashi, Katoh, & Hori, 1999; Hori, Hayashi, & Hibino, 1992; Hori, Hayashi, & Morikawa, 1990; Morikawa, Hayashi, & Hori, 2002; Tamaki, Nittono, Hayashi, & Hori, 2005, p. 195].) The first three stages are defined by varying percentages of alpha waves, which are characterized by a range of 8–13 Hz. (A "hertz" is a measure of the frequency of a wave pattern in studies using the EEG, with one hertz equal to one cycle per second [cps].) Table 2.3 provides an overview of brain-wave patterns during relaxed waking states, sleep onset, and the three sleep stages.) In the detailed sleep-onset studies, the next four onset stages are defined by the presence of theta waves (4–7.5 Hz) and the final stage is defined by the presence of sleep spindles (12–15 Hz) (Hori, Hayashi, & Morikawa, 1994, p. 245; Tamaki et al., 2005, p. 145). In addition to changes in brain-wave patterns during the sleep-onset process, two behavioral changes also occur: declines in reaction time and declines in the subjective experience of feeling awake with each successive stage.

Using the same rating scale for judging the subjective reports that was created for the first laboratory-based sleep-onset study in the United States in the sleep-lab era (Foulkes & Vogel, 1965), which is discussed below, the investigators concluded their results "fundamentally confirmed" the findings from that first study (Hori et al., 1994, p. 250). The reports of what they labeled as "dreamlike imagery" gradually increased from 23.3%, 25.8%, and 33.5% in the three alpha stages to a high of 45.7% during the stages defined by theta, but the percentage declined to 31.9% in the stage defined by sleep spindles (which are generally regarded as the sign that sleep has begun) (Hori et al., 1994, p. 247, table 3). A replication study of the findings on

Table 2.3
Primary brain waves during drowsiness, sleep onset, and sleep stages

Name of pattern	When it occurs	Frequency (Hz)	Illustration
alpha	drowsiness, early sleep onset	8–13	
theta	drowsiness, sleep onset, NREM 2, REM	4–8	
sleep spindles	beginning of sleep, NREM 2	12–15	
K-complexes	NREM 2	n/a (brief peaks of ~100µV)	
delta	slow-wave sleep	0.1–4	

dreamlike reports showed a similar rise and decline: 29.8% for stages characterized by alpha, 36.2% for stages characterized by theta, and 22.8% for the stage with sleep-spindles (Hayashi et al., 1999, p. 677, table 1). The fact that dreamlike reports occur most frequently during the sleep-onset process when theta waves are present is of considerable interest because theta waves are very frequent during REM periods and occur during NREM 2 as well (e.g., Armitage, 1995, pp. 336–338). This finding is important because theta appeared during the decoupling of the dorsal attention network from the anterior default network during waking (Chang et al., 2013). It is also very useful to know that alpha rhythms appear in anterior regions of the default network as part of the decoupling process during both waking and the sleep-onset process.

Dreaming during sleep onset is demonstrated, along with more emphasis on its content, in the aforementioned first large-scale systematic laboratory study of the sleep-onset process in the sleep-lab era (Foulkes & Vogel, 1965). The study, which involved nine participants, used four sleep-onset stages, as defined by differing levels of EEG activation. Eye-movement patterns were used as well. The first two stages had alpha waves, along with wake-like eye movements under the eyelids. The second stage had similar alpha patterns but included slow rolling eye movements. The third stage was defined by the appearance of theta waves and the fourth stage by brief bursts of sleep spindles. Each participant was randomly asked what they were experiencing during the four sleep-onset stages used in this study. The awakenings were followed by a series of questions that included estimations of their drowsiness and their ability to control their thoughts. The researchers found there was increasing drowsiness and a decreasing ability to control thoughts in each stage. There was also an increasing number of dream reports in the first three stages, followed by a slight decline in the sleep-spindles stage. These findings are very similar to the more detailed studies of the sleep-onset process discussed above (Hayashi et al., 1999, p. 677, table 1; Hori et al., 1994, p. 247, table 3).

Importantly, dream reports during the sleep-onset period were similar to those after REM awakenings. They were "no less well organized" than REM-period dream periods, even though there were more reports of "fleeting progressions of visual imagery or dissociated images and thoughts" during sleep onset than during REM (Foulkes & Vogel, 1965, p. 238). Independent judges found it difficult to distinguish sleep-onset reports from REM reports. The researchers concluded that the frequency of "hypnagogic phenomena," which are brief flashes of visual or auditory imagery, are greatly overstated in anecdotal, popular, and clinical accounts of sleep-onset mental activity (Foulkes & Vogel, 1965, p. 242).

Almost all of the participants went through the same general process of increasing drowsiness and a declining sense of volitional control. But there were nonetheless wide individual differences in the frequency with which they dreamed in the different stages. Five participants dreamed during the first two stages, for example, but four participants only dreamed during the last two stages. Six of the nine participants had one or more dreams during periods of "more or less continuous alpha," which is generally thought of as a relaxed waking state (Foulkes & Vogel, 1965, p. 237). Significantly, participants only reported dreams if they also reported a loss of volitional control, whatever the sleep-onset stage they were in. These findings led the researchers to suggest that the loss of volitional control may be the first of the conditions necessary for dreaming to occur. At the content level, participants were found to be a character in 62% of the dream reports during the first stage, with the figure rising to 78%, 83%, and 97% in the next three stages (Foulkes & Vogel, 1965, p. 238).

This study was replicated and extended in a study in which the 32 participants were also awakened during the second, third, and fourth REM periods, leading to 39.1% recall during sleep-onset probes and 69.2% after REM awakenings (Foulkes, Spear, & Symonds, 1966, p. 282). The two types of reports did not differ in their dreamlike quality. There were once again large individual differences in the frequency of reports from sleep-onset awakenings (Foulkes et al., 1966, pp. 282–283). These results were then replicated and further extended in a study that compared sleep-onset, NREM 2, and REM reports from 23 women participants. Using a control for the differing length of reports, the sleep-onset reports had a lower character density and less often included the dreamer as a character, although the dreamer was a character in more than half of the sleep-onset reports (Foulkes & Schmidt, 1983, pp. 267, 271, 276–278).

As shown in this subsection, dreaming does not occur very frequently during sleep onset compared to the amount of dreaming in REM sleep. However, dreaming during the sleep-onset process is important because it demonstrates that dreaming can occur outside of sleep. In addition, the fact that the posterior cingulate cortex decouples from the anterior portions of the default network as part of the sleep-onset process suggests that the sleep-onset process shares some similarities with the transition into mind-wandering, as discussed above (Chang et al., 2013). This point is also indicated by the fact that alpha waves increasingly appear in frontal regions during both the decoupling process within waking and within the sleep-onset process, and that theta rhythms appear in both of these states as well (Chang et al., 2013; De Gennaro, Ferrara, Curcio, & Cristiani, 2001; De Gennaro et al., 2004; Hasan & Broughton, 1994; Hori et al., 1994).

Dreaming in NREM 2 Sleep and in Slow-Wave Sleep
Several of the fMRI-based sleep-onset studies discussed above extended into the first one or two hours of sleep. The default network remains functionally connected during NREM 2 sleep, even as the activation levels slowly decline after the sleep-onset process leads into sleep. One of the larger

fMRI studies of sleep onset, the one based on 25 participants, confirmed the default network is still functionally connected during NREM 2 sleep (Sämann et al., 2011), as did the even larger study of 57 participants (Stevner et al., 2019). At the same time, the default network is more deactivated during slow-wave sleep than it is during NREM 2 and there is also less functional connectivity (Horovitz, 2008; Horovitz et al., 2009; Kaufmann et al., 2006; Larson-Prior et al., 2009; Sämann et al., 2011). Still other fMRI studies more specifically find the default network loses functional connectivity between its anterior and posterior regions during slow-wave sleep, as well as becoming more generally deactivated, which may help account for the isolated and fragmented thoughts that are sometimes reported from slow-wave sleep awakenings (Tagliazucchi et al., 2013; Tarun et al., 2021). These studies uniformly reveal the main factors in dreaming during NREM sleep are the level of activation and the degree of functional connectivity within the default network. In addition, alpha and/or theta waves are often present, as is the case during the sleep-onset process.

The likelihood that NREM 2 sleep can support dreaming is consistent with the findings in numerous earlier studies in sleep-dream labs, although NREM 2 awakenings lead to lower levels of dream recall than do REM awakenings. There are even lower levels of recall, or virtually no recall, after awakenings from slow-wave sleep, and studies that defined dreaming very broadly may have inflated these low numbers (see Nielsen, 2000, for a comprehensive review of the numerous studies of reports after awakenings in the first four decades of laboratory dream research).

The increase in NREM 2 sleep toward morning, at the expense of slow-wave sleep, leads to more dreaming and higher rates of recall. This conclusion was first supported in a study limited by a small sample size (Pivik & Foulkes, 1968). It was then supported in a study with a somewhat larger sample but with fewer controls (Fosse, Stickgold, & Hobson, 2004). A third, well-controlled study with 20 participants found that there were more dream reports from NREM 2 awakenings toward morning and that the reports became longer and more dreamlike (Wamsley et al., 2007, pp. 352–353). Still another study provided new evidence revealing higher activation levels during NREM 2 when dreams are reported (Scarpelli et al., 2017).

The increase in dream reports from NREM 2 as the night progresses gained further credibility in a lab study of spontaneous morning awakenings. In this study, spontaneous morning awakenings led to 144 dream reports from 20 women and 16 men. NREM 2 awakenings accounted for 73% of the spontaneous awakenings, which is similar to what was found in several all-night sleep studies not focused on dreaming (Cicogna, Natale, Occhionero, & Bosinelli, 1998, p. 463, table 1, and pp. 465–466, table 2). There were only small differences in the percentage of dreams recalled after each type of spontaneous awakening: 95.0% from REM awakenings, 90.7% from NREM 2 awakenings. Using a variety of rating scales, the researchers concluded there was only one small difference in the content in the two types of dream reports (Cicogna et al., 1998, p. 467).

There is also evidence that the content of dream reports after REM awakenings does not change throughout the course of the night, even though REM periods increase in length. For example, in a study of 14 participants, who were awakened after both the second and fourth REM periods on two consecutive nights, the dream reports became longer but the degree to which the dream reports were story-like did not change (Cipolli, Mazzetti, Palagini, & Feinberg, 2015). An earlier, methodologically oriented study, reported similar results in relation to dream content. The study was carried out in a remodeled suburban home near the University of Miami, which had been transformed into a sleep-dream laboratory. The study was based on 15 men of varying ages, each of whom slept in the bedroom lab for a month at a time. It first reported that REM dream reports were longer toward morning (Hall, 1966, p. 22). However, the 469 dream reports collected after REM awakenings did not change in substance throughout the night, whether participants had been awakened only once a night to collect reports ($n = 196$) or multiple times per night ($n = 273$) (Hall, 1966, p. 14, table 4, and pp. 22–23). Other findings from this large-scale study are reported at relevant places in later chapters.

Within the context of the neuroimaging findings on the activation of the default network during NREM 2 and the nature of its dream content, two studies of cerebral blood flow in different stages of sleep provide a separate line of evidence for the capacity to support dreaming during NREM 2 sleep. The first study, based on 11 participants, provided a baseline for studying activation levels in NREM sleep. It discovered that cerebral blood flow increased by 4% in secondary visual areas during REM, as compared to waking, and declined by 9% in the anterior prefrontal areas. All of the participants reported dreaming after REM awakenings (P. L. Madsen, Holm, et al., 1991). In a separate study of NREM 2 and slow-wave sleep, based on eight of the participants in the study of cerebral blood flow in REM sleep, there were no statistically significant changes in cerebral blood flow from waking to NREM 2 sleep. This result supports the conclusion that NREM 2 has the general activation capacity to support dreaming, although awakenings were not carried out in this study. On the other hand, cerebral blood flow declined by 25% during slow-wave sleep (P. L. Madsen, Schmidt, et al., 1991, p. 219), which is consistent with the fMRI findings on the relative deactivation of the default network during slow-wave sleep and its loss of functional connectivity (Tagliazucchi et al., 2013; Tarun et al., 2021).

Findings on auditory awakening thresholds supplement and confirm the fMRI and cerebral blood flow studies. An early study of auditory awakening thresholds, based on 319 auditory awakenings of seven participants over the space of four to six nights, found the awakening thresholds were "approximately equal" for REM and NREM 2, but that they were higher during slow-wave sleep (Rechtschaffen, Hauri, & Zeitlin, 1966). These results were replicated and extended in a similar study in the same sleep-dream lab. This study divided the 16 participants into light and deep sleepers on the basis of their auditory waking thresholds and also included awakenings to collect dream reports. The dream reports from REM and NREM 2 did not differ on ratings by independent judges, although some of the participants claimed they were thinking after some NREM 2 awakenings (Zimmerman, 1970, pp. 540, 544, table 1, and p. 547).

The finding that NREM 2 awakenings sometimes lead to reports of thinking, such as in a large-scale study by Kamiya (1961), is supported and refined in another study, which focused on the content of the reports provided after awakenings from NREM 2. The NREM 2 reports that could not be classified as dream reports seemed to be instances of mind-wandering. Decades before mind-wandering became an important topic in neurocognitive psychology due to the discovery of the default network, the coauthors presciently compared these reports to the "large portion of our waking thought which wanders in seemingly disorganized, drifting, non-directed fashion whenever we are not attending to external stimuli or actively working out a problem or a daydream," (Rechtschaffen et al., 1963, p. 411). The fact that there can be either mind-wandering or dreaming during NREM 2 sleep once again shows why it is necessary to have a precise definition of dreaming in order to develop a neurocognitive theory of dreaming within the context of the more general realm of spontaneous thought.

A Note on the Usefulness of High-Density EEG Studies in Studying NREM 2 Dreaming

The neuroimaging findings relating to dreaming during NREM 2 are one of the key issues on which neuroimaging results make it possible to move beyond the study of the brain during sleep by examining EEG patterns (Sämann et al., 2011; Stevner et al., 2019; Tagliazucchi et al., 2013; Tarun et al., 2021). As already noted above, EEG patterns leave the "underlying brain dynamics" very "unclear" in terms of the "spatiotemporal complexity of whole-brain networks" (Stevner et al., 2019, p. 1). This point is demonstrated by examining two high-density EEG studies that attempted to discover the EEG correlates of dreaming by only comparing the brain areas that are active during REM and NREM 2 sleep when dreaming was reported by participants after awakenings (Perogamvros et al., 2017; Siclari et al., 2017). The two studies conclude that the EEG patterns are similar during REM and NREM 2 when dreams are also reported.

These results led the researchers to emphasize the importance of posterior regions in dreaming. These regions are part of the default network, but the concept of a default network is not utilized in their theorizing. The two studies also found some evidence that medial prefrontal and temporal areas are sometimes active, especially during REM periods (Perogamvros et al., 2017, p. 1773; Siclari et al., 2017, pp. 873, 875). As with the posterior regions discussed in these studies, these anterior areas are also part of the default network. Thus, as one of the two studies forthrightly states, their work may be limited by the absence of neuroimaging data (Perogamvros et al., 2017, p. 1773). Nor did the two studies make use of the brain-lesion findings discussed above. Those lesion studies, like neuroimaging studies, show that dreaming also depends on anterior portions of the default network, not simply posterior portions. The EEG-based researchers therefore overlook the evidence that dreaming is based on a complex network, which has anterior and posterior regions in it (Solms, 1997, pp. 47–48). In addition, their results could not be replicated in a similar high-density EEG study, which was focused solely on the claims about NREM brain-wave patterns (Wong et al., 2020).

Due to the emphasis placed on replicated neuroimaging and lesion studies in developing the foundations for the neurocognitive theory of dreaming, the findings from the two high-density EEG studies can be set to one side. It has been evident for over a decade, as demonstrated throughout this chapter, that portions of the default network are the primary basis for the cognitive process of dreaming in REM periods, as well as during sleep onset and NREM 2 sleep (Domhoff, 2011). This new direction has been supported by one new neuroimaging study and two meta-analyses of earlier studies (Eichenlaub et al., 2014; Fox et al., 2013; Uitermarkt et al., 2020).

Dreaming during Waking

The discovery that reports of both dreaming and mind-wandering can occur after NREM 2 awakenings provides a segue to three EEG studies of brief episodes of dreaming during waking. This serendipitous discovery occurred in the context of a laboratory setting in which each of the 16 women participants spent 10 or more minutes alone in a room, and as long as 20–25 minutes in several instances. The purpose of this study was to provide participants with practice for focusing on what they were thinking at the moment they would be awakened in the night sleep-dream lab study in which they were enrolled. As part of the practice sessions, their brain waves were being recorded using the polysomnograph. The reports of dreaming after 24% of the random wake-state probes were not anticipated by the researchers (Foulkes & Scott, 1973). In addition to the reports of dreaming, various forms of waking thought were reported as well. Based on the EEG recordings, there were no instances of EEG patterns indicating that any of the participants had fallen asleep.

In a replication and extension study with 10 women and 10 men as participants, which included a more detailed analysis of the nature of all the post-probe verbal reports, dreaming was reported 19% of the time, mind-wandering or daydreaming 20% of the time, being lost in deep thought 22% of the time, and thinking about the current situation, in full reality contact, 38% of the time (Foulkes & Fleisher, 1975, pp. 68–69). Based on brain-wave and eye-movement patterns, "only 3 of the 120 arousals [i.e., experimental probes] showed anything other than a clearly waking polygraph tracing. No instances of either self-described or objective (EEG stage 1) sleep were observed" (Foulkes & Fleisher, 1975, p. 70). These results were later replicated again in a study that also included probes for dream reports during the sleep-onset process and REM periods (Foulkes, 1985, pp. 70–77, for a summary of the results from all three studies). The results from the first of the two replication studies are presented in figure 2.3.

These findings of brief episodes of dreaming during waking are given more credence by the waking neuroimaging study in which the anterior

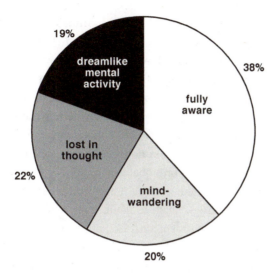

Figure 2.3
The four mental states reported by participants who were asked to "relax, but stay awake," with waking monitored by the EEG. From Foulkes & Fleisher, 1975.

regions of the default network become more connected to each other and more activated when posterior regions in the default network begin to decouple from the dorsal attention network (Chang et al., 2013, p. 230). Recall that the study by Chang et al. (2013) also found that there was an increase in alpha in anterior regions of the brain as decoupling occurred, and evidence of theta as well. In addition, as also noted earlier in the chapter, there have been several other studies suggesting that alpha and theta waves increase when the default network becomes more activated in lower states of attention during waking (see Domhoff & Fox, 2015, p. 348, for a summary). It is within this context that the brief episodes of dreaming during waking very likely occur, much as they do during the sleep-onset process. This hypothesis awaits future studies by means of neuroimaging studies by researchers who wish to find ways to distinguish the brain patterns that underlie dreaming and spontaneous waking thought, but with the added condition that the participants should have their eyes closed and be reclining on a bed (see Domhoff & Fox, 2015, for a discussion of several of the issues that would be involved in carrying out such a study).

The finding that different forms of thinking can occur during drifting waking thought received support in another laboratory study, in which 174 waking-state reports were collected (Reinsel, Antrobus, & Wollman, 1992, p. 165, table 3). The waking reports "were solicited under exactly the same conditions as those of sleep: the subject was lying quietly in a darkened room, wearing electrodes, and was asked to 'tell everything that was going through your mind before I called you'" (Reinsel et al., 1992, p. 160). The transcribed results were then compared to a set of REM dream reports collected in earlier studies from other participants. There were more abrupt topic changes and sudden scene shifts in the waking sample than in the REM dream reports. In addition, there were as many "improbable combinations" (e.g., unusual juxtapositions of objects) in waking thought as in REM dreams, but there were more "improbable identities," such as malformed or blended characters, in the REM dream reports (Reinsel et al., 1992, pp. 169–170, 173).

It therefore seems plausible that there can be brief dreaming or sudden unusual shifts in thoughts during relaxed wakefulness, just as there can be dreaming or mind-wandering during NREM 2. This possibility is further supported by a field study in which 29 college students (16 women, 13 men) were signaled randomly over a period of seven days by means of a pager. Along with evidence that 33% of the participants' 1,425 thoughts were spontaneous, in the sense that their thoughts were not being directed at the moment they were contacted, the study also concluded that 16% of the thought samples had a "trace" of dreamlike thought and another 9% had "more than a trace" of dreamlike thought (Klinger & Cox, 1987–1988, p. 124). Sudden shifts in waking thought and dreaming also have been found by comparisons of discontinuities in both mental states (Horton, 2017).

Based on these varied findings, it would be necessary to include waking samples of drifting waking thought as a control condition in studies of dream content that focus on scene shifts and unusual elements. Otherwise, it would be impossible to assess hypotheses concerning unusual elements and novel thought constructions, which are traditionally assumed to be unique to dreaming.

As in the case of the sleep-onset process, the findings on brief episodes of dreaming, which occur when participants are left alone in a controlled laboratory setting, demonstrate that dreaming is not an inherent part of sleep, even though this unique, but regularly occurring cognitive process happens most frequently during sleep.

Atypical States of Waking Consciousness and Dreaming

There are numerous atypical mental states, during both waking and sleep. Since these atypical states are often compared to dreaming, it is important to discuss neuroimaging findings that reveal their underlying neural substrates. This makes it possible to determine the degree to which these neural substrates are similar to the neural substrates that support dreaming. These findings can lead to a better understanding of dreaming and dream content by eliminating any misleading analogies, which may turn into sidetracks or create distortions. In effect, this section also provides another reason why clinical findings cannot be used in creating a foundational model of dreaming.

Atypical States of Waking Consciousness

Atypical states of waking consciousness are the result of cognitive defects, which may be temporary or permanent. They are supported by unusual neural patterns. To begin with, neuroimaging studies reveal that hallucinating patients have heightened activation in both primary and secondary sensory regions, as well as in premotor and cerebellar areas (E. Allen, Laroi, McGuire, & Aleman, 2008; Ford et al., 2009; Hoffman & Hampson, 2012; Waters et al., 2016; Zmigrod, Garrison, Carr, & Simons, 2016, for a meta-analysis of several studies). The activation patterns that underlie hallucinations are atypical compared to the normal network patterns during waking and highly atypical when compared to the neural substrate that subserves dreaming.

In addition, first-person testimony supports the idea that hallucinations differ from dreaming. An interview study of seven "spontaneously hallucinating patients" on a psychiatric ward, three women and four men, found they "easily differentiate their sleep and dream life from their waking life and hallucinating states." On the other hand, they cannot "effectively differentiate" their hallucinations "from their normal waking perceptions" (Kass, Preiser, & Jenkins, 1970, p. 496). Moreover, clinical studies by neurologists conclude that hallucinations are different from dreaming and that visual hallucinations can occur in otherwise normally functioning individuals (Sacks, 2013, pp. xiii, 26–27, 80, 209, 214). It is therefore not accurate to refer to dreaming as a type of hallucinating. This age-old assumption wrongly medicalizes a generally benign cognitive process that occurs regularly and frequently in the many hundreds of adults who have been studied in sleep-dream labs in numerous different countries.

Drug-altered atypical states, such as those induced by psilocybin and related compounds, suppress and compromise the integrity of the default network, which in many ways makes them the opposite of dreaming, even though they contain heightened sensory imagery (dos Santos, Osório, Crippa, & Hallak, 2016; Fox, Girn, Parro, & Christoff, 2018; M. Madsen et al., 2021). Similarly, studies of meditation states show they are characterized by the relative deactivation of the default network, so they are not at all like dreaming (Fox et al., 2016; Garrison, Zeffiro, Scheinost, Constable, & Brewer, 2015). In all cases of "substance use disorder," a diagnosis which includes a wide range of substances that lead to addictions, the "prominence" of the default network is "transitorily decreased during the intoxication phases." However, it is highly engaged during the "withdrawal and preoccupation phases," along with heightened levels of vigilance on the part of the salience/ventral network (R. Zhang & Volkow, 2019). This neural pattern is very different from the neural substrates that enable dreaming.

As for the hypnotic state, several neuroimaging studies demonstrate its underlying brain patterns reveal an intensely interpersonal state (Landry, Lifshitz, & Raz, 2017). A summary of several studies concludes that studies of hypnosis find "the opposite pattern of brain engagement" than is found during dreaming. This finding leads to the conclusion that the hypnotic state is "deliberate and directed" and dreaming is "spontaneous and involuntarily elicited" (Abraham, 2016, p. 4206). Once again, even though hypnotic states were used to allegedly induce dreams in the past, they are yet another atypical waking state that is different from dreaming.

Taken together, the findings overviewed in this subsection suggest that atypical waking states differ from each other and that dreaming differs from all of them. Studying atypical states of waking consciousness, or making comparisons of them with dreaming, therefore is not useful in developing a neurocognitive theory of dreaming and is in fact misleading.

Atypical States of Dreaming

There are two very different types of atypical dreaming—one of which is frightening and upsetting, the other of which is benign. The first of the two, nightmares, which are defined as frightening dreams that awaken people, may be part of the physical illnesses that everyone experiences or due to temporary chemical imbalances caused by going on or off medications. (To provide an extreme example, Levodopa, a precursor of dopamine that is used to help Parkinson's patients, may cause very vivid, nightmarish dreams until the dosage level is set at a tolerable level [Hartmann, Russ, Oldfield, Falke, & Skoff, 1980].) There also may be a few people who are lifelong nightmare sufferers but score low on tests for stress, anxiety, and hostility and do not express being greatly upset by their nightmares (Hartmann, 1984).

However, adults who begin to suffer upsetting nightmares frequently, and over long periods of time, usually have experienced extreme stress or trauma. Many of them have posttraumatic stress disorder (PTSD). PTSD victims have been studied in the greatest depth with regard to nightmares, in part because a large number of active and retired military personnel, both women and men, develop PTSD (Shapiro & Notowitz, 1988/2016, for a poignant and moving 25-minute video presentation of interviews with four Vietnam veterans in the PTSD unit of the Veterans Administration Hospital in Menlo Park, California).

Based on EEG and neuroimaging studies, those who suffer from frequent nightmares have brain activation patterns during sleep similar to those that support stress and vigilance in waking life (Ebdlahad et al., 2013; Germain, Jeffrey, Salvatore, Herringa, & Mammen, 2013; Mysliwiec et al., 2014; Simor, Horváth, Ujma, Gombos, & Bodizs, 2013). Another study compared 18 participants, who frequently experienced nightmares but had not been diagnosed with PTSD, with 15 control participants. The frequencies were higher for those who experienced nightmares in some EEG bands during waking, REM, NREM 2, and slow-wave sleep. These differences were most clear in frontal and central areas of the brain (Marquis, Paquette, Blanchette-Carrière, Dumel, & Nielsen, 2017). These findings are supported and supplemented by a neuroimaging study that compared 50 active-duty United States soldiers suffering from PTSD with 52 members of the military who were not suffering from PTSD. This study showed that the salience/ventral network remains highly activated at all times, waking or asleep, in PTSD patients (Abdallah et al., 2019). The Abdallah et al. (2019) study is the second of two studies of a psychiatric patient group used in this book. As discussed in chapter 8, it may provide a starting point for the creation of a translational model relating to PTSD nightmares.

The rare reports of self-awareness during dreaming, which is the second type of atypical dreaming, were originally thought to be potentially useful for research purposes. They held out the "promise of an entirely new type of evidence about dreaming: dream reports obtained directly from the dream state, while the subject is still asleep" (Windt, 2015, p. 105). This possibility was first brought to the attention of American dream researchers in an EEG-based study, which used preplanned eye-movement patterns and fist clenches by the participants to signal they were aware they were dreaming. The main participant was at the same time the investigator in the study. He accounted for 14 of the 17 verified signals in the original study. The other three participants each signaled one self-awareness dream in the course of four lab nights (LaBerge, 1980, p. 82). In a slight extension of the original study, which added a few more nights of study, a participant who was employed to help with the project had one verified signal. Another new participant had two verified signals in two nights. Overall, this meant that the experimenter-participant reported 14 of the 20 verified signals in the two parts of the study, which greatly limited the degree to which the results could be generalized. The other six came from the other five participants (LaBerge, Nagel, Dement, & Zarcone, 1981, p. 728, table 1).

The findings in the initial pair of studies were soon followed by two EEG studies that attempted to replicate the original findings. Both found higher percentages of alpha waves during the few instances of REM dreaming that led to post-awakening reports of self-awareness (Ogilvie, Hunt, Tyson, Lucescu, & Jeakins, 1982; Tyson, Ogilvie, & Hunt, 1984). The presence of alpha, as noted above and as shown in table 2.3, is most often found in relaxed waking states and in the early stages of the sleep-onset process, which raised the possibility that at least some of the participants were not asleep.

The early studies of the potential usefulness of this atypical state also seemed to indicate the after-the-fact reports of self-awareness dreams had "a surreal, dreamlike quality," but a thorough review of the literature concluded that "a majority of lucid dreams, it seems, feel very much like standard wakefulness and nonlucid dreams" (Windt, 2015, p. 118). This conclusion fits well with a quantitative comparison of the content of self-aware and non-self-aware dream reports. This content analysis was based on a total of 441 self-aware and non-self-aware dream reports, which the researcher collected in ten different samples from college students and older adults over the space of several years. The comparison found relatively few differences, especially on social interactions. There were more auditory and kinesthetic elements in self-aware dreams and fewer characters and less happiness in the non-self-aware dreams, but overall the two samples *"are more alike than they are different"* (Gackenbach, 1988, pp. 192–193; emphasis in the original). (Although self-aware dreams are often called "lucid dreams," as seen in the quotation earlier in this paragraph, this phrase is not used here because the word "lucidity" frequently has a sense of superiority or insightfulness attached to it.)

There are two small studies of the neural substrate that supports self-awareness during dreaming, which attempt to replicate the initial EEG findings with more recent technologies. Combining the results from the two studies, there were five possible instances of signaling by means of eye movements during REM periods in 5 out of 10 participants, who spent a total of 21 nights in a sleep-dream lab (Dresler et al., 2012, pp. 1018, 1020; Voss, Holzmann, Tuin, & Hobson, 2009, pp. 1191–1192, 1195). In the earliest of the two studies, which used high-density EEG, only three of the six highly motivated college-student participants, who had indicated they remembered three or more self-aware dreams each night at home, reported one instance each of self-awareness during dreaming, through eye-movement signals during a REM period. Their brain patterns were characterized by "wake-like inter-scalp networking, including high-frequency bands," which was labeled as a "hybrid state," and said to be "most pronounced in frontal and frontolateral coherences" (Voss et al., 2009, p. 1196). This finding supports the concerns about wakefulness based on the two EEG studies that attempted to replicate the original findings (Ogilvie, Hunt, Tyson, Lucescu, & Jeakins, 1982; Tyson, Ogilvie, & Hunt, 1984).

The second new study was based on fMRI monitoring of four adult male participants between the ages of 27 and 31, who had actively practiced greater self-awareness during dreaming for several years or more for their own reasons. They were studied over a total of 15 total nights. Despite the numerous REM periods that occurred over the space of 15 nights, only two of the four men were able to provide eye-movement signals (one instance each), both of which were corroborated by post-awakening reports. As in the case of the high-density EEG study, this neuroimaging study found "a reactivation of several areas normally deactivated during REM sleep" (Dresler et al., 2012, p. 1020). These areas are primarily in the frontoparietal control network, including the frontal pole and the dorsolateral prefrontal cortex, which are relatively deactivated in the several studies of REM sleep discussed earlier in this chapter. Thus, the few participants who reported any instances in the course of the two studies were awake or in an atypical sleep state.

When all of the studies over a 40-year period, using either the EEG or fMRI, are taken into consideration, there is good evidence that the reports of preawakening self-awareness in sleep labs do not occur when only the circumscribed neural substrates that support dreaming are activated. Instead, parts of the brain that are activated during waking consciousness are also activated during this self-awareness state. There is also considerable evidence that most people cannot be taught how to experience self-awareness during dreaming (Mota-Rolim, Pavlou, Nascimento, Fontenele-Araujo, & Ribeiro, 2019). This atypical mental state therefore may be limited to those who in some way have atypical neural substrates and association networks. This hypothesis is consistent with the findings in a waking-state fMRI study of 14 participants who were frequent self-awareness dreamers (as defined by three or more such dreams each week). When they were compared with a matched sample of participants, who claimed they had one or less self-awareness dreaming experiences each year, those who reported they frequently experienced a self-aware dream had "increased functional connectivity between anterior areas in the prefrontal cortex and temporoparietal association areas, regions normally deactivated during sleep" (Baird, Castelnovo, Gosseries, & Tononi, 2018, p. 1).

Due to the strong likelihood that the neural substrate that supports self-awareness reports includes regions in the prefrontal cortex that are not active during dreaming, this rare self-awareness may be best characterized as a "hybrid" state (Voss et al., 2009, p. 1196) or as an "intermediate" state between mind-wandering and dreaming (Windt, 2015, pp. 441–442). As a hybrid state it is not useful in developing a foundational neurocognitive theory of dreaming, but it might provide the basis for a translational model that would be helpful in understanding waking consciousness.

The "How" of Dreams: Dreaming as a Form of Embodied Simulation

The path to understanding simulation as an important type of thinking began with pioneering studies by experimental cognitive psychologists studying categories and concepts. This work showed that categories are built around natural prototypes (Barsalou, 1982, 1991; Rosch, 1973; Rosch

& Mervis, 1975; Rosch, Mervis, Gray, Johnson, & Boyes-Braem, 1976). Natural categories do not have a defining property; instead, it is the category members themselves that have properties. Categories are therefore based on family resemblances.

Shortly thereafter, studies of simulation led to the hypothesis that categorizing involves the simulation of perception and action, which brought simulation to the fore in studies of waking thought (Schacter et al., 2008, for a detailed historical account). This conception of simulation was introduced into cognitively oriented dream research in the mid-1980s through the claim that dreams are "organized mental acts," which are "generally plausible simulations of the texture of waking life experience"; more specifically, dreams are an "imaginal life simulation" (Foulkes, 1985, pp. 32, 176). This view was subsequently incorporated into an early version of the neurocognitive theory of dreaming (Domhoff, 2003, p. 31).

By the early 2000s, a pioneer researcher on concepts in waking life argued that "conceptual processing uses reenactments of sensory-motor states—simulations—to represent categories" (Barsalou, 2003, p. 521). The concept of simulation also was expanded to include the possibility that at least some simulations are "embodied"; this concept begins with the general notion that "cognitive processes are deeply rooted in the body's interactions with the world" (Margaret Wilson, 2002, p. 625). The concept of embodied simulation soon had six different meanings. For example, the half-dozen meanings include the idea that cognition is shaped by the context ("situated") and time-constrained by the pressures of the situation (Margaret Wilson, 2002, pp. 625, 626). Most importantly for the development of a neurocognitive theory of dreaming, embodiment also has a strictly psychological meaning. It is a form of simulation in which "sensory and motor resources are brought to bear on mental tasks whose referents are distant in time and space or are altogether imaginary" (Margaret Wilson, 2002, p. 635). This fully psychological conceptualization of embodied simulation includes mental imagery, imagination, episodic memory, reasoning, and problem-solving.

The idea of embodied simulation as a strictly psychological concept received strong support from a compelling analysis of studies of visual imagery. It employed a combination of carefully designed psychological experiments, lesion studies, and PET-scan studies (Kosslyn, 1994). The analysis claimed to provide a resolution to the imagery debate with artificial intelligence researchers and their allies within psychology. These computationally oriented theorists insisted that alleged mental "imagery" of all kinds relied upon propositional algorithms. At about the same time, an edited book on auditory imagery made a strong case for its reality as a form of mental imagery (Reisberg, 1992). Twelve years after the 407-page book on visual imagery appeared, the case for visual imagery became even stronger on the basis of 58 neuroimaging studies, nearly four times as many as were available earlier (Kosslyn, Thompson, & Ganis, 2006, pp. 186–194, table A-1, for the full list of neuroimaging studies relating to visual imagery up to that point).

During the next decade, neuroimaging evidence for auditory, olfactory, gustatory, and tactile imagery began to appear (Moulton & Kosslyn, 2011). In an fMRI study of auditory imagery, for example, the researchers concluded that "Auditory imagery of the same sounds evokes similar overall activity in auditory cortex as perception" (Linke & Cusack, 2015, p. 1322). A unique neuroimaging study took advantage of implanted electrodes in an epilepsy patient, who also was an excellent keyboard musician. It demonstrated the reality of auditory imagery with greater precision than was possible in the past (see Martin et al., 2018, for this study as well as an overview of past studies).

Still other researchers found evidence that embodied simulations are distinctive because they are subjectively experienced as the body in action (Calvo-Merino, Jones, Haggard, & Bettina, 2017; Gibbs, 2006; Landau, Meier, & Keefer, 2010; Niedenthal, Winkielman, Mondillon, & Vermeulen, 2009). In the case of the tool-use network discovered in human brains, this network, including its sensorimotor areas, is activated merely by viewing tools. This is true even if the person has no intension to act (MacDonald & Culham, 2015). In the case of recently learned procedures for tying a knot, the researchers were "able to reliably identify when participants were planning to tie knots before they physically tied them." This finding demonstrates that the brain regions involved during the learning of the task provide a "neural representation" of what the participant intended to do (R. Mason & Just, 2020, p. 729). Based on over 30 years of research, the case for embodied sensory imagery was well established by 2020.

The research on embodied simulation took on added importance for the development of a neurocognitive theory of dreaming when fMRI studies demonstrated that mental imagery and imagination were both supported in good part by neural substrates in the default network (Addis, Pan, Vu,

Laiser, & Schacter, 2009; Addis, Wong, & Schacter, 2007; Andrews-Hanna, Reidler, Huang, et al., 2010; Andrews-Hanna, Reidler, Sepulcre, et al., 2010; Buckner et al., 2008; Christoff et al., 2009; Domhoff, 2011, for a summary of these studies and related work in relation to the neurocognitive theory of dreaming). Dreaming therefore can be understood as a form of simulation, defined as "imaginative constructions of hypothetical events or scenarios" (Buckner et al., 2008, p. 20). It also can be understood as "a particular kind or subset of thinking, which involves imaginatively placing oneself in a hypothetical scenario and exploring possible outcomes" (Schacter et al., 2008, p. 42). In addition, dreaming is a form of "episodic simulation," a cognitive process that allows people "to mentally 'try out' different ways upcoming events might play out without engaging in actual behaviors" (Spreng et al., 2018, p. 23). These definitions of simulation inadvertently read as if they were written with dreaming in mind.

Due to the sense of subjectively experiencing the body in motion during dreaming, the neurocognitive theory of dreaming is premised on the idea that dreaming is an especially dramatic version of embodied simulation. According to this view, dreaming is a form of embodied simulation that leads dreamers to experience themselves and their bodies as being involved in hypothetical scenarios that include a vivid sensory environment. This vivid sensory environment often involves interpersonal interactions. In terms of a neurocognitive theory of dreaming, the defining feature of dreaming is therefore the sense of being a participant in—or more rarely, an observer of—an ongoing event. This event is experienced as "real" while it is happening. It often lasts for as long as 15–30 minutes and is sometimes temporarily remembered upon awakening as an actual experience, even while realizing that it didn't "really" happen.

From the perspective of the neurocognitive theory of dreaming, this sense of being part of a highly engaging event is the key factor in distinguishing dreaming from other forms of thinking during sleep. These other forms of "sleep mentation" are usually thought-like or fragmented, or actually occur in brief micro-awakenings, as briefly noted in chapter 1. They also may be instances of mind-wandering during NREM 2 (Rechtschaffen et al., 1963). Based on these findings, it is no longer useful to assume that all "sleep-mentation" reports after awakenings are a form of dreaming. Nor is it useful to include instances of mind-wandering during NREM 2, which is not dreaming.

The Importance of Cognitive Factors in Dream Recall

Despite the considerable amount of dreaming that can occur between the early stages of the sleep-onset process and the moment of a spontaneous morning awakening, most people forget almost all of their dreams. This lack of recall is primarily due to cognitive factors. The influence of personality factors seems to be secondary, although they may help explain why some people remember a few more of their dreams than do others.

The fact that people do not recall 95–99% of their dreams can be explained first in terms of the importance of memory consolidation in retaining memories of recent experiences (i.e., the process of conserving a memory once an event is experienced) (McGaugh, 2000). The importance of memory consolidation is first demonstrated in the laboratory finding that the "percentage of recall drops dramatically if the awakening is delayed for even a few minutes after the end of REM sleep." "Even a gradual awakening" from a REM period reduces recall "substantially, particularly among people who rarely recall dreaming in the morning under everyday conditions" (Goodenough, 1991, p. 144).

Other long-standing findings in studies of waking memory also contribute to the understanding of dream recall. The four most important factors are recency, length, salience (defined primarily in terms of novelty and the intensity of an experience), and motivation to remember, which is usually indexed by a person's degree of interest in dreams (Beaulieu-Prevost & Zadra, 2005a, 2007; D. Cohen, 1974, 1979; Goodenough, 1991; Horton, 2009; Schredl & Göritz, 2017; Tonay, 1993). The importance of recency is best seen in the study of dream recall after spontaneous morning awakenings in the laboratory after a night of uninterrupted sleep. In this study, as noted above, dreams were recalled after 95.0% of the spontaneous REM awakenings and 90.7% of the spontaneous NREM 2 awakenings (Cicogna et al., 1998).

Then, too, cued recall during the day, which in the case of dreams means seeing a particular person, object, or setting that triggers recall, accounts for a small portion of dream recall as well. It also accounts for the even smaller percentage of unexpected recall during mind-wandering, which participants report to be the result of a connection to what they were thinking (Cipolli, Calasso, Maccolini, Pani, & Salzarulo, 1984; Domhoff, 1969; Strauch & Meier, 1996, pp. 59–60). On the other hand, numerous studies related to personality variables since the 1960s have found little or no relationship between dream recall and various personality variables (Blagrove, 2007; Blagrove & Akehurst, 2000; D. Cohen, 1979; D. Cohen & Wolfe, 1973; Domhoff & Gerson, 1967; Levin, Fireman, & Rackley, 2003; Tonay, 1993). Similarly, a meta-analysis related to absorption and psychological boundaries did not find any relationship to dream recall (Beaulieu-Prevost & Zadra, 2007). However, a high score on openness to new experiences, which is one of the dimensions on the Five Factor Personality Inventory, may help explain why some individuals recall a few more of their dreams than most people do (Schredl & Göritz, 2017).

Due to the secondary role of personality factors in understanding the small percentage of dream recall, it is likely that the importance of memory factors may lead to reasonably representative samples of nonlaboratory dream reports. This unexpected conclusion is plausible for two reasons. The fact that the narrative structure and content of REM dream reports does not change throughout the night, along with the greater similarity of NREM 2 dream reports to REM dream reports late in the sleep period, means that recency alone may contribute a very large portion of representative dream reports in spontaneous morning recall. More specifically, recency is very likely the explanation for as many as 70–75% of the dreams that are recalled (D. Cohen, 1974, 1979; Domhoff, 1969; C. Meier, Ruef, Ziegler, & Hall, 1968). Nor is it likely that the relatively mundane dream reports recalled because of externally cued recall or mind-wandering would lead to biased samples. Motivation also may contribute to representative samples because of the evidence, which will be presented in chapter 4, that a few people are motivated to recall as many dreams as possible for completely different reasons. However, the salience factor very likely introduces some degree of bias into nonlab samples of dream reports, as shown in studies in which samples of lab and nonlab dream reports from the same participants are compared (e.g., Domhoff & Schneider, 1999; Weisz & Foulkes, 1970). The differences in dream content between lab and nonlab samples are discussed early in chapter 3.

Cognitive Insufficiencies in Dreaming

Reasonably alert waking cognition, based on the interactions of the five association networks, provides a starting point for a consideration of "cognitive insufficiencies" during dreaming. In the past, aspects of the concept of cognitive insufficiencies have been called "deficits," such as in neurological studies of patients who suffer from inherited brain abnormalities or lesions caused by injuries or illnesses. Other aspects of the concept have been used in past studies of "temporary deficits" in executive functions in children, which are due to the gradual maturation of the frontal lobe and the gradual development of cognitive capacities (Welsh, Pennington, & Groisser, 1993). For purposes of a neurocognitive theory of dreaming, the concept of "cognitive insufficiencies" provides a more encompassing way to understand all types of long-term and temporary deficits.

Most of all, it can incorporate the major factor in creating cognitive insufficiencies during dreaming: the smaller and more limited neural substrate that subserves dreaming. In other words, four of the five association networks that support the complex nature of waking thought are relatively deactivated in all stages of sleep. It is this basic fact, discovered through neuroimaging studies, that may explain much of what is "different" about dreaming. Two straightforward examples can serve as quick examples here. First, the obvious fact that dreamers are not aware of their current environment (in bed, sleeping) is very likely explained by the gradual deactivation of the most anterior and lateral regions of the prefrontal cortex, such as the frontal pole and the dorsolateral prefrontal cortex, along with the deactivation of the dorsal attention network, during the sleep-onset process. Second, the rarity of waking episodic memories in dream content has often been noted (e.g., Baylor & Cavallero, 2001; Fosse, Hobson, & Stickgold, 2003; Malinowski & Horton, 2014). This rarity may be due to the deactivation during sleep of several regions in the subsystem that supports this form of memory (e.g., Cooper & Simons, 2019, for a summary of past studies of this subsystem during waking, and S. Zhang & Li, 2012).

The concept of cognitive insufficiency leads to several "predictions" about the limits that may be imposed on the complexity of dreaming and the nature of dream content by one or another insufficiency. These predictions can be "tested," in a loose sense of that term, by examining some of the unexpected and anomalous results of the 60+ years of systematic studies of dreaming and dream content, inside and outside the sleep-dream laboratory. Such possibilities are examined in several of the following chapters. Some of the less coherent and incongruous elements sometimes found in dream reports, which are discussed at various points in later chapters, also may be due to cognitive insufficiencies.

Under What Conditions Does Dreaming Occur?

Based on studies of dreaming during REM, NREM 2, the sleep-onset process, and brief episodes of dreaming during long periods of drifting waking thought, it can be concluded that dreaming is a form of spontaneous and undirected thinking, which occurs in adults if and when the following four conditions are met:

First, the neural substrate that subserves dreaming has to be intact. This qualification allows for the demonstrated impact of lesions on dreaming (Bischof & Basset, 2004; Jus, Jus, Villeneuve, et al., 1973; Kerr et al., 1978; Poza & Marti-Masso, 2006; Solms, 1997).

Second, there has to be an adequate level of cortical activation, which is provided by thalamic and extra-thalamic subcortical ascending pathways, as well as by hypocretin neurons in the hypothalamus (Edlow et al., 2012; B. E. Jones, 2020; McGinty & Szymusiak, 2017; J. Siegel, 2017a).

Third, there has to be an occlusion of external stimuli, which is provided by gating mechanisms in the thalamus (Chow, Horovitz, Picchioni, Balkin, & Braun, 2013; Picchioni et al., 2014; Tarun et al., 2021).

Finally, there has to be a loss of conscious self-control. This step-by-step process involves the gradual deactivation of the frontoparietal control network, the dorsal attention network, and the salience/ventral network (De Gennaro et al., 2004; Domhoff, 2018a, pp. 194–196; Marzano, Moroni, Gorgoni, Nobili, & De Gennaro, 2013; Sämann et al., 2011; Stevner et al., 2019). The fifth and sixth conditions necessary for dreaming, which concern the development of dreaming in children, are added at the end of chapter 6. As shown in that chapter, the default network is not mature enough to support adultlike dreaming in all its aspects until ages 9–13, and most of the cognitive processes that are necessary to produce dreams develop only gradually between the ages of 5 and 11.

Conclusions and Implications

There is a relatively well-defined neural substrate that subserves dreaming. It is centered in portions of the default network. The early neuroimaging studies were focused on REM sleep, but later studies showed this neural substrate is active and functionally connected to varying degrees during the sleep-onset process and NREM 2 sleep. The neuroimaging studies are supported by the lesion studies and electrical brain stimulation studies, which provide convergent validity for the foundational starting point for the neurocognitive theory of dreaming. The fact that dreaming can occur during sleep onset and for brief periods during waking in a controlled setting, demonstrates dreaming is a cognitive process, independent of any stage of sleep, even though it occurs most frequently during sleep, and in particular REM sleep.

In addition, a series of step-by-step advances in cognitive psychology over the past 70 years has led to the conclusion that dreaming is an intensified version of mind-wandering and daydreaming, based on embodied simulation. Dreaming is different from mind-wandering or daydreaming because the frontoparietal, dorsal attention, and salience/ventral networks are relatively deactivated while dreaming is occurring. At the same time, the differences between waking imagination and dreaming suggest the truncated nature of the neural substrate that enables dreaming may lead to cognitive insufficiencies during dreaming. These cognitive insufficiencies may explain several of the differences between dreaming and waking thought. Still other studies suggest several specific conditions have to be present for dreaming to occur. Finally, dreaming differs from all other forms of mental activity during sleep and from the various atypical states of waking thought. In developing a foundational model of dreaming, the dreaming state therefore cannot be usefully compared to the various types of atypical waking thought, nor with any of the other forms of mental activity during sleep.

3 Dream Content as Revealed by Quantitative Content Analysis

Based on large samples of dream reports provided by college students and older adults, this chapter provides a detailed account of the main findings on dream content. For the most part these samples were collected in either exclusively laboratory or nonlaboratory settings, but six of the studies involved comparisons of samples collected from the same participants inside and outside the laboratory setting. Those six studies demonstrate that dream reports differ relatively little in content with regard to the context within which they were collected. (Aggression in dream reports is an exception.)

The chapter demonstrates that five categories of embodied simulations can be detected in the overall findings from a normative quantitative study of dream content. It also presents evidence that the dream reports in some of these categories differ in distinctive ways from an overall normative sample. Building on these classifications, the chapter then discusses the percentage of dream reports that include familiar settings and characters, and the frequencies of various thought processes and novel elements in dream reports. A section on age, cross-national, and cross-cultural studies of adult dream reports reveals age, nationality, and cross-cultural differences are relatively minor, with the exception of variations in aggressive elements. In addition, an analysis of four relatively large gender differences in dream content provides evidence there is continuity between the personal concerns that appear most frequently in dream reports and the degree to which those same personal concerns preoccupy the dreamers in waking life.

One important general conclusion drawn from these findings is that most dream reports are focused on personal and interpersonal issues. Perhaps for this reason, the dream reports of women, men, young adults, older adults, and people in different nation-states and indigenous societies are more similar than they are different in several content categories. Finally, these various studies demonstrate dream reports are in most instances a reasonable simulation of the waking world.

Before focusing on what has been found in dream reports, however, it is first important to discuss the methodological and statistical strategies on which the findings are based. Just as studies of the neural substrate that supports dreaming were made possible by new neuroimaging technologies, so too, the advent of powerful desktop computers has made quantitative studies of large samples of dream reports far faster and easier to analyze with new search algorithms. It is also now possible to use approximate randomization strategies to examine or bypass the many assumptions that underlie established inferential statistical tests.

The benchmarks set by the large sample sizes, along with the methodological controls built into these replicated studies, provide baselines for assessing the usefulness of many smaller and unreplicated studies in later chapters.

Methodological and Statistical Issues in the Study of Dream Content

The findings presented in this and several of the following chapters are based on quantitative content analysis, a methodology that is widely used in the social sciences. It has made systematic studies possible for such widely differing texts as personal diaries, speeches, newspaper articles, novels, poetry, personal letters between correspondents, and psychotherapy sessions (e.g., Gerbner, 1969, for the classic compendium; Krippendorff, 2004; Charles Smith, 2000). Past experience demonstrates that no one general coding system is applicable to these many different kinds of texts. Novels are one thing, diaries are another, and dream reports are still another.

In fact, even though dream reports are similar to texts based on waking-state thinking in that they have a narrative-like structure, they are unique in several ways, starting with the fact that they are neither "instrumental" nor "representational." In other words, they are not meant as communications to one or more individuals or as allegorical representations. Further, most of the dream reports used in research studies would not have been reported if dream researchers or anthropologists had not asked for them. Then, too, dream reports are unique in that they are verbal or written reports of a form of thinking that occurred in an atypical neurocognitive state. The uniqueness of dream reports is also demonstrated by the repeated lack of success in finding anything new or useful by applying one or another coding system for waking texts to the study of dream reports. What is dreamed can only be known to others through such reports. Nor is there any independent way to assess the completeness and accuracy of dream reports. Dream content therefore has to be studied with systems of content analysis developed on the basis of detailed examinations of large samples of dream reports.

This conclusion is supported by an exhaustive compilation of the waking content methods used in the first 25 years of scientific dream research, most of which did not lead to replicable results and were never used again (Winget & Kramer, 1979). Nor did a comparison of 5,208 dream reports with many types of waking texts (from the Linguistic Inquiry and Word Count analytical system) yield any useful findings. It concluded that dream reports are somewhat similar to novels and expressive writing, and have higher frequencies for first-person singular terms, along with more terms for motion, space, and home (Bulkeley & Graves, 2018, pp. 43, 52–53). Similarly, a comparison of the Linguistic Inquiry and Word Count methodology with the system of quantitative analysis discussed in this chapter found that it can explain only 25.1% of the variance found with the system developed for studying dream reports (Zheng & Schweickert, 2021, p. 219).

The methodology of content analysis used in the study of dream content is best described as "the categorization of units of qualitative material in order to obtain frequencies, which can be subjected to statistical operations and tests of significance" (Hall, 1969a, p. 175). It includes four basic steps: (1) creating clearly defined categories that can be understood and applied in a reliable way by researchers who were not part of the effort to create them; (2) tabulating frequencies; (3) using percentages, ratios, or other statistics to transform raw frequencies into meaningful data; and (4) making comparisons with control groups or normative samples (Hall, 1969a, 1969b; Hall & Van de Castle, 1966; Van de Castle, 1969).

Ordinal-level rating scales for studying dream content, usually resembling 4- to 5-point Likert scales, have produced several useful findings, some of which are discussed in this and later chapters. However, they also have limitations in the study of dream content, as discussed in a summary fashion in chapter 8, and more fully elsewhere (Domhoff, 2003, pp. 58–60). Most of the findings in this and several subsequent chapters are therefore based on a detailed and comprehensive coding system that rests on the categorical (nominal) level of measurement (Hall & Van de Castle, 1966, pp. 144–149). The 10 categories in the Hall/Van de Castle (HVdC) coding system are divided into two or more subcategories. They cover every type of element that appears in dream reports, including settings, objects, characters, activities, social interactions, good fortunes, misfortunes, successes, failures, and descriptive elements. There are also four categories for coding emotions, along with five categories for food and eating elements, and seven for elements from the past. The HVdC coding system has been used by investigators in several dozen studies in at least 11countries, including Canada, Germany, Greece, India, Iran, Italy, Japan, Switzerland, Spain, the Netherlands, and the United States.

The most frequently used categories concern settings, characters, social interactions, and misfortunes. In the case of the social interaction categories, they include codings for who was the initiator and who was the recipient in the social interactions, along with codings for mutual interactions and reciprocated interactions. The codings of the dream reports are best analyzed using percentages and ratios ("content indicators") to correct for differences in the length of dream reports. Percentages and ratios also correct for the varying densities of characters and social interactions in different samples of dream reports. Two different studies—one using the HVdC male norms, the other using the HVdC female norms—confirm that the content indicators are successful in eliminating any biases created by report lengths that vary from 50 to 300 words by comparing dream reports with 50 to 175 words to dream reports with 176 to 300 words. Similarly, a large sample of dream reports from one person, which ranged in length from two to 800 words, found that the content indicators are able to correct for the longer length of dream reports, which in general have more of everything. But dream reports with less than 50 words are not useful for quantitative studies (see Domhoff, 2003, pp. 80–84, for both of these studies). The corrections for length based on percentages and ratios are important because there is evidence that word frequencies may not be adequate to control for differences in report length (Foulkes & Schmidt, 1983, p. 274; Hall, 1969a, pp. 151–152). In addition, reports of findings based on a metric of "characters per 100 words" or "aggressions per line of text" are often difficult to understand and do not readily relate to information presented in more conventional ways.

To repeat a frequently used example of a content indicator, the "Animal Percent" is simply the total number of animals found in a sample of dream reports, divided by the total number of all types of characters. The use of the Animal Percent makes it possible to compare samples of dream reports from groups of all ages from all over the world. This is because the Animal Percent is independent of report length, character density, or sample size. In addition, findings presented as percentages are readily communicated and comprehended. People immediately understand what is meant if it is reported that the Animal Percent in dream reports declines from 10–20% in childhood to 4–6% in adulthood. It also "makes sense" to people if they are told the Animal Percent is higher in hunting and gathering societies than it is in industrialized nations (Domhoff, 1996, pp. 89–94, 117–126; Van de Castle, 1983).

In a parallel fashion, the total number of friendly interactions in a sample of dream reports, when divided by the total number of characters, provides a ratio called the "Friendliness per Character Index," or "F/C Index." This ratio controls for report length and the varying density of characters in dream samples. Furthermore, the F/C Index can be calculated for specific characters or types of characters in dream reports, leading to the possibilities of an F/C Index with, for example, parents, friends, children, or strangers. A list of the content indicators most frequently used in this book, and explanations for how they are calculated, is presented in table 3.1. The definitions of these indicators are also mentioned again the first time they appear in the text.

In addition, the categories can be used to make fine-grained analyses of one or another form of social interaction. For example, the HVdC categories for aggression (defined as hostile thoughts toward another character or deliberate, intentional acts by one character to annoy or harm some other character) demonstrate this point. Each aggression in a dream report is tabulated into one of eight categories: (1) covert feelings of hostility; (2) verbal criticism; (3) rejection or coercion; (4) verbal threat of harm; (5) theft or destruction of a person's possessions; (6) chasing, capturing, or confining; (7) attempts to do physical harm; (8) murder. The frequencies for each category can be compared with normative findings with large samples. Categories 1 through 4, the nonphysical types of aggression, can be summed, as can categories 5 through 8, the physical aggressions. These two figures lead to a content indicator called the Physical Aggression Percent (all physical

Table 3.1

The formulas for calculating the Hall/Van de Castle content indicators

Characters	
Male/Female %	Males ÷ (Males + Females)
Familiarity %	Familiar ÷ (Familiar + Unfamiliar)
Friends %	Friends ÷ All humans
Family %	(Family + Relatives) ÷ All humans
Animal %	Animals ÷ All characters

Social interactions	
Aggression/Friendliness %	Dreamer-involved aggression ÷ (D-inv. aggression + D-inv. friendliness)
Befriender %	Dreamer as Befriender ÷ (D as Befriender + D as Befriended)
Aggressor %	Dreamer as Aggressor ÷ (D as Aggressor + D as Victim)
Physical Aggression %	Physical aggressions ÷ All aggressions

Social interaction indexes	
Aggression/Character	All aggressions ÷ All characters
Friendliness/Character	All friendliness ÷ All characters

Percentage of dreams with at least one:	
Aggression	Dreams with aggression ÷ Number of dreams
Friendliness	Dreams with friendliness ÷ Number of dreams
Sexuality	Dreams with sexuality ÷ Number of dreams
Misfortune	Dreams with misfortune ÷ Number of dreams

Adapted from Domhoff, 2003.

aggressions divided by all physical aggressions + all nonphysical aggressions). This indicator reveals age, gender, and cross-cultural similarities and differences, as discussed later in this chapter. Finally, all categories of aggressions can be totaled for an overall aggression score, which can be expressed by an Aggression/Character Index (A/C Index). It also can be expressed as the percentage of the dream reports in the sample with "at least one aggression," or "at least one physical aggression." No information is lost in this system, and there are no assumptions about how much "stronger" one aggression is compared to another, which is one of the problematic assumptions that underlie rating scales for dream content (see Van de Castle, 1969, for several examples). However, the "at least one" indicators do not control for dream length for reports that contain over 300 words, so they need to be used with caution, or not used at all, with samples that contain numerous lengthy dream reports.

The coding system includes normative findings, based on 500 dream reports from 100 men and 500 dream reports from 100 women. These dream reports were collected from predominantly white middle-class students at Case Western Reserve University and Baldwin-Wallace College in Cleveland, Ohio, in the late 1940s and early 1950s (Hall & Van de Castle, 1966). The normative findings were subsequently replicated three times for selected coding categories (Dudley & Fungaroli, 1987; Dudley & Swank, 1990; Reichers, Kramer, & Trinder, 1970). They were also replicated for the major categories in studies of men and women's dream reports at the University of Richmond (Hall, Domhoff, Blick, & Weesner, 1982), women at the University of California, Berkeley (Tonay, 1990/1991), and women at the University of California, Santa Cruz, based on codings by Veronica Tonay, who also did the final codings on her Berkeley project (Domhoff, 1996, p. 67).

Determining Reliability in Studies of Dream Content

The percentage of perfect agreement method is a standard reliability measure for all types of content-analysis studies in the social sciences (Charles Smith, 2000). (This reliability measure is based on the number of agreements between two coders, divided by the total number of agreements and disagreements). It is the ideal method for determining reliability with the HVdC system. This conclusion was demonstrated empirically in a study that compared various methods of determining reliability (Hall & Van de Castle, 1966, chap. 13). In particular, this study demonstrated the risks of using a correlation coefficient with the HVdC coding system, because it does not answer the question of how often the two judges agreed exactly on their codings (Hall & Van de Castle, 1966, pp. 148–151, 154–155).

Within this empirically based context, using Cohen's kappa coefficient (κ) for reporting reliability is highly problematic. Cohen's kappa was created to correct for chance agreement by two judges making "yes" or "no" judgments of clinical protocols, who may be *guessing* at least some percentage of the time (J. Cohen, 1960). However, well-trained HVdC coders are making informed choices on the basis of explicit coding rules, in which two coders usually only disagree in terms of which subcategory to use, so a correction for chance makes no conceptual sense. Nor is kappa actually chance-corrected because such a correction implies a model of how judges are making their decisions and can vary depending on whether few or many codings are made (see Uebersax, 1987, 2014, for a summary and a full bibliography relating to the weaknesses of kappa).

Determining Statistical Significance and the Magnitude of Effect Sizes

The percentage and ratio indicators used in the HVdC system are best analyzed for both statistical significance and the magnitude of effect sizes with the test for the significance of the difference between two independent proportions (hereafter simply called "the proportions test"). The issues involved in making this claim, and the claims in the next few paragraphs, may seem to be arcane or minor matters, but the use of statistical procedures premised on the ordinal, interval, or ratio levels of measurement, as found in readily available statistical packages, can lead to inaccurate outcomes.

The deceptively simple proportions test for nominal data is in fact a type of mean for which all the values in the distribution are either zero or one. A proportion only seems simpler than a mean because it is familiar to everyone from an early age in the form of a "percentage," which is simply a proportion multiplied by 100. Since a proportion is a type of mean, the same inferential issues are involved with proportions as with means in general (J. Cohen, 1977, p. 179; Ferguson, 1981, p. 185). Another frequently used statistic, chi-square, provides the same results as the proportions test for 2×2 categorical tables, which are frequently used with the HVdC system (Ferguson, 1981, pp. 211–213).

Moreover, the proportional difference between two samples is exactly equal to the Pearson r between the two samples. For example, a difference of .13 between two samples can be understood as an r of .13 between two dichotomous variables (Rosenthal & Rubin, 1982). In the case of studies with the HVdC system that analyze three or more variables, it is possible to use other nonparametric statistical tests often deployed with categorical data. They include chi-square, the Wilcoxon signed-rank test, Krusak-Wallis one-way analysis of variance, and Friedman two-way analysis of variance (Hall, 1966; Strauch, 2004, 2005; Strauch & Lederbogen, 1999).

The accuracy of proportions testing in determining p values has been demonstrated empirically for all but small differences between two small proportions at the extremes of a distribution (proportions below .10 to .15 or over .85 to .90). In those cases, the proportions test is sometimes wrong.

This point has been demonstrated by comparing estimates based on the proportions test with those based on the computationally intensive randomization strategy called *approximate randomization* (Domhoff, 2003, pp. 84–87; 2018b; Domhoff & Schneider, 2008a). The use of approximate randomization obviates the need for random samples, similar sample sizes, or a normal distribution of scores, through the use of random resampling. It is carried out by randomly drawing thousands of pairs of samples from a common pool made up of values from *both* samples (R. Franklin, Allison, & Gorman, 1997; Noreen, 1989). Approximate randomization thereby provides exact p values, not approximations. In a simulation study, randomization statistics were more powerful than either the nonparametric Wilcoxon signed-rank test or the parametric t test when distributions were skewed, which they usually are in studies of dream content (Keller, 2012).

The use of the proportions test with ratios, such as the A/C and F/C indexes, may at first glance seem inappropriate given that the relationship between a ratio and a proportion is seldom clearly articulated. However, ratios are the more general and encompassing category. They are defined as a comparison of any one quantity to any other quantity, and they can vary between zero and infinity. A proportion therefore is simply one type of ratio, by definition, but it is a ratio that only can vary between zero and one. It is thus possible to use the proportions test with a ratio as long as it varies between zero and one, as the HVdC social-interaction indicators invariably do. The fact that proportions testing is accurate with the A/C and F/C indexes has been demonstrated empirically through the use of approximate randomization testing, which returns the same results (Domhoff, 2018b).

The use of the statistic for the significance of the difference between two proportions segues smoothly to the use of an effect-size measure called h. It is similar in its logic to the better-known d statistic for determining effect sizes based on means. Both of these statistics were created by the same statistical psychologist (J. Cohen, 1977). Although h is calculated with a mathematical formula, as a rule the effect size is simply the difference between the two proportions, multiplied by two. For a more exact measure, there are easily used look-up tables, which make it possible to derive h from the two proportions (J. Cohen, 1977, p. 180; Domhoff, 1996, p. 315). However, the h statistic is best determined by the use of the DreamSAT spreadsheet, which is available on dreamresearch.net for doing all the statistical calculations for a full HVdC analysis (SAT stands for "statistical analysis tool"). As simple as any of these avenues to determining h are in practice, they are in fact based on complex mathematical issues that have been resolved by Jacob Cohen (1977; see p. 180 for the statistical rationale) and summarized in other sources (Domhoff, 1996, p. 315; 2018b).

The h statistic has the added value that it is equal to phi and lambda, the two statistics used to determine effect sizes with chi-square (Ferguson, 1981; Reynolds, 1984). Then too, the magnitude of the difference between two proportions is equal to the Pearson r for dichotomous variables, so nothing would be gained by using a correlational approach instead of percentages (Rosenthal & Rubin, 1982). More importantly, it is comparable to Cohen's more widely known "d" statistic for determining effect sizes between two means. The relative sizes of h in various research areas in psychology can be readily compared with the findings with d. As shown in one study of the usual size of d in different fields within psychology, the h in HVdC studies of dream content, which are generally between .20 and .40, are higher than those for studies of laboratory interviews and reaction times (.14 and .18). They are lower than those for learning studies and person-perception studies (.52 and .54) (Domhoff, 2003, p. 89, table 3.6; Rosnow & Rosenthal, 1997).

Cohen (1977, p. 184) suggests that what is considered to be a "small," "medium," or "large" effect size is best determined by "theory or experience" in each research area. Based on numerous HVdC studies, effect sizes up to .20 should be considered small, effect sizes from .21 to .40 are best thought of as medium, and effect sizes above .40 are large. Effect sizes of .50 or above have been extremely rare in studies of dream content based on representative samples (Domhoff, 1996, chap. 8; 2003, chap. 5; Domhoff & Schneider, 2008a).

Necessary Sample Sizes in the Study of Dream Content

Sample sizes have to be large in most psychological studies to conclude anything with confidence. This point has been demonstrated empirically as part of the concern that many results in psychology, the neurosciences, and medicine cannot be replicated (Nosek, 2015). This point may be even more important in studies of dream content for two reasons. First, the fact that effect sizes are relatively small in dream research makes statistically significant differences more difficult to detect. Second, the units of analysis in studies of dream content are not the individual dream reports. Instead, the units of analysis are specific categories of content within the dream report, such as characters or social interactions. These categories have varying frequencies; most of them appear in less than half of dream reports, although there are a few exceptions. Even in the case of characters, which appear in most dream reports, there may be far fewer *specific* characters, such as one or another of the members of the dreamer's family, or the dreamer's best friend (Hall, 1969a, 1969b). The traditional minimal sample size of 30 to 40 observations in psychological studies is therefore reduced to an effective sample size of 15 to 20 observations or less in a sample with only 30 to 40 dream reports.

According to Cohen's (1977, p. 205) detailed work on the sample sizes necessary for attaining statistical significance with varying magnitudes of differences in *proportions* between samples, it takes a large number of observations to detect small differences with any degree of accuracy. For example, with a real difference in proportions of .20, which is roughly equivalent to an h of .40, 125 observations are needed to have an 80% chance of attaining statistical significance at the .05 level. For proportional differences of .10, it takes a sample of 502 observations to have an 80% chance of attaining significance at the .05 level.

Three different empirical studies of this issue using the HVdC coding system led to the conclusion that Cohen's (1977) calculations have to be taken very seriously in dream research (Domhoff, 1996, pp. 64–67; 2003, pp. 92–94; Domhoff & Schneider, 2008a). For all but a few of the content indicators, it takes at least 125 dream reports to find statistically significant differences and accurate effect sizes. For example, examining the codings of 500 dream reports from women and 500 dream reports from men, which led to the HVdC normative findings, the results are replicated exactly for either gender with random samples of 250 and 125 dream reports, at the .05 level of significance. However, many of the significance levels and effect sizes in the two normative samples cannot be detected with any reliability with a subset of 100 dream reports, and the results become progressively weaker with 75 or 50 dream reports. The exceptions involve a few dream elements that appear frequently (Domhoff, 1996, pp. 64–67; Domhoff & Schneider, 2008a). It therefore is essential to use large sample sizes in building a sound scientific foundation for the understanding of dream content.

The Importance of Replication Studies

Although the evidence for the usefulness of the various statistical tests discussed in this section is very strong when sample sizes are adequate, virtually all statisticians agree there is no substitute for replication studies in psychology, no matter how large the sample size or how sophisticated the statistical analysis. It is now recommended that studies be replicated at least once (F. Schmidt, 1996; S. Schmidt, 2009). Replications are necessary because no statistical test is perfect (J. Cohen, 1990, 1994; Fife, 2020; S. Schmidt, 2009). Moreover, there are bound to be a few later studies based on small samples in which the random draw does not lead to the statistical significance that has been established in replicated studies. Although there are statisticians who suggest ignoring such small-sample studies, other statisticians suggest that any failure to replicate a finding that has been replicated several times should itself be replicated by the researchers and by independent investigators before it is taken seriously (Hedges & Schauer, 2019; S. Schmidt, 2009).

The Issue of Multiple Testing of the Same Sample

Multiple tests of the same pair of samples can greatly increase the probability of finding at least one or two statistically significant differences by chance (Fife, 2020, pp. 1056–1057). For example, if 10 comparisons are made, which is a realistic number when using the HVdC system, there is a 40% probability of at least one statistically significant difference ($p < .05$). To correct for multiple testing, the HVdC system has used the "false discovery rate" since 2015, which controls for false positives by focusing on the comparisons that yielded statistically significant differences (Benjamini & Hochberg, 1995). It does not suffer nearly as much from the large loss of statistical power experienced when the entire list of p values is used in making the correction, even when some comparisons are not significant, as in the case of the earlier Holm-Bonferroni correction (for further analyses, see Benjamini & Hochberg, 1995; Benjamini & Yekutieli, 2001; for a comparison of the two methods, see Domhoff & Schneider, 2015b; for the Holm-Bonferroni rationale, see Holm, 1979).

As a result of its low statistical power, the Holm-Bonferroni correction has been widely criticized by statisticians and by practitioners in epidemiology, ecology, and medicine on the grounds that it stifles the further exploration of unexpected findings, particularly in fields that are primarily at a descriptive stage in the theory-building process (Ellis, 2010; Garcia, 2004; Moran, 2003; Perneger, 1998). Dream research, which has mostly remained at a descriptive level, can be placed in the same category as ecology, epidemiology, and medicine.

When the Benjamini-Hochberg correction for multiple testing was used in a comparison of the women's and men's normative samples for 22 HVdC content indicators, it did not result in any changes in the original p values that were calculated (Domhoff & Schneider, 2015b).

Do Dream Reports Collected in Laboratory and Nonlaboratory Settings Differ?

The most methodologically rigorous comparison of lab and nonlaboratory reports, based on 38 dream reports from 12 young male participants, revealed only two differences, both of which related to aggressive interactions (Weisz & Foulkes, 1970). The participants slept in the lab on two nonconsecutive nights over a two-week period and contributed nonlab reports from two morning awakenings at home, using voice-recorded reports in both conditions. There were no statistically significant differences between the lab and nonlab settings in the percentage of recall or the length of the reports, or on content indicators other than aggression (see Weisz & Foulkes, 1970, pp. 590–593, for details). Since the same participants were used in both conditions, a smaller sample size than is usually recommended could be used. Moreover, these findings were replicated in a study of 53 lab and 56 nonlab dream reports from five young male participants in a study at the University of Zurich, all of which were voice-recorded (Gross, 1988; Strauch & Meier, 1996, p. 107). Using the HVdC coding system and corrections for report length, there were no differences in settings, characters, activities, social interactions, or successful strivings (Gross, 1988).

The largest and most detailed comparison of lab and nonlab dream reports was carried out in the Miami sleep-dream lab, which had once been a suburban home. The study used voice-recorded reports in the laboratory and written reports in the nonlab condition. This was the least-controlled condition in the study, although the differences in the length of the reports from the two different conditions were not large and there were controls for length of the reports (see Domhoff & Schneider, 1999, for an accessible summary, and Hall, 1966). The original HVdC codings, which showed few or no differences in the initial analysis, were reanalyzed three decades later using the new content indicators, which have excellent controls for word length with 50 to 500 words, as already mentioned above. This second analysis also included the effect size h.

Based on 120 nonlab dream reports and 272 lab dreams, which came from 8 of the 11 young male participants in the original study, there were only four statistically significant differences in the comparisons of 21 content indicators. The nonlab dream reports scored higher on three aggression indicators and the Animal Percent. However, the effect sizes were relatively small for two of the indicators but large in the case of two of the aggression indicators. Very importantly, this study also showed that the laboratory dream reports include only about half as many codable elements as do the nonlab dream reports in terms of the social interaction, striving, and misfortune categories (Domhoff & Schneider, 1999). It therefore seems there may be less of everything in at least some samples of lab dream reports, which has to be kept in mind for some type of studies. However, the basic results in terms of the content indicators remain the same.

Generally speaking, then, these three rigorous studies demonstrate there are few or no substantive differences between lab and nonlab dream reports, with the important exception of more frequent aggression in nonlab dream reports. There are also two studies that are less well controlled, which used the same participants in both conditions, and came to the same conclusions (Domhoff & Kamiya, 1964; Zepelin, 1972).

Although nonlab samples are useful when they are large and contain dream reports with 50 words or more, the differences on aggression have to be kept in mind when making major generalizations. Moreover, the differences between the two types of samples raise theoretical issues as to why nonlab dream reports would be similar to laboratory dream reports on some issues but not on others. More specifically, and as discussed in chapter 2, it may be that some of the well-known factors involved in the recall of dreams, such as recency and cued recall, may lead to reasonably representative samples of dream content, while saliency may introduce some biases into nonlab samples.

Five Categories of Embodied Simulations in Dreams

To provide a general picture of what people dream about, it is useful to begin with analyses of the codings for characters, social interactions, and activities in the HVdC normative dream reports. These analyses reveal there are five main categories of embodied simulations in dreams. Three of these categories involve social interactions with, or thoughts about, other human beings. A fourth involves dream reports in which the only characters are the dreamer and one or more animals. The fifth concerns dream reports in which the dreamer is the only character. The results are derived from a spreadsheet containing the original codings for 991 dream reports, 491 of which are from 100 women and 500 from 100 men. Although nine of the original coding cards for women's dreams could not be located when these materials came into the author's possession, there were no differences on any content indicators when the original normative findings were compared with those from the extant normative samples. Within the context of the five categories extracted from the extant database, it is also possible to present the distinctive features that mark some of these categories and to mention any gender differences as well.

Aggression, Friendliness, Sexuality (A/F/S)

The first of the five categories includes three types of social interactions that occur in dream reports: *aggression, friendliness,* and *sexuality*. They have been analyzed in many dozens of investigations relating to age, gender, culture, and individual differences, some of which are summarized later in this chapter. These results are discussed in more detail in earlier books (Domhoff, 1996, chaps. 4–6; 2018a, chaps. 2–4). Aggressive, friendly, and sexual interactions can be combined to create a content indicator that expresses the percentage of dream reports with "at least one A/F/S" interaction involving humans. In the combined normative sample, 67.2% of dream reports have at least one aggression, friendliness, or sexuality interaction involving humans. There is a small gender difference (65.6% for women, 68.8% for men; $h = .07$, *ns*). Further, the dreamer is part of these social interactions about 80% of the time and an observer of them in the remaining 20%. These results demonstrate that most dreams are highly social in nature. The findings concerning the frequency with which aggression, friendliness, and sexuality occur in the combined normative sample, separately or together, are presented in figure 3.1.

Despite the large percentage of dream reports from both women and men involving aggressive, friendly, or sexual interactions, it is striking that only about 10% of the social interactions, whether initiated by the

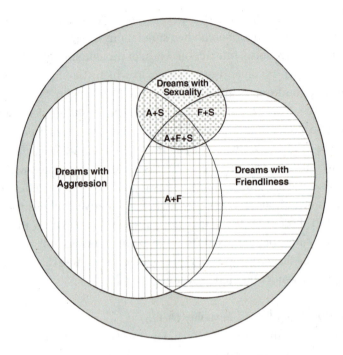

Figure 3.1
Euler diagram of social interactions (aggression, friendliness, and sexuality) between
humans in the Hall/Van de Castle normative dreams.
Note: The enclosing gray circle represents the complete set of dreams.

dreamer or one of the other characters, are *reciprocated* by the other character involved. More specifically, a reciprocation concerns a response to
a social interaction with a social interaction of the same type. Based on
that index, the "social interactions" in dream reports do not involve much
interaction. Omitting aggressive thoughts and murders, neither of which
can be responded to, fewer than one in six aggressions is met with any type
of aggressive response. After omitting friendly thoughts (which cannot be
responded to), only 4.3% of the friendly acts in the remaining six categories
lead to a response. These two results are the same whether the dreamer or
another dream character initiates the aggression or friendliness. There are
no significant gender differences on this issue. According to classic socio-
logical understandings of what constitutes a "social interaction," it is necessary for the two or more parties involved in the potential interaction to be

aware of the context and the likely thoughts of the other party or parties involved. Since so few friendly or aggressive acts in dream reports even involve a reciprocation, it may be that this lack is another example of a cognitive insufficiency during dreaming. The general results on reciprocation to aggression and friendliness are provided in table 3.2.

Directed and Joint Activities

In addition to A/F/S social interactions, social interactions often occur in the context of *activities* in dream reports, such as talking, walking, looking, listening, and thinking. There are eight subcategories of the overall activities category, all of which focus on "what characters *do* in dreams" (Hall & Van de Castle, 1966, p. 87; emphasis in the original). The activity subcategories include three that involve the body: *physical* (voluntary movements of the body or body parts while staying in about the same place); *movement* (self-propelled changes in location); and *location change* (changes in locale via a vehicle). The activities category also includes five sensory and cognitive categories, which are for the most part self-explanatory on the basis of their labels: *verbal, visual, auditory, expressive communication* (primarily laughing and crying), and *thinking/cognition*. Overall, activities are very

Table 3.2

Reciprocated aggression and friendliness in the Hall/Van de Castle norms

	Female norms (491 dreams)	Male norms (500 dreams)
Aggression		
Number of directed aggressions, subclasses 2–7	244	274
Percent that are reciprocated	13.1	17.9
Directed aggressions where the dreamer is the recipient	144	151
Percent that are reciprocated by the dreamer	11.1	17.9
Friendliness		
Number of directed friendly acts, subclasses 2–7	237	210
Percent that are reciprocated	5.5	2.9
Directed friendly acts where the dreamer is the recipient	112	101
Percent that are reciprocated by the dreamer	9.8	1.0

frequent in the HVdC normative dream reports, averaging 4.9 per dream report for women and 4.7 for men; 25% of women's dream reports and 22% of men's reports contain from 6 to 20 activities (Hall & Van de Castle, 1966, pp. 182–185, for the highly detailed findings on activities).

Although not all activities in dream reports are part of social interactions, many activities are indeed carried out in "interactions between characters" or in "conjunction with other characters" (Hall & Van de Castle, 1966, p. 87). The characters' interactions with each other (called "directed" activities) or in conjunction with each other (called "joint" activities) are used in defining the second of the five categories that can be found in dream reports. Directed and joint activities may or may not involve the dreamer. Sometimes they simply involve a group of unnamed people carrying out an activity together.

These directed and joint activities include characters talking with, or moving to another locale with, another character or characters, or two or more characters seeing or listening to someone together. They also can include such joint activities as lifting an object together or working to repair a vehicle together. There are even a few instances of characters laughing or crying together, or thinking together, as in "deciding" or "plotting" some activity together. This type of embodied simulation is analyzed by means of a content indicator expressed as the percentage of dream reports with "at least one directed/joint activity" involving another human character. In the normative sample, 77.1% of the dream reports have at least one directed or joint activity.

When the 77.1% of dream reports with at least one directed/joint activity are combined with the 67.2% of dream reports that have at least one A/F/S, then fully 86.9% of dream reports have at least one A/F/S interaction *or* at least one directed/joint activity. Women have a slightly higher percentage of dream reports with a directed or joint activity (79.2% for women, 75.0% for men). However, both women and men end up with approximately the same percentages when the A/F/S and directed/joint activity categories are combined (87.4% for women, 86.4% for men). The overall distribution, as displayed in table 3.3, demonstrates the overwhelmingly social nature of dreams even better than the two categories considered separately. Note that a majority of dream reports (57.4%) contain both a social interaction *and* a directed or joint activity. Note also that table 3.3 contains findings on three categories of embodied simulation discussed below.

Table 3.3

Categories of "social simulations" (or lack thereof) in the combined Hall/Van de Castle normative sample of dream reports

Type of social simulation	Percent of dream reports	Cumulative percent
Social interactions (aggression, friendliness, or sexuality)*	67.2	
Activities (joint/communal or directed)*	77.1	
Social interactions OR joint/directed activities	86.9	86.9
Human characters but no interactions	6.6	93.4
Only animal characters	2.2	95.7
Dreamer only	4.3	100.0

* 57.4% of dream reports had both social interactions and joint/directed activities.

Perception/Cognition of Other Characters

In addition to A/F/S interactions and directed/joint activities, there is a type of embodied simulation that includes *only* visual, auditory, or thinking activities by the dreamer but in which another character may be seen, heard, or thought about (e.g., "I saw/heard Joe across the room;" "There were hundreds of people along a fence and I thought of a woman I know"). This category also includes dream reports in which one or more human characters are reported to be in the same setting with the dreamer but are not interacting with the dreamer or with each other (e.g., "I was standing in a railroad station where there was a great crowd of people"). These dream reports are labeled "perception/cognition" simulations because they only include the dreamer's perceptions and/or thoughts in relation to some other character or characters. By definition, perception/cognition dream reports do not include *any* type of social interaction or any directed or joint activities. Dream reports within the perception/cognition category are analyzed by means of a content indicator expressed as the percentage of dream reports with "at least one perception/cognition activity" involving another human character. The perception/cognition percent is 6.7, which means that 93.5% of dream reports include a social simulation of some type that involves other people.

Dreamer and Animals Only

A fourth type of embodied simulation in dreams is one in which only the dreamer and one or more animals or creatures are present. ("Creatures"—the stuff of science fiction, fairy tales, and cartoons—appear only twice in the combined norms.) For all intents and purposes, then, this category involves only the dreamer and one or more animals, but it is by definition a "dreamer-plus-animals-or-creatures" category. This category allows for future studies in which there may be more creatures. In any case, only 2.1% of dream reports fall into this category. Although this percentage of all the dream reports is small, these dreams are distinctive because they are very high on aggressive interactions. The dreamers are often being chased or physically attacked by an animal or dangerous insect. The Physical Aggression Percent in animal-only dream reports is 72.2, compared to 42.4 in the rest of the dream reports. This difference leads to an unusually large effect size ($h = .62$, $p < .01$).

Dreamer Only

Fifth, and finally, there is a type of embodied simulation in dreams in which the dreamer has no social interactions, directed activities, joint activities, or social/perception activities. That is, there are no other characters, human or animal, in the dream report. Nor is there even any *thinking* of other characters in these "dreamer-only" dream reports. In these dream reports, the person is carrying out some activity alone, such as listening to music, shaving, driving, trying to fix an automobile, walking alone on a street or in a forest, or simply observing his or her surroundings. These "dreamer-only" dreams comprise the final 4.3% of the dream reports in the normative sample.

Dreamer-only dream reports are distinctive in that they are higher than the normative baselines for the presence of at least one of several types of "misfortune." Misfortunes are defined as bad outcomes that happen to the dreamer, or any other dream character, but are not due to actions by any dream character. Misfortunes are analyzed in terms of five nominal subcategories. The categories range from annoying to difficult to very sad outcomes: a dream character is facing an obstacle, such as a locked door; is falling or in danger of falling; is threatened by something in the environment, such as a falling tree limb; is involved in a minor accident; is injured or ill; or dies or is already dead (Hall & Van de Castle, 1966, pp. 102–104). In dreamer-only dream reports, misfortunes are almost twice as likely to occur than in the remainder of the normative sample. In addition, the effect size is quite large (62.8% vs. 33.5%; $h = .59$, $p < .001$). For example, the dreamers are not only shaving, they also see blood from a cut. The dreamers are not only walking in the forest, they are also becoming lost. They are not simply driving alone; they are also worried about experiencing car failure.

As this general analysis based on five separate categories documents, dreams are overwhelmingly social in nature. About two-thirds of dream reports contain at least one aggressive, friendly, or sexual interaction. A little over three-fourths of dream reports involve a shared activity with another human character. Another 6.7% involve thoughts or perceptions of other people. Finally, there are the 2.2% of dream reports that involve the dreamer and one or more animals, and the 4.3% that only involve the dreamer. As these percentages suggest, no one generalization captures the full range of dream content.

These findings can be used to examine various cultural stereotypes about dreams. This diversity and complexity are further supported by the findings in the rest of this chapter and in chapter 4. The wide individual differences in the content of the dream reports discussed in chapter 4 also add to this general portrait of dream content.

Age, Cross-National, and Cross-Cultural Differences in Dream Reports

Numerous studies of adults inside and outside of lab settings have examined the issue of age differences. Age is not only of interest in and of itself but also has implications concerning the degree to which normative findings based on young adults in university settings can be used with older adults. Similarly, there are both substantive and methodological reasons for determining if there are differences in dream content between dream reports collected in the United States and other industrialized countries (called "cross-national differences" in this book), as well as any differences between nation-states and small indigenous societies (called "cross-cultural" differences in this book).

Age Differences in Dream Reports in Three Different Countries

Although age is usually one of the first issues most people assess when they meet an individual, there are few age differences in dream reports. Nor is it very often obvious what the dreamer's age is, based on the settings, characters, and social interactions in dream reports. However, phrases such as "my children" or "my aging father" sometimes provide some hints. In one of the first large-scale studies of age differences in nonlab dream reports, which compared aggressive and friendly interactions in 281 dream reports from women ages 30 to 80, as well as aggression and friendliness in a similar number of dream reports from men in the same age range, the main differences concerned a possible decline in the frequency of aggression (Hall & Domhoff, 1963a, p. 260, table 2; 1964, p. 310, table 1). Similar age differences on aggression were reported in a study using a different coding system. It compared 148 dream reports from 38 college women, ages 18 to 26, with 185 reports from 43 women, ages 40 to 86, all but two of whom had attended college (Brenneis, 1975, p. 433). A study of age differences between 58 males ages 27 to 64, which compared laboratory and nonlab dream reports from the same participants, made use of several HVdC coding categories. One of the few differences involved the frequency of aggressions. Family-related content was most prominent from ages 35 to 55, which is consistent with the focus on raising children for many adults in that age range (Zepelin, 1980, 1981).

Perhaps the most convincing study on age changes in dream content involved a unique longitudinal design in which the researchers collected two-week, follow-up dream diaries from Canadian women. They had originally written down their dreams 10, 15, or 17 years earlier. At the time they were asked if they would keep a second dream diary, seven of the women were between ages 30 and 39, 10 were between 40 and 49, and four were between 50 and 55. Using a large number of HVdC content indicators, as well as other measures, the research team found no statistically significant differences (Lortie-Lussier, Cote, & Vachon, 2000, pp. 71–72, tables 1, 2, and 3).

Similarly, in a cross-sectional study of 47 French-Canadian women ages 25–35, 36–45, and 46–56, each of whom contributed two dream reports, the main difference, using various HVdC coding categories, concerned more pleasant outcomes in the oldest group. There was also a slight decline in the frequency of emotions for the older age group (Côté, Lortie-Lussier, Roy, & De Koninck, 1996). However, two other cross-sectional studies with heterogenous samples found mixed evidence for age consistency for both women and men. The first study was based on 375 dream reports from each of 375 Canadian women, who varied in age from 12 to 17, 19 to 24, 15 to 39, 40 to 65, and 65 to 85. The second study used dream reports from 231 Canadian men, who were divided into roughly the same five age categories. Although several categories remained stable for the women, there was a slight decline in the Aggressor Percent and larger declines in familiar characters, total number of activities, aggression, and the F/C Index. In the case of the men, the changes involved a decline in the rate of aggressions from the adolescent group to the young adults. It then remained relatively stable thereafter. On the other hand, the F/C Index and the number of activities in the dream reports gradually increased in the three older groups of men (Dale, Lortie-Lussier, & De Koninck, 2015, 2016).

Two large studies of the elderly at the University of Zurich made use of HVdC coding categories. In the first study, 253 dream reports from 15 women and 9 men between the ages of 66 and 78 were collected in a lab setting. Then dream reports were collected from the same women and men through the use of voice recorders in each individual's home setting for two weeks. The reports were coded for characters, settings, social interactions, and activities. The only two differences between the elderly and a control sample from younger Swiss adults involved a higher percentage of unfamiliar characters and unfamiliar settings in the dream reports of the elderly (Strauch, 2003). In a second study, a recent dream report was collected from each of the 106 women and 45 men, ranging in age from 65 to 89, who attended one of the two repeated lectures on dreams provided for senior citizens in Zurich. Once again, the dream reports were compared with dream reports from younger Swiss adults, and once again there was a higher percentage of unfamiliar characters and unfamiliar settings in the elderly sample. There was also less aggression in this sample of dream reports from the elderly (Strauch, 2014).

Substantively speaking, then, there are few age differences among dreamers in several samples from three different countries. However, there are one or two relatively consistent differences between the youngest and most elderly of adults, especially on indicators related to aggressive interactions. In terms of methodological issues, these findings suggest normative findings based on dream reports from young adults have to be used selectively with samples of elderly dreamers.

Cross-National Studies of Dream Content

The most detailed study of dream content in a European country compared the dream reports of young adult Swiss women and men, collected at a University of Zurich sleep-dream lab, to the HVdC normative findings. Based on HVdC codings, the Swiss and American men and women were very similar in terms of the mean number of settings, the mean number of characters, the Male/Female Percent, Animal Percent, and the Familiarity Percent. However, the Swiss women had a higher percentage of strangers and the Swiss men had a higher percentage of outdoor settings (Domhoff, 1996, p. 202, table 6.3). The differences between the Swiss and American samples are greater in the categories of aggression and friendliness. The HVdC norms, based on nonlab samples, therefore are not ideal for a cross-national comparison (Domhoff & Schneider, 1999). Swiss dreamers of both genders were also much more likely than their American counterparts to initiate friendly interactions.

Similar findings emerge in two different studies of the same database of 246 dream reports from 106 German college women and 95 reports from 39 German college men (Domhoff, Meyer-Gomes, and Schredl, 2005–2006; Schredl, Petra, Bishop, Golitz, & Buschtons, 2003). In the second of the two studies, for example, the German women differed from the HVdC norma-tive sample of American women on 5 of 17 comparisons based on HVdC indicators, and the German men differed from the HVdC normative sample of American men on 6 of 17 comparisons (Domhoff, Meyer-Gomes, and Schredl, 2005–2006, pp. 274–275, tables 1 and 2). The most unexpected large difference, on the Male/Female Percent in German men, is discussed later in the chapter.

A collection of 218 single dream reports from 115 men and 103 women at the University of Tehran found many similarities with the HVdC norms, along with lower levels of aggression and sexuality and a higher percentage of family members and other familiar characters (Mazandarani, Aguilar-Vafaie, & Domhoff, 2013, p. 167). The findings on family members and known char-acters are similar to findings in a study in India (Prasad, 1982). The results from a lab and nonlab comparison study at the University of Tokushima in Japan, which resulted in a total of 193 dream reports from women and 104 dream reports from men, provided findings that were almost identical to the HVdC norms for women and men for the objects and activities catego-ries (Yamanaka, Morita, & Matsumoto, 1982, pp. 34–35, 38). However, there were more characters in Japanese reports than in the HVdC norms, and far more of the characters in Japanese reports were familiar to the dreamer. Both women and men in the Japanese study had very low percentages of dream reports with at least one aggression, compared to the American norms. The figure was 26% for women, as compared to 44% for American women, and 14% for men, compared to 47% for American men.

By and large, dream reports are more similar than they are different across a number of cases, although the United States is far higher than the others on content indicators related to aggression. In addition, there are differences on the degree to which the characters in dream reports are familiar to the dreamer. Although the available samples from a wide range of countries makes it possible to find commonalities, further studies in specific countries would be necessary to examine the differences from the HVdC normative samples in more detail.

Cross-Cultural Studies of Dream Content

Using dream reports brought together from a variety of anthropological studies, dream reports from 10 indigenous societies have been analyzed with HVdC categories. Two of the societies were based on hunting and gathering economies, and the rest of these indigenous societies largely relied on one or another type of agriculture. These dream reports were collected under many different conditions and many of the reports were relatively brief, so the most important conclusions are somewhat general (Domhoff, 1996, pp. 99, 115–120, for a more detailed discussion). Most impressively, the dream reports from these 10 societies are similar to dream reports collected in the United States and other countries in that there are always more single than plural characters, more humans than animals, and more familiar than unfamiliar characters. As in the United States, the A/C Index is higher than the F/C Index in these societies, with the exception of Hopi women. Dreamers in indigenous societies are also more often victims of aggression, with two exceptions. There is usually more physical than non-physical aggression. Finally, the Physical Aggression Percent is higher than it is in the United States in all but one of the indigenous societies, namely the Hopi, who had been forced to live on reservations almost 100 years before their dream reports were collected (Domhoff, 1996, pp. 119–120, tables 6.17 and 6.18, and p. 128). Once again, aggression emerges as a key variable between societies.

The Familiar and the Unfamiliar in Dream Reports

Most dream reports include familiar settings and familiar characters, but some dream reports include neither. In the HVdC women's normative sample, 33.6% of the dream reports include at least one family member, 59.1%

include at least one friend, and 39.7% have at least one familiar setting. More generally, either a family member, a friend, or a familiar setting are present in 82.7% of the women's normative dream reports. In contrast, only 18.2% of the dream reports in the men's normative sample include at least one family member, only 44.8% include at least one friend, and only 33.6% have at least one familiar setting. As a result, only 65.0% of the men's dream reports have at least one family member, friend, or familiar setting, so there is a large difference (82.7% vs. 65.0%, $h = .41$, $p < .001$). Gender differences on this issue aside, the dream reports with varying degrees of familiarity in terms of settings and characters also can be examined more closely using several of the HVdC coding categories.

When characterized in terms of degrees of familiarity and unfamiliarity, the dream reports differ on several content variables. In particular, dream reports that involve the dreamer interacting with unknown people, within unknown settings, have more physical aggression, less friendliness, more misfortune, and a higher Animal Percent. These dreams may be more "narrative-driven," which means they are more like sagas or adventures than enactments of personal concerns (Foulkes, 1999, p. 136). The results from the various analyses in this paragraph are displayed for women and men separately in table 3.4. The table also can be examined to show there is only one gender difference, which concerns the A/C Index.

The differences between familiar and unfamiliar dream reports also have been studied using four general ad hoc categories, which have some overlaps with HVdC categories. These categories concern "familiar characters," "familiar activities," "school/work/politics," and "nothing familiar." They were used to study six different sets of dream reports 50 to 300 words in length. The samples consisted of 246 dream reports from 106 German college women; 95 reports from 39 German men; random samples of 100 HVdC norm dream reports from 100 women and 100 men; and 100 dream reports collected at the University of Miami in a lab setting and 100 dream reports from the same male participants outside of the lab setting (Hall, 1966). The same bilingual research assistant who coded the German dream reports, Katrin Meyer-Gomes, also coded the other five sets of dream reports (Domhoff, Meyer-Gomes, & Schredl, 2005–2006). She did so on two separate occasions several months apart (a temporal measure of reliability). She then reconciled the few differences between her two sets of codings. In every sample, the percentage of dream reports that included one or more

Table 3.4

Differences in content in the familiar and unfamiliar dream reports of women and men

	Female norms			Male norms		
	Familiar ($n=168$)	Unfamiliar ($n=80$)	h	Familiar ($n=127$)	Unfamiliar ($n=164$)	h
Aggression/ Friendliness %	51	57	+.12	57	65	+.15
Physical Aggression %	32	66	+.70**	33	63	+.60**
A/C Index	.24	.24	+.01	.29	.44	+.33**
F/C Index	.22	.17	−.13	.21	.20	−.02
At least one misfortune %	35	46	+.23	31	47	+.32*
Animal %	3	13	+.39**	3	11	+.32**

Note: "familiar" dreams are those with both familiar characters and familiar settings. "Unfamiliar" dreams have neither familiar characters nor familiar settings.
* $p<.01$; ** $p<.001$.

of the "familiar" elements was 70% or higher. The results from the samples of women from Germany and the United States were very similar (87% and 82% familiar), as were all three samples from the American men (70–74% familiar). The German men had a Familiarity Percent of 80, which is more similar to the results for the two samples from women than it is to the results for American men.

These results with the six samples from Germany and the United States support the HVdC findings above, which are based on a comparison of familiar and unfamiliar dream reports with the normative dream reports for women and men. More generally, the results from the study using four ad hoc categories and the study using HVdC codings suggest there may be gender and nation-state differences on the degree to which samples of dream reports can be characterized in terms of the Familiarity Percent.

Cognition and Cognitive Appraisals in Dreams

Cognition, as defined in the HVdC system, involves a "deliberate continued mental effort," such as "concentrate," "deliberate," and "think about." But it does not include "brief, transient mental activities," such as "I forgot my coat," "I remember the room," or "I think it was blue" (Hall & Van de Castle, 1966, p. 90). Defined in this way, 113 cognition codings appeared in 13.8% of the women's normative dream sample and 75 such codings appeared in 11.6% of the men's sample dream reports. On the other hand, a category for cognitive appraisals, which concerns the assessment of a puzzling or possibly concerning situation or event, appears more frequently (21.0% of dream reports for women, 13.4% for men). Since there is a small overlap in the two categories at the juncture between confusion and puzzlement, 29.3% of the women's dream reports contain at least one cognitive element and 23.0% of the men's reports include at least one cognitive element. Both categories are useful due to the concern with cognition in the neurocognitive theory of dreaming.

However, the focus in this book is on cognitive appraisals. They have important theoretical implications in terms of the development of an adequate theoretical understanding of emotions in dreams, as explained in chapter 8. Cognitive appraisals, which also can be defined as the "process of evaluating the affective significance of an event" (Dixon, Thiruchselvam, et al., 2017, p. 1034), often occur as part of a process that may or may not lead to the expression of one or another type of emotions. In the HVdC system, the concept of cognitive appraisals is indexed by terms such as "astonishment," "bewilderment," "confusion," "puzzlement," and "surprise." This point is captured in Hall and Van de Castle's (1966, p. 112) discussion of the category as one of "cognitive ambiguity," which is due to an "unexpected event," or to an "inability to choose among available alternatives." They further note this state is also expressed by terms such as "amazed, conflicted, mystified, perplexed, uncertain, and undecided" (Hall & Van de Castle, 1966, p. 112). However, the two researchers were uncertain as to whether to put this category under the general rubric of cognition or emotion. After noting "it may be debatable as to whether confusion is a condition possessing the same degree of autonomic involvement," they mention what came to be the key point concerning the later development of the concept of cognitive appraisals: "the feeling state accompanying uncertainty may begin to shade toward a type of free-floating anxiety, toward frustration, or toward depression" (Hall & Van de Castle, 1966, p. 112). They then concluded that confusion was "emotionlike."

Due to advances in cognitive psychology, it now makes more conceptual sense, especially within the context of a neurocognitive theory of dreaming, to categorize confusion, surprise, and puzzlement as "cognitive appraisals." Cognitive appraisals arise in the face of new situations or sudden events (I. Roseman & Evdokas, 2004; I. Roseman & Smith, 2001). To repeat, a cognitive appraisal is the first step in what may or may not lead to an emotion. It is for these reasons that confusion, surprise, and closely related cognitive states are now an important subcategory within the general category of cognition.

The Continuity between Dreaming and Waking Thought

The neurocognitive theory of dreaming puts a strong emphasis on the "continuity" between what people dream about and their waking personal concerns. This section provides good preliminary evidence for the validity of this hypothesis by means of studies based on three very different topics. The first compares the dream reports from a hunting and gathering society with the HVdC norms. The second study compares the degree to which the "negativity bias" detected in waking thought in numerous psychological studies also appears in dream reports. The third and most detailed analysis compares dream reports from women and men in the United States on the basis of a large and unexpected gender difference. This large difference in turn relates to other relatively large gender differences in dream reports and to gender differences in waking personal concerns.

The dream reports from the Yir Yoront, a hunting and gathering society in Australia, were collected in the 1930s by a cultural anthropologist (see D. Schneider, 1969, for details). Asking about dreams seemed useful to this field researcher, even though the Yir Yoront put no special emphasis on dreams and made no use of them in ceremonies. This is because he soon realized that talking about dreams provided a way to "establish and maintain rapport" on "a neutral subject that could be exploited without offending anyone" (D. Schneider, 1969, p. 15). The Yir Yoront sample consists of 140 dream reports from 43 men, one of whom contributed 14 dream reports. Another man contributed 11 and the rest contributed from one to five each.

There were relatively few differences between Yir Yoront men and American men, except that the Yir Yoront men dreamed far more about animals, had a higher proportion of aggression with animals, and had a very high percentage of physical aggressions with animals. They also had a higher frequency of friendly encounters with humans, particularly with familiar female characters. These friendly interactions with women mostly involved the Yir Yoront men sharing meat from the animals they had killed. When compared with the men's normative samples and five indigenous agricultural societies, the findings suggest continuity between the dream lives and waking lives of Yir Yoront males (Domhoff, 1996, p. 120, table 6.18). They thought constantly about the animals they hunt and kill, and feared them. They looked forward to increasing their stature within the group by sharing the meat from the animals they killed with relatives and friends, including women relatives. And they had ceremonies to propitiate the spirits of the dead animals they feared and respected. In general, then, this comparison of animals in three extremely different samples of dream reports provides solid evidence for the concept of continuity.

Numerous waking studies over many decades have led to the conclusion that there is a negativity bias in people's thinking, which is a tendency to think more often of their fears, worries, failures, and bad experiences than of positive thoughts and events in their lives (see Baumeister, Bratslavsky, Finkenauer, & Vohs, 2001, and Rozin & Royzman, 2001, for reviews and summaries of several decades of studies on this topic by research psychologists). This work has been supplemented and extended by research in social neuroscience (see Norris, 2021, for an analysis of neuroimaging and other types of neuroscience findings on the negativity bias from a social-neuroscience perspective). Based on an examination of the HVdC normative findings, the negativity bias is present in dream reports as well. When the total number of aggressions, misfortunes, and failures is compared to the total number of friendly acts, good fortunes, and successes, women have a negativity bias of 68.2%/48.7%, which can be expressed as a negativity ratio of 1.40, and men have a negativity bias of 73.6%/51.2%, which is a negativity ratio of 1.65.

Turning to the third study that is relevant to the issue of continuity between the concerns expressed in dream reports and waking thoughts, it began with one large gender difference in dream content in the HVdC normative sample. The analysis next used the findings in other HVdC categories to understand that initial difference. The results of these analyses were then related to gender differences found in a variety of waking studies. The gender difference that provided the starting point for the study is simply that women dream far more equally about both genders than do men. More exactly, 48% of the human characters in women's dream reports are men and 52% are women. Men, on the other hand, dream twice as often about other men as they do about women, 67% vs. 33%. This difference on the "Male/Female Percent" is determined by dividing the total number of men by the total number of men plus women, which leads to an h of .38 and a p value of .0001 As noted above, but worth mentioning again, this is a fairly large effect size for studies of dream content. It is also one of the four largest differences between the normative dream reports of American women and men.

The gender difference on the Male/Female Percent has been found at all ages in many different countries and cultures, including Argentina, Switzerland, and many of the small traditional cultures studied in the past by cultural anthropologists (Domhoff, 1996, chap. 6; Hall, 1984). However, it is by no means a "universal" difference, in the sense that it is invariably found in every group. In fact, the dream reports of African American male college students at a community college in Chicago in the late 1960s had a Male/Female Percent of 53/47, which was very similar to the Male/Female Percent for the African American women in the study (Domhoff, 1996, p. 75). Nor was the usual difference found in studies of Mexican and Peruvian teenagers and young adults, where the men tended to dream equally of men and women and the women dreamed more frequently of men (Domhoff, 1996, p. 106). It also was absent in the study of the dream content in several hundred reports from German college students, in which the Male/Female Percent was 56/44 for women and 58/42 for men (Domhoff et al., 2005–2006; Schredl, Petra, Bishop, Golitz, & Buschtons, 2003). In the cross-cultural study of indigenous societies discussed above, there was no male group in which the Male/Female Percent was below 59/41. However, two of five female groups had a high Male/Female Percent—66/34 in one of them and 58/42 in the other. These two differences perhaps indicate they lived in extremely patriarchal societies (Domhoff, 1996, p. 119, table 6.17).

The unexpected general finding on the Male/Female Percent for white Americans, as well as people in several other societies, is a discovery that is the product of the coding system. There is no immediately obvious reason for this difference. However, it provides an opening for using other findings, based on the HVdC content indicators, to show how this finding led to further discoveries, which relate to gender differences in waking personal concerns. More specifically, if it is hypothesized that the quantitative analysis of dream reports reveals the same personal concerns people think about in waking life, then the findings on the Male/Female Percent suggest women are equally concerned about both men and women, and that men are more concerned about other men than they are about women.

As a starting point for examining this hypothesis, there is another gender difference that provides an initial clue. It is based on the frequency of appearances by familiar characters, who are defined as family members, friends, and acquaintances. Women dream more often of characters who are familiar to them than do men. They have a Familiarity Percent of 58, compared to 45 for men ($h=.26$, $p<.001$). A closer examination of this difference reveals that women dream more often about familiar females than men do (29% vs. 16%, $h=.31$, $p<.001$). On the other hand, men dream more often about unfamiliar men than do women (28% vs. 15%, $h=.32$, $p<.001$). Within this context, it is relevant that women and men dream equally about familiar men (23% vs. 25%) and also equally about unfamiliar women (11% vs. 10%). The gender difference on the Familiarity Percent is therefore created by the presence of more familiar women in women's dream reports and more unfamiliar men in men's dreams. This finding suggests that familiar women (such as mothers, sisters, and women friends) are of greater concern to women and that unfamiliar men are of greater concern to men.

These results take on further interest when the nature of the aggressions in dream reports, along with the patterns of aggressive and friendly interactions, are compared in the normative dream samples. At a very general level, the dream reports of American women and men are similar on aggression. About the same percentage of the reports have at least one aggressive interaction (44% for women, 47% for men), as well as a similar number of reports with at least one friendly interaction (42% for women, 38% for men). However, women have a lower A/C Index than men (.24 vs .34, $h=.26$, $p<.0001$). By way of contrast, women and men have the same rate of friendliness per character, .22 for women and .21 for men. If the aggressions are categorized as physical or nonphysical, then a very big difference appears. The Physical Aggression Percent for women is far lower than it is for men (34% vs. 50%, $h=.34$, $p<.0001$). Conversely, the aggression in women's dream reports is twice as likely to involve rejections and exclusions than in men's dream reports (36% vs. 18%, $h=.41$, $p<.0001$). This is the largest gender difference that has been found.

The findings on gender differences on physical aggression and relational aggression lead to studies of the patterns of aggressive and friendly interactions with specific categories of characters. This comparison is made by determining the Aggression/Friendliness Percent (A/F%), which is calculated by dividing the total number of aggressions with characters in any given category by the total number of aggressions plus friendly interactions with characters in that category. A percentage under 50% means the dreamer has more friendly than aggressive interactions with that character. A percentage over 50% means the dreamer has more aggressive than friendly interactions with that character. To sharpen the analysis, "friends" in dream reports can be defined as characters with an A/F% of 40% or lower and "enemies" in dream reports can be defined as characters with an A/F% of 60% or greater. Characters in dream reports with an A/F% between 41 and 59 are neither friends nor enemies.

Based on this metric, known men are neutral for women dreamers, at 41%. Unknown men, though, are enemies, 62%. For men, unknown men are enemies to an even greater extent than for women, 73%, while known men are neutral at 50%. In women's dreams, neither known women (49%) nor unknown women (48%) are friends *or* enemies. In men's dreams, however, both known women (37%) and unknown women (36%) are friends. These findings are displayed in table 3.5.

In a nutshell, then, women find both their friends and their enemies among male characters. Men are friends in women's eyes if they are known and enemies if they are not known. Further, women have mixed relationships with both known and unknown women characters. On the other hand, men have a very different pattern. They have clear friends (all women,

Table 3.5

Aggression/Friendliness percent for known and unknown male and female characters in the Hall/Van de Castle normative sample of the dream reports of women and men

	Unknown A/F %	Known A/F %	h
Female norms			
Male characters	62	41	−.43*
Female characters	48	49	+.03
	Unknown A/F %	Known A/F %	H
Male norms			
Male characters	73	50	−.47*
Female characters	36	37	+.02

* *p* < 0.01.

whether known or unknown) and clear enemies (unknown men) in their dream reports. Known men are neither friends nor enemies in men's dreams.

Waking studies of American women and men support the hypothesis that the largest gender differences in dream content are continuous with waking thoughts and concerns. In a study of short stories by well-known authors of the late 1950s and early 1960s, 18 written by women and nine written by men, the Male/Female Percent in the stories by women authors was 54%/46% and the Male/Female Percent in the men's stories was 80%/20% (Hall & Domhoff, 1963b, p. 280). This finding is similar to what is found in dream reports using the Male/Female Percent. A second study asked 40 college women and 35 college men to write down the initials of people they liked for one minute and to write down the initials of people they did not like for another minute. Half the participants started with the initials of those they liked, half began with the initials of those they disliked. The Male/Female Percent, based on the total number of initials for women and men that were written down, was 41%/59% for women and 61%/39% for men ($h=.44$, $p=.08$) (Hall & Domhoff, 1963b, p. 280). This difference is very similar to the difference found in dream reports.

Studies of waking gender differences by research psychologists, many of whom have a focus on developmental psychology, provide further support for the hypothesis that the concerns expressed in dreams are continuous with waking concerns. Men's inclination to be involved in physical aggression is perhaps the largest of the few gender differences in studies of women and men in the United States, with an effect size of .55 (Hyde, 2014, p. 385). Moreover, just as in dreaming, women make greater use of relational aggression in waking life, such as exclusion, rejection, and criticism. On this issue, the difference is not as large in waking studies as it is in dreams (Bussey, 2013, pp. 87–88; Leaper & Farkas, 2015, p. 830). Then, too, it seems likely from varying types of evidence that men react differently to threatening male bodily postures and faces, even while focusing more often on women's faces in nonthreatening situations (Kret & de Gelder, 2012; Kret, Pichon, Grèzes, & de Gelder, 2011). Men also commit 85–90% of homicides, and mostly kill other men. To the degree that women kill anyone, it is usually an abusive male partner (e.g., Kellerman & Mercy, 1992). In a classic anthropological study, which compared highly male-dominant societies with those in which the women had considerably more power, it was nonetheless the case that men in the latter societies would organize to kill women. They did so if they thought it was necessary to change the social system in the face of challenges from other cultural groups. On the other hand, women were unwilling to band together to kill men (Sanday, 1981, pp. 210–211).

Overall, this third analysis of continuity between the concerns expressed in dream reports and waking life demonstrates how an unanticipated gender difference on the Male/Female Percent leads to further findings in dream reports. Those differences involve more unfamiliar men in men's dream reports, along with a higher Physical Aggression Percent for men. There are also differences in the patterns of friends and enemies in women and men's dream reports. These various differences between women and men in dreams in turn lead to continuities with differences in the waking personal concerns of women and men.

More generally, this section has used three different avenues—one based on dream reports collected in a hunting and gathering society over 80 years ago, one based on the normative findings on the negativity bias in the contents of white American women and men's dream reports, and one based on the small handful of medium and large gender differences in dream content—to provide solid evidence for an important tenet in the neurocognitive theory of dreaming. The conceptions and personal concerns discovered in dream reports are similar to those in waking life. The evidence for this claim is expanded in chapters 4 and 7.

Conclusions and Implications

This chapter has placed considerable emphasis on sample sizes, methods of analysis, and statistical tests. These issues are emphasized because numerous studies show there are no substitutes for well-designed studies with large sample sizes, which can be and have been replicated. Most of the content-oriented studies of dreams used in this book therefore have very large sample sizes and only are used with confidence when the p value is .01 or less in both the original and replication studies. Moreover, most of the basic studies used in developing the neurocognitive theory of dreaming have been replicated twice or more.

Based on the importance that should be attached to sample sizes, statistical procedures, and replication studies, it seems likely that many past studies are not useful in developing a neurocognitive theory of dreaming or any other type of dream theory. Some of these studies have been critiqued in earlier analyses (e.g., Domhoff, 1999, pp. 120–124, 129–134; 2003, chap. 2; 2005; 2017, pp. 25–33). Still others are discussed in terms of their specific shortcomings in relevant contexts in chapters 8, 9, and 10. Such studies are often mentioned in review essays because they put forth possible implications. These implications generate endless speculations that are never tested. If small and unreplicated studies are given the same weight as large-scale replicated studies, it hinders the process of altering or abandoning theories that have no other evidentiary basis.

In terms of substantive findings, this chapter shows dream content differs relatively little by age, nation-state, or culture. The most important of the few exceptions concerns the frequency and nature of aggressive actions. These studies of dream reports from widely varying societies further show there is nothing in dream reports that cannot be encompassed by the comprehensive HVdC coding system. The differences are in degree and in clusters of indicators (such as those concerning aggression) but not in kind. However, there are just enough differences from society to society, especially in the case of non-Anglophone nation-states, and even more so with indigenous societies, to conclude it would not be methodologically sound to use the HVdC norms for women or men to study groups within those societies. Although culture is not a prominent variable in dream reports, culture does matter in shaping societies and relationships between societies. It would not be helpful to overlook culture in studies of dream content before many more results are available.

The chapter also demonstrates that there is a pattern of relatively small gender differences, but the differences are larger on aggression, and especially on physical aggression. Then too, there are differences among dreams in terms of their content. About 70–75% of the dream reports of Americans include personal concerns, with the remaining 25–30% in the realm of adventure dreams. The great majority of dream reports include aggression, friendliness, or sexuality, and also directed or joint activities. A small minority include only the dreamer (4.3%) or only the dreamer and one or more animals (2.2%). In addition, there are differences in the degree to which dream reports contain familiar or unfamiliar settings and characters. Dream reports with only unfamiliar settings and characters also contain more aggressions and misfortunes.

When all of the dream reports in the HVdC normative samples are analyzed, in women's dream reports there are 1.40 times as many "negative" elements (aggressions, misfortunes, and failures) as there are "positive" elements (friendly acts, good fortunes, and successes), and in men's dream reports there are 1.65 times as many negative elements as there are positive elements. These findings are consistent with the negativity bias found in waking thought and therefore provide good evidence for the continuity between the concerns that are expressed in both dream reports and waking thought.

These and other findings discussed in this chapter are important in and of themselves, especially in narrowing the theoretical focus of attention to what actually appears in dreams. Then, too, the findings also are important for laying the foundations for many of the analyses presented in later chapters. Those analyses begin with the large individual differences in what people dream about, which are discussed in chapters 4 and 7. The numerous findings on dream content overviewed in this chapter also provide a good basis for examining the adequacy of a variety of theories in terms of their ability to incorporate and explain replicated descriptive empirical findings, as shown in chapters 9–11.

4 Findings on Dream Content in Individual Dream Series

Dream journals, which usually consist of hundreds to thousands of dream reports, are kept by a very small but diverse set of individuals in many parts of the world. These individuals keep a record of their dreams for their own separate reasons, which only rarely have anything to do with psychotherapy or an interest in any dream theory (A. Bell & Hall, 1971; Domhoff, 2003, chap. 5; 2015; Hall & Lind, 1970; M. Smith & Hall, 1964). The dream series contained in dream journals are a form of unobtrusive, nonreactive archival data. They are uninfluenced by demand characteristics, expectancy effects, and social desirability factors, which can arise as subtle confounds in experimental settings, whether using human participants or other animals (Kihlstrom, 2002; Orne, 1962; Rosenthal, 1976; Rosenthal & Ambady, 1995; Webb, Campbell, Schwartz, Sechrest, & Grove, 1981).

Unobtrusive measures have enjoyed a respected methodological pedigree in psychology and related fields since the early 1940s. They have led to new insights in the study of personality, social psychology, and community psychology (Allport, 1942; Webb et al., 1981; Whitley & Kite, 2013). Nonetheless, they are not frequently used because most researchers prefer to do experimental and observational studies whenever it is possible. In addition, there are relatively few sources of good nonreactive data for most substantive issues. Perhaps the biggest exception to the lack of good nonreactive archival data is in the field of dream research. It is also the area of study that is most limited in terms of the possibilities of experimentation and in which it is impossible to obtain dream content until after the fact. When the difficulties of obtaining even four or five dream reports from volunteers over a one-week or two-week time span are added to the picture, the possibility that dream series might provide useful data becomes of greater interest.

Most analyses of individual dream series are based on Hall/Van de Castle (HVdC) content categories, which make it possible to compare the results with the replicated group norms for American men and women (Hall et al., 1982; Hall & Van de Castle, 1966; Tonay, 1990/1991). These comparisons with the HVdC normative findings also lead to inferences about the dreamers' waking thoughts, which are based on statistically significant deviations in any given dream series from the norms. In that regard, the analysis of multiple content indicators in a dream series can be placed in the same realm as standardized personality tests, which are based on numerous questions and also rely on group norms. Thus, studies of dream series are not similar to "case studies." Case studies usually do not include the same detailed, systematic quantitative information from individual to individual and therefore cannot be compared with quantitative, group-based norms.

Quantitative studies of about two dozen dream series, using a variety of HVdC coding categories, have shown that the main characters, avocations, and personal preoccupations in them are consistent over months, years, or decades (Domhoff, 1996, chap. 7; 2003, chap. 5; Lortie-Lussier et al., 2000). This past research reveals three types of consistency are possible: absolute, relative, and developmental. Absolute consistency, which also can be called constancy, means the frequency of a dream element varies very little over time periods of several months or years. Relative consistency is defined in terms of one or another dream element always exceeding the frequency of a related element, although one or both of them may increase or decrease in frequency. Developmental consistency is defined as a continuous increase or decrease in any dream element.

Moreover, dream series provide a good opportunity for studies that examine the degree to which the conceptions and concerns found in dream reports also might be present in waking thought. This search for continuities between dreaming and waking thought can be carried out by asking participants to confirm or reject a series of inferences that are based on blind quantitative analyses of a dream series. In these studies, the researcher knows little or nothing about the dreamer at the outset. Studies of roughly a dozen dream series based on this strategy have shown that there is usually considerable continuity between the frequency with which specific people and avocations appear in a dream series and the intensity of waking personal concerns about those people and avocations. These studies thereby establish a general rule: frequencies in dream series reveal the intensity of concerns in waking life (Bulkeley, 2012, 2014; Domhoff, 1996, chap. 8; 2003, chap. 5; 2015).

Finally, studies of ongoing dream series provide an excellent way to test claims based on questionnaire studies or two-week dream diaries. This is particularly the case when researchers allude to some recent event as part of their questionnaire studies. By asking people if they have had a dream about a recent dramatic event, such as an earthquake, researchers may inadvertently introduce implied demand characteristics. This may be especially the case for participants who follow the media closely or have a strong desire to tell researchers what they think the researchers want to hear. This possibility is examined in this chapter in a study of six ongoing dream series in regard to the concern about the coronavirus pandemic and COVID-19. All of these participants had been keeping a record of their dream reports for their own personal reasons, well before the investigator thought about doing such a study.

Most of the findings in this chapter have been presented in detail in published sources, so they can be presented in a more summary form here. The emphasis in this chapter is therefore on the theoretical implications of the various studies. In addition, most of the dream series discussed in this chapter can be examined by readers on DreamBank.net, an online resource, which is discussed early in the chapter. Before turning to the results of these varied studies, however, it is essential to provide the methodological and statistical rationales that underlie these studies. If these rationales make sense to readers, then the study of dream series can contribute to the creation of a plausible neurocognitive theory of dreaming and dream content.

Safeguards, Methodological Strategies, and Statistical Rationales

The consideration of a dream series for possible use begins with efforts to determine its authenticity and accuracy. Dream reports posted on the Internet are never used. The veracity of such reports and the motives for posting them cannot be known with any certainty. Dream reports written down from memory months or years after they were thought to have occurred are not useful. They are not part of any dream series that has been studied. It is also necessary to ask the person why they began to write down or voice-record their dreams to begin with, and also to ask why they decided to offer their dream reports to dream researchers at the time they did so. People are usually very candid in response to these questions. Their answers often lead researchers to write tactful expressions of regret that it is not possible to study their dream journal. It is also necessary to ask for photocopies of pages from handwritten dream series or to examine Word files for information on the dates on which the first entries were made. It is useful to search the Internet for possible information on the actual origins of dream series that seem to be questionable as to their authenticity.

If a decision is made to examine the dream series, it is essential to read through it before deciding to study it. There is no certain way to know that any given dream report collected in any setting, whether a sleep-dream laboratory or otherwise, is fabricated, slightly altered, or authentic. Two past studies found some content differences in the two types of reports, but one did not. Using rating scales, one found that it was possible for independent judges to distinguish made-up and authentic dream reports, but the other did not (see Domhoff, 1996, pp. 47–48, for a summary). In the Berkeley normative study of women, a later post-collection information questionnaire found that 3.8% of the participants said they made up one or more of the reports. Fully 43% of the students in the class, many of whom did not participate in the study, reported they probably would have made up dream reports if participation had been required (Tonay, 1990/1991).

Methodological and Statistical Issues

As noted and demonstrated in chapter 3, the first crucial methodological issue confronting dream research concerns the length of dream reports, which can vary greatly from sample to sample. If there is no control for dream length, results often cannot be replicated. To examine this issue in detail, a random sample of 250 dream reports from the first 3,116 dream reports in a lengthy dream series were studied. This series carries the pseudonym Barb Sanders. These dream reports, which ranged in length from 50 to 300 words, as in the HVdC normative samples, first were coded for the most important and length-sensitive coding categories—characters, social interactions, and misfortunes. The analysis showed the HVdC content indicators control for length within this range, including for the at-least-one indicators. In addition, random samples of 200 shorter and 200 longer dream reports from this series were coded for the same content categories. This analysis revealed that the content indicators, whether they involve percentages or ratios, cannot adequately correct for length in dream reports with less than 50 words. However, they controlled for length perfectly for dream reports with more than 300 words. These results support the decision by Hall and Van de Castle (1966) to exclude dream reports shorter than 50 words, which was based on their unpublished results.

However, the various at-least-one indicators cannot control for dream length above 300 words because they do not include controls for length. Since dream reports with more than 300 words have increasingly more of several types of dream content in them as they grow longer, at-least-one indicators should not be used with the relative handful of very long dream reports in a dream series (Domhoff, 2018b, for details).

Sample Sizes

As also noted and demonstrated in chapter 3, sample size is another major methodological issue that faces all dream researchers. A study was therefore carried out using the HVdC codings for the same random sample of 250 dream reports from the Barb Sanders series, using approximate randomization. Based on 10,000 resamplings, *any* sample of 125 of the 250 dream reports replicates the overall results almost exactly. However, as was the case with the studies of the HVdC normative sample discussed in chapter 3, there are many deviations from the overall results with subsamples of 100 dream reports. The drop-off is large at 75 and 50 dream reports, demonstrating once again how risky it is to accept or reject hypotheses based on inadequate samples sizes (J. Cohen, 1977, 1990, 1994; Domhoff, 1996, pp. 65–66; Domhoff & Schneider, 2008a).

The Issue of Autocorrelation

The issue of autocorrelation is a concern in any psychology study that includes repeated responses from the same participant. In dream research, the possible lack of independence among a series of responses is unlikely in terms of finding a false consistency. This is because autocorrelation is thought to be increasingly unlikely as the time between responses increases, from a few seconds to a minute to several minutes, and so on. Since dreams usually are recalled only three to five times a week by many of those who keep dream series, autocorrelation is less of a possibility. However, the issue nonetheless needs to be addressed statistically.

The Wald and Wolfowitz (1940) runs test, a statistical technique designed to test for randomness in categorical time-series data, can detect the presence of any form of dependency within a dataset. This generalization includes monthly or seasonal cyclical patterns. It is the best-suited test for categorical data because the frequently used Durbin-Watson test assumes at least an interval level of measurement (Klingenberg, 2008). The Wald-Wolfowitz test determines whether there is any pattern in the "runs" that appear. (A run is defined as a series of similar responses, followed by a dissimilar observation, which in turn may or may not become the first instance in another run.) The application of the runs test begins by counting the observed number of runs and comparing them with the expected number of runs. The expected number of runs is determined by a formula that accounts for the number of runs and the overall frequency of occurrence for each type of observation. Using a formula based on the difference between observed and expected proportions, a Z score and p value are calculated.

In a study of five different dream series, 125 runs tests were carried out on several hundred dream reports. These dream reports had been previously coded for various HVdC categories as part of substantive studies (Domhoff & Schneider, 2015a). The analysis of these 125 different sets of codings, all of which are longer than most of the sets of repeated responses analyzed in other types of psychological studies, ranged from 86 to 171 dream reports. The analysis resulted in only six statistically significant results, five at the .05 level and one at the .01 level. The percentage of statistically significant p values, 4.8% at the .05 level and 0.8% at the .01 level, is very close to what would be expected by chance for each significance level (Domhoff & Schneider, 2015a, pp. 73–74, table 1). Since the failure to reject the null hypothesis of randomness also can be due to very weak autocorrelations, a one-period lagged phi coefficient also was computed for each dream series in order to examine the results more rigorously. These first-order phi coefficients were extremely small, which suggests the results of the Wald-Wolfowitz test are accurate (Domhoff & Schneider, 2015a, p. 73).

Based on these results, it is very unlikely autocorrelation will be found in any dream series that has been carefully checked for the authenticity of the dream reports to avoid hoaxers. However, if there are any doubts about any given dream series, it can be checked for autocorrelation with the aforementioned utility, which can be used for studies of HVdC categories.

The Issue of Multiple Tests on the Same Sample

The problem of generating false positives by carrying out numerous tests on the same sample is as potentially serious for the study of a dream series as it is for studying representative samples of dream content, as discussed in the previous chapter and in more detail elsewhere (Domhoff & Schneider, 2015b). For purposes of examining this issue in dream series, the codings for the Barb Sanders sample of 250 random dream reports was studied, using the Benjamini-Hochberg statistic for ascertaining the false discovery rate (Benjamini & Hochberg, 1995; Benjamini & Yekutieli, 2001). Nineteen content indicators were calculated and compared with the HVdC female norms. Twelve of the 19 indicators were originally statistically significant at the .05 level; 5 of those 12 were also significant at the .01 level. After applying the Benjamini-Hochberg correction, 10 of the 12 previously significant p values remained significant at below .05, with four of five remaining significant below .01. The two measures that crossed over from significant ($p < .05$) to nonsignificant ($p \geq .05$) had small effect sizes ($h = .11$, $h = .16$) (Domhoff & Schneider, 2015b). The similarity of these results to those for the HVdC normative samples for both men and women strengthens the case for the findings with dream series.

In addition, there are good reasons to assume approximate randomization works well with a dream series from one individual, mainly because there is no assumption of independence built into this approach (R. Franklin et al., 1997). Further, there is evidence that the h statistic for determining effect sizes can be used with a dream series. This evidence is based on a simulation study, which made use of approximate randomization with 10,000 resamples. The simulation study demonstrated that Cohen's d statistic for determining the effect size between two independent means is as accurate with repeated measures from one individual as it is with representative samples based on numerous individuals (Dunlap, Cortina, Vaslow, & Burke, 1996). Since proportions are a special case of the mean, this result holds for h as well.

DreamBank.net: An Online Resource for Carrying Out Content Analyses

The online and publicly accessible website, DreamBank.net, provides a way to carry out all four steps of a content analysis in a matter of seconds or minutes. It does so through an algorithm that makes the use of word strings possible. (Simple examples of word strings might include father|mother|

sister|brother in the case of characters or living_room|dining_room|kitchen| bathroom|bedroom in the case of indoor settings.) The word strings entered into the search program are also the definition of the category's breadth and limits. Since the same results are guaranteed each time the sample is searched, the categories have perfect reliability. The search engine also provides instant frequency counts. It also creates analyses in the form of percentages, which are "at-least-one" indicators in the parlance of the HVdC coding system. The fourth step of a content analysis, a comparison with a normative or control group, can be carried out by using any given word string to analyze the extant 491 normative dream reports from women and the 500 normative dream reports from men. Then, too, relevant dream series available on DreamBank.net can be used as control groups. Dream-Bank.net contains roughly 34,000 dream reports, 26,000 in English and 8,000 in German. Most of these dream reports are part of several lengthy dream series.

Two types of word strings, generic and individually tailored, can be created for use on DreamBank.net. They are constructed with a slightly modified version of the Perl language for "regular expressions," which are codes used for pattern matching in computer programming (Domhoff & Schneider, 2008b; A. Schneider, 1999). Generic word strings concern topics likely to be relevant in the lives of most individuals. In principle, they can be used on a wide range of representative sets of dream reports as well as on individual dream series. Individually tailored word strings are separately created for each new dream series.

Generic word strings have to be based on a relatively circumscribed number of words commonly used by most people. Even though generic word strings rely on commonly understood words, they are susceptible to both false positives and false negatives, mostly because many words have multiple uses. They also have to be assessed for statements that deny the presence of an element (e.g., there were no trees in the backyard, there were no books on the shelf). This is especially the case if the word strings are not constructed using the language of regular expressions, which can eliminate some problems. Then the results with generic word strings usually have to be searched for false positives and false negatives, using the "show dreams" utility on DreamBank.net. Such searches can be carried out in a reasonable amount of time. The search terms appear in boldface when the relevant dream reports are displayed.

There are several simple generic word strings, including one for sensory terms, which was very useful in studying dream reports from people who were blind from birth or became blind later in their lives (Hurovitz, Dunn, Domhoff, & Fiss, 1999). (The results from this study are utilized in chapter 6 as part of an analysis of the development of visual mental imagery in children.) The others were primarily developed for demonstration purposes (Domhoff, 2003, chap. 4; Domhoff & Schneider, 2008b). However, there also is a useful word string for finding the degree to which religious elements are present in a series of dream reports.

Religious Elements in Dream Series and in the HVdC Normative Dream Reports

The religion word strings are built on coding categories created by other dream researchers on the basis of the same breakdown of dream content into settings, characters, objects, and activities used in the HVdC coding system (Krippner, Jaeger, & Faith, 2001). Such an adaption was possible because the terms used in relation to religious and spiritual concerns are limited in number. Religious settings, such as churches, synagogues, temples, and mosques, are relatively few in number, and the characters in religions, such as gods, angels, and spiritual leaders, are also few in number. Religious activities involve such matters as praying and worshipping. Religions usually have only a relatively few sacred objects, such as crosses, rosaries, and altars. Nor do any of various terms in this word string usually have additional, nonreligious meanings, except when used figuratively or in analogies (e.g., it was a "church-like" situation). It is therefore necessary to use the "show dreams" utility on DreamBank.net to search for false positives.

The religion keyword searches initially focused on the two normative samples, which found there were few religious terms once false positives were eliminated. After eliminating false positives, only 3.6% of the dream reports had at least one of these terms. Specifically, 2.5% of the dream reports mentioned places of worship. Only 0.8% of the dream reports mentioned specific religions or religious denominations. Only 0.7% mentioned religious objects. The percentages were even smaller for religious concepts, such as "worship," "divine," and "nirvana" (0.5%), and for religious leaders, such as deities, ministers, and priests (0.4%). The most prominent religious object in the dream reports was the altar, which sometimes appeared in dreams about weddings. When 10 dream series from adult men and 10 dream series from adult women were searched, similar results emerged. However, there were some individual differences, and the dream series with the highest frequencies for each of these categories, as well as overall, were significantly different from both the HVdC normative dream reports for women, as well as from the cumulative results for the other nine dream series from women.

In addition to the generic word strings for sensory terms and religious elements, word strings have been created for other topics including perception, cognition, culture, natural elements, and emotions, along with numerous subcategories. They have been used to study consistency and continuity in several dream series (see Bulkeley, 2014 and 2018, for findings and references). These word strings are part of the Sleep and Dream database at http://sleepanddreamdatabase.org/.

Individually Tailored Word Strings

As stated above, individually tailored word strings have to be created for each new dream series. They are therefore less likely to have defects. The process of creating them begins with a reading of several dozen dream reports in a new series. During this reading, as many repeated personal names for people and places as possible are entered into a Word file, along with repeated words and phrases. Separate word strings for characters, activities, places, avocations, and other common and well-understood topics, which are designed to parallel the HVdC coding categories, are then created from the initial compilation of words. The initial word strings are next extended and refined into revised word strings. This step includes reading through the dream reports found with the original string. At this point the revised word strings are again refined and some new word strings may be created as well. Through repeated iterations of this simple but sometimes tedious process, many different word strings can be developed for the study of any dream series. The updated strings and any new word strings added on the basis of the later iterations are then used in searches of the entire series or subsets of it. Individually tailored word strings can consist of a few simple separate words (e.g., ^my_mother^ ^my_father^ ^my_older_sister^ ^my_younger_brother^) or slightly more complex formulations (e.g., my_mother|my_father|my_brother|my_sister).

With these methodological and statistical considerations serving as a backdrop, the chapter turns to the substantive issues of consistency and continuity in dream series.

Consistency in Two Very Different Dream Series

The dream series discussed in this section come from different eras and the opposite ends of the adult age spectrum. The first one was kept by Jason, out of his own personal curiosity. He wrote down his dreams from age 37 until the time of his death in his 70s. He was born early in the twentieth century, earned a Ph.D. at a major university, and was employed as a professor at several different universities for the rest of his working life. Since Calvin S. Hall came to know him slightly as a colleague at one point in their careers when they taught at the same university for several years, there is every reason to accept its authenticity. The two professors then stayed in touch about the dream series as they both moved on to several different universities. This dream series is not available on DreamBank.net. Jason, who was a very private and strait-laced person, stipulated before he shared the dream series that his dream reports could not be published (Domhoff, 1996, pp. 133–143).

Kenneth, on the other hand, wrote down every dream he remembered during his first three years of college in the late 1990s, which adds up to 2,222 dream reports. He sometimes documented two or more dream reports from the same night, which led to a mean of almost two dream reports a night. After he finished his undergraduate years, he went to graduate school in a physiological field. After a long conversation with this author about the quantitative content analysis of dream reports when he happened to be attending a meeting of sleep researchers, Kenneth unexpectedly said he had a dream journal. He then offered it for quantitative studies, which was well after he had stopped documenting his dreams. He earned a Ph.D. in physiology and later worked in the private sector. Kenneth's dream reports are available on DreamBank.net.

Jason's Dream Reports at Ages 37–54, 55–59, and 70–73

Jason was not a frequent recaller, and he made no special effort to improve his recall. During one four-year period, which began eight years after he started keeping a record of his dreams, he wrote down only 15 of his dreams. His yearly recall totals for the eight years before the four-year lull ranged from 11 to 103, for a mean of 49.5. For the five years after the years of low recall, his recall ranged from 13 to 70, with a mean of 37. 8. The fact that the dream series is sparse necessitated the use of only six subsets of 100 dream reports to analyze it for the age period 37 to 54.

To begin with, there was excellent absolute consistency on the Male/Female Percent, as has been found with most later dream series, with a few exceptions (Hall, 1984). There was good absolute consistency on several other general character categories, such as known adult men and women strangers. There was good absolute consistency for his Friendliness per Character Index (F/C Index) with all characters. His Aggression/Character Index (A/C Index) showed adequate absolute consistency but was slightly higher in the last two subsets. His A/C Index was always higher with women than with men, which is the opposite of the case with his F/C Index (Domhoff, 1996, p. 139). This is an example of relative consistency.

There was one developmental regularity. In each of the six subsets there is an increasing A/C Index with his wife and daughter, and the increase is very large in the last two subsets. This development regularity is displayed in table 4.1 as part of a larger table, which also provides instances of absolute constancy and relative consistency in the Jason dream series. Unfortunately, it was not possible to determine the degree to which these changes were continuous with changes in his waking thoughts about his wife and daughter. This was because he did not want to divulge any information about his waking life.

Table 4.1
Types of consistency in the Jason dream series, ages 37–54

	Set (100 dreams each)					
	I	II	III	IV	V	VI
Absolute constancy						
Male/Female %	63/37	61/39	57/43	63/37	62/38	60/40
Relative consistency						
F/C with male characters	.35	.37	.39	.47	.31	.35
F/C with female characters	.22	.15	.10	.27	.19	.28
Developmental regularity with female family members						
A/C with wife and daughter	.17	.19	.21	.25	.33	.68

Consistency and Change between Ages 37–42 and 48–53

The general consistency in Jason's dream content, despite major life changes, was examined more closely by comparing his dream reports from ages 37 to 42 to those when he was between 48 and 53. Jason was married and a parent of a child while recording the first subset, but by the time of the second subset he was legally separated from his wife and retired. He traveled often. He lived in two foreign countries for several months each, and eventually settled far from the area in the United States in which he spent his career. In addition, his daughter was a college graduate and lived in a state distant from where he lived.

Jason's A/C Index and F/C Index indicated absolute consistency between the two subsets. There was a relatively new cast of characters, with the exception of his immediate family members, a high-school friend, and a favorite professor from his college years. All those holdover characters appeared less frequently than they did in the first five-year subset. The known men in the second subset were mostly his friends, as in the first subset analyzed in this subsection. They were not the same friends that appeared in the first subset, who had almost entirely disappeared from his waking life as well as his dream life by age 53. However, despite this turnover in specific individuals, his percentage of known characters remained consistent. This finding shows that absolute consistency can occur within general character categories.

Consistency in the Jason Series at Ages 40+ and 70+

Finally, consistency in the Jason series was analyzed by comparing the first 100 dream reports from his early 40s with the first 100 from his early 70s. There were few or no changes on the main content indicators. For example, there were three acts of physical aggression in the first subset, two in the second subset. Nor were there any differences in mentions of injury, illness, or death. Based on the elements from the past scale in the HVdC coding system, the present-mindedness of his dream life is shown by the fact that he dreamed of the past in 10% of the dream reports in both subsets.

The present-mindedness of Jason's dream reports in both subsets is further demonstrated by the finding that he once again dreamed most frequently of his friends at the time, just as he did in the comparison of two subsets in the previous subsection. There was very little carryover of friends from his early 40s to his early 70s, a 30-year gap. The few characters who overlapped in the two subsets were once again his former wife, his daughter, his father, a male friend from the past who he still saw on occasion, and the same favorite professor from his college years. By then, these five important people from his youth and young adulthood had appeared as characters in his dream reports over a 33-year time span (Domhoff, 1996, pp. 142–143).

An Extensive Three-Year Dream Series from a College Student

Kenneth, as already mentioned above, recalled dreams frequently, at least during his first three years of college. He claimed he wrote down every dream he remembered. The issue of consistency is therefore packed into a much shorter time span in this dream series, and it involves nearly four times as many dream reports. For a HVdC study of consistency, the Aggression/Friendliness Percent and the percentages for at least one aggression, friendliness, and sexuality were utilized. This study focused on his first 100 dream reports from 1996, the first 100 for 1997, the first 100 for 1998, and the last 100 dream reports he recorded, which were written down in the fall of 1998. There was absolute consistency on all four indicators in the first three sets. In the fourth set, the A/F%, at least one aggression, and at least one friendliness declined slightly, and there was a slight increase in sexuality. There was relative consistency among the four indicators as well. These results are shown in table 4.2.

For purposes of word string studies, the full series of 2,022 dream reports was divided into four subsets of 506 dream reports per subset (or 504 in the case of the last subset, to be exact). Kenneth's consistency on words relating to religion and sexuality was examined using the search engine

Table 4.2

Consistency in social interactions among four sets of 100 dream reports from Kenneth

	Set (100 dreams each)			
	I	II	III	IV
Aggression/Friendliness Percent	66	68	67	64
At least one aggression (%)	57	53	55	42
At least one friendliness (%)	37	39	37	33
At least one sexuality (%)	9	9	8	11

on DreamBank.net. There was considerable consistency in sexual content, but the findings with the religious word string were markedly higher in the third subset than in the other three subsets.

Kenneth's series was also studied for consistency using several individually tailored word strings that were created using the procedure discussed above. Since Kenneth appeared upon inspection to dream frequently of his family, his friends, food and eating, cars and trucks, taking part in outdoor activities, and playing sports, six search patterns were created. His parents appeared consistently across the four sets, ranging from 17.0% to 21.1% of dream reports. His friends appeared 47% of the time overall, with a slightly wider range of 42.9% to 51.4% for the four subsets. When combined into a single search term, either a parent or a friend appeared in 56.3% of Kenneth's dream reports. The other four categories (food–eating, cars–trucks, the outdoors, and sports) showed varying levels of consistency, which is shown in table 4.3. Overall, food-related terms appeared in 17.2% of the dream reports; cars and trucks were mentioned in 25.3%; outdoor activities (hunting, fishing, and boats) appeared in 16.0%; and sports showed up in 12.0%. When all six of the individually tailored word strings are combined, 75.7% of the dream reports mentioned at least one of these terms. The specific figures and trends for both the generic and individually tailored word strings over four subsets are presented in table 4.3.

As the following section shows, there are other examples of consistency in adult dream series, using either HVdC codings or word strings. More

Table 4.3

Frequency of search terms in four subsets of the Kenneth dream series

	I ($n = 506$)	II ($n = 506$)	III ($n = 506$)	IV ($n = 504$)	total ($n = 2{,}022$)
Religion (%)	3.8	3.6	7.1	4.8	4.8
Sex (%)	18.2	15.2	19.4	16.1	17.2
Parents (%)	21.1	20.2	17.0	18.1	19.1
Friends (%)	49.4	51.4	44.3	42.9	47.0
Food & eating (%)	17.8	17.4	19.0	14.7	17.2
Cars & trucks (%)	27.1	26.9	24.7	22.6	25.3
Hunting/fishing/ boats (%)	8.5	20.0	19.6	16.1	16.0
Sports (%)	14.4	12.5	11.7	9.3	12.0

examples could be added (Domhoff, 1996, chap. 7; M. Smith & Hall, 1964). However, the two very different series from two very different people analyzed in this section, with one series covering adulthood and old age and the other documented in the first three years of young adulthood, provide strong replicated evidence for consistency in dream series. This unanticipated finding was first discovered in the Jason series many decades ago but remained unpublished until consistency was documented in other dream series (Domhoff, 1996, chap. 7).

Consistency and Continuity in Three Unusual Dream Series of Varying Lengths

This section discusses three different dream series of varying length from different eras, one from a woman and two from men. These three dream series not only differ in length and era but also have unique features that present potential challenges to the concept of consistency and, to some extent, to the concept of continuity. The lengthy dream series contributed by the adult woman (Barb Sanders) began when she was in her late 20s. It contains over 4,000 dream reports overall, which were documented in the last quarter of the twentieth century. It provided the opportunity for the most detailed study that has been done of any dream series using HVdC codings.

The 173 dream reports written down by an entomologist in a summer in the 1930s is an unusual document from an unexpected source. The section concludes with an analysis of the 143 dream reports a bereaved widower documented about his deceased wife over a 22-year period. It is highly atypical in its own poignant way. The three series are analyzed with slightly differing subsets of the HVdC coding categories and also with two individually tailored word strings in the case of Barb Sanders.

The Dream Reports of Barb Sanders

Barb Sanders grew up in a small town in the West. She is the oldest of four children. She has a brother about three years younger, a second brother about five years younger, and a sister about six years younger. Her parents both had college degrees from a small denominational college. They worked all their lives in education and social work. Both of her parents played musical instruments, as did Barb Sanders and her three siblings, which led to frequent household music and singing sessions.

Barb Sanders married after one year of college, had three daughters in about a five-year period, and completed a B.A. degree at a nearby state college when she was in her mid-20s. She left her husband, a technical employee for a natural resources corporation, when she was 30. She then returned to her home state, where she earned an M.A. in a helping profession and worked in a community college as a counselor for several years. Her three daughters, ages 7, 4+, and 2+ at the time of the divorce, were raised by her ex-husband, although her middle daughter came to live with her about 10 years after the divorce. Barb Sanders had a long-standing personal interest in dreams. However, she did not write any dreams down until a few years after her divorce, at a time when she was having disturbing dreams, including some relating to the divorce and its aftermath. She hoped to gain insight about herself by keeping a dream journal and perhaps use that insight to enhance her efforts at creative writing. She became involved in local theater productions as an actress, writer, and director, and participated in dream groups. She had several boyfriends after her divorce and never remarried.

Twenty years after she started documenting her dreams, she unexpectedly offered them to the author, who had not met her before. She was attending, out of her own curiosity, a meeting that included a panel of dream researchers. Her dream series originally consisted of 3,116 dream reports. It was first analyzed by drawing the earlier-mentioned random sample of 250 dream reports. This sample, which was in effect a "normative" sample of the 3,116 dream reports, was coded for five major HVdC categories (characters, social interactions, misfortunes, and successes and failures). The codings were carried out by three research assistants, who knew nothing about the dreamer. In addition, all 3,116 dream reports were coded by another research assistant for social interactions with family members, along with a few of her friends, her high-school sweetheart, and a younger man she developed a crush on two decades after her divorce. After all of the codings were completed, the author interviewed the dreamer and also did separate individual interviews with four of her close women friends, during a two-day visit to her hometown.

Barb Sanders later provided another 1,138 dream reports that were documented, again for her own reasons, while the initial study was carried out. Subsequent analyses of this second subset showed no appearances by the author or his research assistants, or any evidence that her dream reports were influenced because she provided the first subset for study. This second subset was therefore used in later studies using word strings. All 4,254 dream reports and the HVdC codings for the baseline-250 sample can be found on DreamBank.net, along with partial codings of several hundred other dream reports, and transcripts of the interviews with Barb Sanders and her four friends.

Consistency in the Barb Sanders series Consistency in the Barb Sanders series was initially studied by comparing the first 125 dream reports in the baseline-250 sample with the second 125. It showed the two subsets are consistent within five or six percentage points on 12 content indicators, ranging from the Familiarity Percent, Aggression/Friendliness Percent, Befriender Percent, and Aggressor Percent, to the A/C and F/C indexes and the "at-least-one" percentages for aggression, friendliness, and sexuality. Her Male/Female Percent fell from 58/42 to 49/51, placing her very close to the female norms of 48/52. In the second half of the baseline-250 sample, she dreamed less of friends and more of family members (see Domhoff, 2003, pp. 114–115, table 5.3).

These findings were supplemented by making word and phrase searches using DreamBank.net In this analysis, the first and second subsets within the full series were compared on the frequency of appearances by at least one of the 13 main people in her life—parents, ex-husband, three siblings, three children, granddaughter, and three best women friends. They appeared in 33.6% of the dream reports in the first set and 35.1% of the second set. Her interest in theatrical productions, as a writer, actor, and producer, is reflected in the fact that 4.9% of the dream reports in the first set contained one of several terms related to this activity, as compared to 5.2% for the second set. This strong evidence for consistency over a period of 35+ years in the Barb Sanders series replicates the findings with the Jason and Kenneth series.

Further evidence for consistency for all the characters in the full Barb Sanders series was provided in a network analysis by mathematical psychologists (Han, 2014; Han & Schweickert, 2016). This analysis is discussed later in the chapter.

Continuity in the Barb Sanders series The codings for the baseline 250 made it possible to determine how she did and did not differ from the HVdC norms. At the same time, it established an overall self-norm, which could be used for comparisons with the findings with individual characters. Although she is normatively typical on her Aggression/Friendliness Percent and her Befriender Percent, she is above the norms on the A/C Index (.33 vs. .24, $h = .21$, $p = .000$), the F/C Index (.32 vs. .22, $h = .24$, $p = .000$), and the Aggressor Percent (50 vs. .33, $h = .36$, $p = .000$). These results establish that she is very socially active in her dream reports. This is especially the case for aggressive interactions, which in her case usually involve angry thoughts toward a person, critical comments, or rejections. Moreover, the additional codings for her interactions with several important people in her life made it possible to study continuity in more detail than was possible in past or subsequent studies of any dream series.

Barb Sanders's mother was the most frequent character and also the central figure in her overall network of characters, even though her waking circumstances and her relative lack of contact with her mother would not have predicted it (Domhoff, 2003, chap. 5; Han, 2014). Her mother appears in more dream reports than any other character. Barb Sanders's A/C Index with her is .70, which is above her normative figure in the baseline 250 for all characters (.33). Her F/C Index is a much lower .27, although it is well within her normative range. The A/F%, which is consistent through the first subset, is 72%, well above the dreamer's normative figure of 49%. Barb Sanders reported in the interview that she conceived of her mother as a cold and rejecting person. She said her relationship with her mother remained the number one concern in her life and a constant preoccupation for her. Her friends made similar observations about Sanders's relationship with her mother in the individual interviews with them (Domhoff, 2003, chap. 5).

Sanders's relationship with her middle daughter is almost as problematic for her as her relationship with her mother. As noted above, this daughter was 4+ years old at the time of the divorce. She was also the daughter most upset by the divorce. At age 14, she ran away from her father's home and came to live with Barb Sanders. She did poorly in school, could not hold a job, and suffers from severe psychological problems. She had a daughter when she was in high school and often left her to Barb Sanders to raise. She returned to live with Barb Sanders from time to time. There is a high level of tension between them, and Barb Sanders constantly worried about her. Barb Sanders's A/C Index with this daughter, .92, is far higher than it is with her mother or any other character. The F/C Index is also relatively high at .52, because Barb Sanders is always helping this daughter. Barb Sanders initiates 79% of the aggressive interactions and 70% of the friendly interactions, which are both above her normative figure. By contrast, Barb Sanders dreams only half as often of her oldest and youngest daughters and has more friendly than aggressive interactions with both of them. Sanders is once again more likely to initiate both aggressive and friendly interactions with them.

The fact that Barb Sanders had a primary role in raising her granddaughter, who lived with her, provides an opportunity to examine the way in which changing conceptions and concerns lead to changes in content in the 205 reports with her in them. When the granddaughter was a cute and charming baby, there was a predominance of friendly interactions in the 14 dream reports about her, with an A/F% of 29%. Barb Sanders took the initiative in all the interactions. The predominance of friendly interactions when the granddaughter was a young child is even greater in the 33 dream reports from this period, with an A/F% of 20%, and Barb Sanders initiates 88% of these interactions. However, this friendly conception of her relationship with her granddaughter declined greatly thereafter. The A/F% rose to 69% in the 93 dream reports during her late childhood and to 75% in the 65 dream reports during her adolescent years. Barb Sanders told the author she became increasingly frustrated and annoyed by her granddaughter's many deficiencies as a person and student. Barb Sanders was often highly critical of her in waking thoughts and in social interactions with her (Domhoff, 2018a, p. 103, table 3.4).

Barb Sanders's 164 dream reports involving her ex-husband enact the same negative interactions with him for the first 15 years after they were divorced. This finding is best reflected in the A/F percentages of 57%, 59%, and 61%, in the first three segments of this subseries. All of these figures are above her baseline with all characters, which is 49%. However, her conception of him began to change at the beginning of the fourth segment. Then he unexpectedly died of a heart attack in his early 50s, without any history of serious illness. At this point the dream reports that include him become more reflective and sometimes include the awareness that he is dead, even though she is interacting with him in the dream. The changes in this fourth segment are captured in the dramatic decline of the A/F% from 61% in the third segment to 34% ($h = -.53$, $p < 0.005$).

Still another dimension of Barb Sanders's dream life is demonstrated in the first 43 dream reports she documented about a man she became infatuated with well after she was divorced. These dreams are in the realm of pure fantasy, but they are continuous with waking personal concerns expressed in daydreams. She also talked about him all the time with her friends, they said in the interviews with them. These dream reports portray an arc of passionate initial involvement and her eventual angry rejection of him, even though they knew each other only slightly through occasional involvement in the same social occasions. Nor did he know of her infatuation with him. Thirteen of the first 16 dream reports about him contain sexual or intimate physical interactions, but they also express a fear he does not care about her. A few months later, the dream reports rarely include sexual interactions. Then she documents dreams in which she is angry with him or jealous because he is having sex with someone else. At that point, she broke off her real-life acquaintanceship with him, because he brought a woman with him to a small social gathering she also attended. After appearing in 43 of the 334 dream reports she recorded after her first dream about him, he appeared in only four of the next 461 and then disappeared from her dream life. In complete contrast with her dream reports about her ex-husband, which continued off and on for decades, the dream reports about her infatuation declined greatly in frequency as the infatuation turned to disappointment. They then turned to angry thoughts of rejecting him, and then he disappeared from her dream life (see Domhoff, 2003, pp. 122–126, for a detailed account).

The dream reports relating to the man with whom she became infatuated share some similarities with the 53 dream reports concerning the boy she fell deeply in love with in her senior year in high school. However, these dream reports appeared sporadically throughout the dream series, and they were consistently more positive in terms of the A/F% and the F/C indexes. They often included sexual interactions as well. Her dreams of him primarily express the love she felt for him, along with their intense sexual interactions when they were a couple, which stopped just short of sexual intercourse. In the interview with her, she said the high school sweetheart was similar to the man she developed a crush on decades later in that they were the two men she had loved the most. They were also the two men in her love life with whom she never had sexual intercourse.

However, several of the dream reports about this first love also drama- tize her bitter memories of a relationship he had while he was in naval training. This intimate relationship deeply angered her, even though they had agreed they would see other people while they were separated. As she explained in the interview in March of 2000: "I was so angry and felt so betrayed that at that point." "I said, 'That's it, we're done. . . . I was out, and very, very angry and was not willing to trust him again'" (Domhoff, 2018a, pp. 105–106).

Despite all the negative notes in her relationships with the characters discussed up to this point, the dream series also captures her positive rela- tionships with the favorite people in her life. She paints a more positive picture of her relationship with her father, who is the second-most frequent character in the dream series. She said in the interview she had a more positive attitude toward him in waking life than she did with her mother, as also stated by her friends. In addition, she has great affection for the brother closest to her in age. He appears in 97 dream reports, which is one more than the total for her other two siblings combined. The A/F% with him is 25, which is nearly the mirror opposite of her interaction pattern with her mother.

Even more, she has an extremely positive relationship with her closest friend of long standing, whom she met when she returned to college for her M.A., even after this friend later moved to a city 100 miles away and raised a family. She appears in 96 dream reports and has an A/F% of 23, the most positive balance with any known character, along with a high F/C Index of .89 and a low A/C Index of .26. The comradely nature of their relationship is seen in the fact that they are equally likely to initiate friendly or aggressive interactions.

Barb Sanders's dream reports about another close friend reveal a very different pattern. She takes an initiatory role with this friend, who is many years younger. This younger friend had been a student at the community college where Barb Sanders once worked, giving her instructions, help- ing her, and becoming annoyed when she was late or resisting direction. This pattern is reflected in the fact that Sanders initiates 78% of the many friendly interactions between them, as well as 78% of the relatively few aggressive interactions. Her A/C Index with this friend is a low .39, and her F/C Index is a high .63.

The findings included in these thumbnail sketches of how Barb Sanders's dream reports enact her differing waking conceptions of the significant people in her life, were found in analyses of several other dream series as well. However, those analyses were sometimes limited in scope by the sample size or the lack of full information on the dreamers' waking-life concerns. The analyses of the Barb Sanders series, on the other hand, do not suffer in the slightest from either of these defects. It is highly likely that similar detailed analyses of any large sample of dream reports from an individual would provide similar detailed portraits. Still, the dream series discussed next demonstrate this point well for a more limited set of characters. In conjunction with past analyses and with the two analyses that follow, the studies of dream series overviewed in this chapter fully establish continuity between the nature of dream enactments in a dream series and the dreamer's thoughts about those people in waking life.

The Dream Reports of an Entomologist in 1939

The person given the pseudonym "Natural Scientist" kept a dream journal in the summer of 1939, at age 46 He did so in part because he remembered dreaming "copiously" and having nightmares as a youngster. He also wanted "to learn whether the nature, course, and speed of a dream can be controlled; and I wish to test Freud's belief that the subconscious mind is largely a cesspool." He was also careful to state his method: "My method is to place a notebook at my bedside every night and, on wakening in the morning, or during the night, to jot down the 'high spots' and 'a rough outline' of each dream remembered." He would then write a more complete account.

The Natural Scientist was born in 1893 in a small farming town in the Midwest. He graduated from a state university in Iowa in 1916. Except for nine months in the US Army in 1918–1919, he worked as an entomologist in an agency of the Department of Agriculture in Washington from the time of his college graduation until the late 1940s, when he took an early retirement due to declining health. His work over a period of 30 years focused on trying to construct an accurate taxonomy of the 400,000 species of beetles. He was a lifelong bachelor and did not have any direct descendants. Nor does he have any other traceable descendants. His dream journal eventually was sold to an antiquarian book dealer. It was purchased in the early 1980s by a dream researcher, who generously shared it (see Hobson,

1988, chaps. 11–14). (The Natural Scientist's dream reports, along with his introduction to the dream journal, are available on DreamBank.net.)

Even in the space of just three months, the 234 dream reports provide a large enough sample size to determine there was a very great degree of consistency. Using a range of HVdC coding categories, the first 93 reports with 50 or more words were compared with the second 93 reports with 50 or more words. The details do not need to be repeated by this point in the chapter, because the empirical fact of consistency has been replicated many times (Domhoff, 1996, p. 148, table 7.9). A blind analysis of the characters, social interactions, settings, and emotions in the dream reports provides a revealing picture of the dreamer's waking concerns and interests. This analysis was made possible by a four-page obituary in an entomology journal, which contains personal information supplied by one of his sisters (Domhoff, 2003, pp. 154–157). The dreamer is low on aggressiveness and even lower on dreamer-involved aggression and dreamer-involved friendliness. He is in good part an observer of social interactions and of other people's activities, which is atypical. These findings are continuous with his quiet personality and his focus on observation in waking life, including in his work. There are no sexual interactions in his dream reports, which fits with his bachelor status.

On the other hand, he is above the norms on Animal Percent, which fits with the fact that he liked to fish and hunt and became an entomologist. The animals in his dream reports are primarily birds, barnyard animals, and the insects he studied in his professional life. They often appear when he dreams that he is on the farm on which he was raised and to which he returned occasionally in summers. They also appear in dream reports in which he is working with beetle specimens in his office. He is also somewhat above the norms on familiar settings, which are most often his home, his office, or the family homestead (Domhoff, 1996, pp. 147–150). There is great attention to detail, which parallels the large taxonomic task he had undertaken with beetles. There are also 13 dream reports concerning golf, which he enjoyed in waking life as part of a lifelong involvement in competitive sports.

This quantitative content analysis is also supported by what the Natural Scientist wrote in the introduction to his dream journal: "At the time this is being written, in my 46th year, the majority of my dreams are of the neutral sort (emotionless), neither particularly pleasant nor unpleasant, like the prosaic episodes of everyday life." He further explained he had reached this point as his activities gradually changed over the previous 20 years:

> One by one I have eliminated a good many of my earlier activities, for one reason or another; partly from the ever-strengthening conviction that it is only a simple life that holds for me any chance for a reasonable degree of contentment. The gradual simplification (which of course goes in the face of the much advertised and praised "more abundant life") of my mode of existence is beginning to give me a certain amount of leisure, much of which has been spent in reading miscellaneous literature, and in reflection.

This dream journal, coming as it does from an unlikely source and kept over one summer in the late 1930s, out of his own curiosity as a careful scientific observer and thinker, is an ideal example of an unobtrusive, nonreactive dream series. It strengthens the case that there is both consistency and continuity on most issues in most people's dream lives.

A Widower's Dreams of His Deceased Wife

For 22 years after his wife's death, a bereaved husband ("Ed") wrote down 143 dream reports that included his deceased wife. He did so because it gave him solace and comfort to see her in his dreams. He sometimes included the memories and thoughts the dream had occasioned for him after he documented the dream. Because Ed only recorded dreams that included his deceased wife, Mary, the full range of his dream life cannot be known. Comparisons between dream reports including his deceased wife and his other dream reports cannot be made. Nevertheless, his dream series provides a unique opportunity to study the memories dramatized in relation to his love for her and a marriage of 32 years, during which Ed and Mary raised three children.

After he had been writing down dreams about Mary for 14 years, Ed, by then retired for several years, wrote a reminiscence of his life with Mary. It was part of a tentative plan to publish a pamphlet or book for other grieving spouses. This project was in keeping with the help-oriented occupation he had been involved in for decades after his service in World War II. (However, to quell some likely speculations about his profession, he was neither a clinical psychologist nor a psychiatrist.) The draft manuscript included a commentary on how his dreams about Mary had impacted his life after she died. Although his plan never materialized, his commentary on how his dreams affected his life, along with his reflections on some of his dream reports shortly after writing them down, provides a very useful waking-life perspective for dream researchers. He later offered a copy of his dream series to a psychologist and her colleagues at a nearby university for research purposes (Belicki, Gulko, Ruzycki, & Aristotle, 2003). The dream reports and the HVdC codings of them, along with his reminiscences, are available on DreamBank.net. The full results of the detailed study of this series are reported elsewhere (Domhoff, 2015).

When Mary was 45, she began to suffer bloating and stomach pain, which turned out to be ovarian cancer. Following a radical hysterectomy, she was assured she was cured. She was indeed symptom-free for six years, but the cancer returned during the seventh year after the operation. Medication suppressed the tumors, and there seemed to be a chance she would live normally for some time—or so Ed and Mary thought. The cancer came back a third time during the next year. She had a long and painful terminal illness that ended with her death one year later. Ed was 57, and their children were 27, 24, and 19 when she died at age 54.

The bereaved widower reports that he suffered great agony, loneliness, and confusion during her final illness and after her death. He met another woman, Bonnie, about four months after Mary died, and married her about a year later. However, this marriage did not work out at all well. He and Bonnie often argued, and they separated five years after they married. He remained single for the final decades of his long life. He lived alone in an apartment until he had to move into a home for the elderly. In his reflections, he wrote that his second marriage was a great mistake, which caused shock to his children. He wondered in retrospect how he had ever gotten involved with someone so different from Mary. Mary had been his first and only girlfriend, he wrote. He added that he was a shy person who did not find it easy to take the initiative in relations with women (Domhoff, 2015, pp. 7, 20).

Consistency and continuity in Ed's dream series   Ed's dreams about Mary were relatively infrequent, between four and eight in most years, although he recalled 14 in the 16th year after she died. There was a 14-month hiatus in recall about nine years after she died, which provided a natural dividing point for comparing the content of the early and later parts of the series. There are 62 dream reports in the first part and 81 in the second.

A comparison of these two subsets showed there was considerable consistency over the 22 years of documenting his dreams. There was no change in the Befriender Percent or the Aggressor Percent, nor in the indicators for at least one aggression, at least one friendliness, or at least one sexuality (Domhoff, 2015, p. 243, table 4). There was a decline in the number of dream reports in which Mary is ill or dead, from 39% of the dream reports in the first half of the series, to 23% in the second half. The percentage of dream reports with at least one misfortune also declined, primarily because Mary is less often portrayed as ill or dying in the second subset. On the other hand, the A/F% rises from 17% in the first subset to 28% in the second subset, which suggests the social interactions between Ed and Mary become somewhat more aggressive. However, these trends, which had medium effect sizes, did not reach statistical significance at the .01 level, nor were they statistically significant even at the .05 level after the false discovery rate correction for multiple testing was applied.

All of the 10 dream reports in which Ed was surprised to see Mary was alive again, which are coded as a "good fortune" in the HVdC coding system, are in the first subset. (A good fortune is defined as something good that happens out of the blue.) However, the percentage of dream reports with at least one good fortune was the same overall in both subsets. Ed did have one dream report toward the end of the second subset, nearly 20 years after her death, in which she returns home unexpectedly due to the success of a last-minute surgery. But there is no implication in this dream report that she has been dead, and he is not surprised to see her alive.

In addition to good consistency, there is very strong continuity with his waking thoughts and concerns. This conclusion is based on a comparison of the findings from his dream series with his written comments after documenting some dreams and with his later commentary on his dream reports. It is also based on his response to a question the author sent him via email, when he was in the home for the elderly. First, the Bodily Misfortunes Percent in Ed's dream reports is far higher than in the HVdC normative sample for men (100% vs. 30%, $h = +1.99$, $p < .001$). (The Bodily Misfortunes Percent is defined as the misfortunes to a human body, divided by the total number of misfortunes.) These dream reports embody Ed's upsetting memories about Mary's illness. For the most part, they are realistic portrayals of Ed's thoughts concerning Mary's illness, hospitalization, or impending death during various stages of her illness. In many of these dream reports Mary looks very good or they are embracing, but he is aware that she is dying.

She still looks very beautiful in other dream reports, but there is one or another telltale sign of her illness, such as paleness. Sometimes she is overweight or bloated, but in another dream report she is thin and gaunt, and in two others she is bald. These illness dream reports also portray his concerns about her treatment, as reported in his waking reflections. However, there is no sequence or progression in the dream reports with bodily misfortunes in terms of the increasing severity of her illness and her imminent death. Once again, dreams seem to emerge through a random draw from the dreamer's cognitive rolodex of semantic memories.

In addition, Ed's dreams about Mary are characterized by a large number of friendly interactions (208 in 101 dreams) and a small number of aggressive interactions (64 in 41 dreams). This finding is most clearly demonstrated by the Aggression/Friendliness Percent. The A/F% is 36.4% with women characters in the norms for men, but it is only 22.1% for Ed's interactions with Mary ($h = .32$, adjusted $p = .004$). There is also more friendliness than would be expected in terms of the percentage of dream reports with at least one friendly interaction, and less aggression in terms of dream reports with at least one aggressive interaction. This pattern of findings suggests Ed's dreams act out his predominantly friendly feelings toward Mary, which are evident in his reflections. There are also 25 sexual interactions in 15 of the dream reports, which is close to what would be expected based on the men's norms.

The findings on social interactions become more revealing when they are analyzed in terms of who initiates the three types of social interactions. Ed initiates most of the friendly interactions, 81.2% vs. 18.8%, which is higher than the 67% vs. 33% difference in the men's normative sample ($h = .32$, adjusted $p = .034$). Similarly, Ed initiates 76% of the 21 sexual interactions that do not begin with mutual overtures. In striking contrast, Ed is the aggressor in only 25% of their relatively few aggressive interactions, whereas men initiate 50% of the aggressive interactions with women in the men's normative sample ($h = .52$, adjusted $p = .034$). Put another way, Ed is the victim of Mary's aggressions in 75% of their aggressive interactions. As demonstrated in the following paragraphs, the interaction patterns concerning aggression simulate Ed's conceptions of his relationship with Mary. They are therefore excellent evidence for continuity between dream portrayals and waking thoughts.

There are 64 aggressive interactions between Ed and Mary in the dream series. They are primarily nonphysical in nature, which include hostile thoughts, critical remarks, rejections, and accusations or threats. There are four physical aggressions, in all of which Mary is the aggressor. In one dream, she throws "a ham or a roast" at him. In two, she pushes him away. In another, she takes something away from him. Several other aggression dreams seem to embody the annoyances, tensions, and rejections that often typify married life, such as arguments and criticisms. The most dramatic of the dream reports with aggression in them involve arguments about their sexual life.

In the first of these dream reports, Mary angrily criticizes Ed for not understanding her. In two of these dream reports Mary is critical of Ed about an affair she says that he had with another woman. These accusations, which first appear 16 years after her death, give the impression that Ed is simulating an actual event. In the second dream about an affair, which occurred roughly eight months after the previous one, Ed is in a play or a business venture (he is not sure which). In this dream he must have an association with a woman, and perhaps even kiss her. Friends warn him that people will gossip, and that Mary will hear about it. When this author asked him via email if there was a waking behavioral basis for this accusation, he said that such a relationship did occur. He then recounted a brief encounter, a few years before Mary's final illness began. He said he had thought about this affair quite often over the decades and had deeply regretted it ever after.

In perhaps the most cerebral and revealing of the dream reports in which they argue about sexuality, Mary blames him for their lack of sexual relations. This accusation leaves Ed unable to reply, even though a litany of answers is running through his mind as she is talking. In the long comments Ed writes after documenting this dream, he says he was "very conscious of the argument put forth" by Mary. He felt she was "unfair in equating infrequent sex with my not loving her." He "went over in my mind," while lying in bed after waking up, "what our sex life had been during the years preceding her death." He adds that even though "this dream prompted me, now, to think about this aspect of our relationship," he had "thought about it frequently when Mary was alive" (Domhoff, 2015, p. 20).

Ed then recalls a discussion they once had about her often falling asleep on the sofa at night. At that time, Mary told him that she felt he didn't want sex with her because he didn't urge her to leave the sofa and join him in bed. He then recalls replying he did not come out of the bedroom to the sofa in the living room because he thought she wanted to avoid having sex with him. He then characterizes himself as a person who does not assert himself when he feels he has been rejected: "I bowed out of the act and did not try to get this straightened out." He also comments "This dream had a strong impact on me in reviewing, yet again, our sexual relationship" (Domhoff, 2015, p. 20).

Very clearly, the overall findings from Ed's dream series support the idea that dreams are embodied simulations, which express waking concerns. The main concerns in these dream reports—his wife's health and his relationship with her—enact concerns his written reflections attest to in waking life. In his dreams, he seeks Mary's help or assurance, on occasion enacts fond memories of the good times they had together, often puzzles over the sexual tensions in their marriage, and relives the horrible events of the terminal illness that ended her life. The dream reports express and embody the conflicted thoughts and feelings on the dreamer's mind and in his semantic memory bank during the 22 years after she died. His answer to the author's email query about a possible affair in the distant past suggests that he continued to think about his brief affair and its impact on his marriage long after he no longer recalled any dreams about Mary.

Character Networks in Dream Series Are Small-World Networks

A mathematical psychologist and his graduate students, who turned to the study of dream series as part of their work on memory, made a major discovery. They determined that the network of characters in a dream series has the same small-world properties as waking-memory networks. Building on earlier studies by the leader of this research team, Richard Schweickert (2007a, 2007b, 2007c), their studies of five different dream series extended his earlier work. They did so by assuming that the characters appearing in the same dream report are linked in the dreamer's cognitive social network. The series range in length from 208 to 423 dream reports, three from women and two from men (Han, Schweickert, Xi, & Viau-Quesnela, 2015). Both the waking and dreaming social networks differ from random networks, which are discussed in chapter 2 in relation to brain networks. They include short paths to other people, a tendency for the people who appear frequently to appear together, and a strong tendency for a large number of characters to be connected in a large general component (Han et al., 2015). Based on this work, it was further demonstrated that the distribution of characters in a dream series can be understood, by means of the power law (Zipf's law), as a lawful draw from a dreamer's semantic memory network (Schweickert, Zhuangzhuang, Viau-Quesnel, & Zheng, 2020).

The findings with the first five dream series were replicated and extended through a unique comparison, the first and only of its kind. It focused on both the character network in the 4,254 dream reports from Barb Sanders and on the people in her social network in waking life. Her waking-life social network was constructed on the basis of a questionnaire in which Barb Sanders rated how well each possible pair of individuals in her dream reports actually knew each other in waking life. She also rated how emotionally close they were to each other on a 1 (low) to 5 (high) scale. She also rated her own emotional closeness to each person on the same 5-point scale (Han, 2014, p. 36; Han & Schweickert, 2016). In all, 120 characters known personally by the dreamer appeared in her dream reports. They included nine immediate family members, 55 other relatives, 28 friends, 14 coworkers, nine boyfriends, and five miscellaneous characters who appeared once or twice. The number of dream reports including at least one of these 120 characters was 2,048, which is 48.1% of all the dream reports.

The first finding of interest in terms of social networks is that the 120 characters in the dream reports are close to the estimate of the number of relationships among people (150), which can be cognitively managed in waking life (Han, 2014, p. 38). Then, too, the more a pair of people appeared together in various dream reports, the more likely they were to be connected in waking life (Han, 2014, p. 50). The two networks were also similar in several other important ways. The "density" of the networks, defined as the proportion of all possible connections that actually appeared, was .16 in the waking network and .14 in the dreaming network (Han, 2014, p. 47). A centrality measure, based on how connected a person is to other well-connected people, revealed that all nine of her immediate family members, including the granddaughter she helped to raise, were among the 10 most central characters in both networks. Her mother and father had the highest centrality scores in both networks. When the character networks in the dream reports were separated into four chronological subsets, they were found to be very consistent over time (Han, 2014, p. 68). However, her granddaughter was the most central figure in the fourth subset, and Barb Sanders's mother became the second-highest in centrality (Han, 2014, p. 83).

Unexpectedly, the dreaming and waking networks differed somewhat. The people in Barb Sanders's waking-life network were sorted into separate "communities" of family, friends, and coworkers, with the boyfriends usually not connected to any other character. However, her network of dream characters more often brought together immediate family members, other relatives, and friends into a common network, even though they were not necessarily in the same social networks in waking life (see Han, 2014, pp. 48–49, figs. 5 and 6, for graphics of this difference). This tendency increased for the people she rated as extremely close to her in waking life (Han, 2014, pp. 50, 67). The coappearances of pairs of characters in dream reports who are closest to Barb Sanders in her waking life reinforce a key concept in the neurocognitive theory of dreaming: the frequency of appearances in dream reports reveals the intensity of personal concerns in waking life.

Dream Reports from 2020 Relating to COVID-19

The potential for a worldwide coronavirus pandemic became apparent to many experts and to journalists in the mass media in the United States in late February 2020. The rapid spread of the virus soon led dream researchers in the United States and Italy to do online questionnaire studies (e.g., D. Barrett, 2020; Iorio, Sommantico, & Parrello, 2020; S. Scarpelli et al., 2021; Schredl & Bulkeley, 2020). The pandemic also led to articles in the popular press concerning COVID-19's influence on dreams. In early October, for example, an article on the science pages of the *New York Times* appeared under the headline "Pandemic Escape? Maybe Not in Your Dreams." It began with the claim that people were dreaming of swarms "of insects—sometimes gnats, sometimes wasps or flying ants; being caught in a crowd, naked and mask-less; of meeting men in white coats who declared, 'We dispose of the elders'" (Carey, 2020, p. 7).

However, there are several difficulties with questionnaire studies. Since very few people recall their dreams and soon forget the few that they do recall, they rarely have any detailed knowledge of the content of their dreams (Beaulieu-Prevost & Zadra, 2005a, 2005b, 2007, 2015; Bernstein & Belicki, 1995–1996; Bernstein & Roberts, 1995; Domhoff, 2017, pp. 25–33). Even if people feel certain they can remember at least one dream of the type asked about, they usually are not able to provide any reliable information concerning content or frequency. The findings from such studies therefore may be based on a combination of people's beliefs about dreams and their general waking thoughts, not the actual recall of dream content. People may feel an implicit pressure to be helpful to researchers by answering questions about a widely discussed and frightening event, such as the COVID-19 pandemic. In that type of situation, they may provide answers on the basis of what they hear and read about other people's dreams. Such studies may have demand characteristics in them, and people also might answer on the basis of the wake-state bias (Windt, 2015). These factors add up to the fact that the retrospective recall of dreams in response to a questionnaire is not likely to be accurate in regard to the frequency or content of what people dream about, perhaps especially in the midst of a pandemic.

In the specific instance of the COVID-19 pandemic, these inadequacies in questionnaire studies can be examined by studying ongoing dream series. Six dream series from American women, which began long before the first year of the pandemic, provide an unobtrusive baseline for studying these issues. Four of the six series were volunteered in 2020 or 2021, so they have not been analyzed for other dimensions in their dream content. However, they have been authenticated, primarily through their use of voice recordings that document their dreams, through voice recordings on a smartphone app, or through the timestamps other participants use in digitally entering each report soon after it is recalled. Two of the series were already known to the author on the basis of earlier studies of them. The analysis begins with the dream series provided by the youngest participants.

Rebecca, Penelope, and Zelda: Ages 18–23

Rebecca was 18 years old and in her first year of college in a southwestern state when she volunteered her dream journal in early 2021. She has been documenting her dreams since her mid-teens. She entered 53 dream reports into her computer from March through December. Two of them related to the pandemic, which is 3.8% of the dream reports. On March 26, after the campus had been closed for three weeks due to the pandemic, she dreamed that the students were asked to return for just one day. She was sitting with her best friend and an acquaintance at their usual lunch table. On March 29, she dreamed that she and her friends were wearing masks, which was the day after her mother had dropped off a large pack of disposable masks for her and her close friends.

Penelope, who volunteered her dream series in early 2021, is in her early 20s. She lives in a southern state, which has an ongoing high incidence of COVID-19 infections. She was raised in a highly religious family, which belongs to a mainstream Protestant nomination. She has been writing down her dream reports since she was in her mid-teens. Penelope documented 93 dream reports between March 1 and the end of 2020. Four of them clearly relate to the pandemic and another could conceivably relate to it. On March 30, she dreamed that "some parasite was infecting everyone else around me." She did not think the parasite related to the coronavirus, but when she sent the dream reports, she "thought it worth mentioning." On May 22, she dreamed she was working again in a store she had left in 2019. In her dream "The place had changed because of COVID-19 and the new policies implemented."

On June 13, she dreamed that her sister, who "has a compromised immune system in real life, died of a car crash, when I expected it to be the coronavirus." On June 20, she was on a "horrible movie date, so I bring up COVID-19 violation to get out of sitting next to him." On September 29, she walked up to the house she and her parents once lived in and gave her "mother a hug." But the young person she was walking with, who did not know her very well, "seemed shocked I would hug a random lady possibly giving her COVID-19 or vice-versa." Depending on how the first dream is counted, 4.3% or 5.4% of her documented dream reports related to the pandemic.

Zelda is also a recent college graduate, only a little older than Penelope. She was born and raised in a large city in Western Europe and went to college in the United States. She works as an assistant in a scientific laboratory in a large city in the United States. The city is located in a region that has tried to control the spread of the coronavirus. Due to her scientific training and the type of cell-oriented basic research lab she worked in, she clearly understood the potential for a pandemic from the outset. She volunteered her dream reports in late 2019 and sent what she had up to that time, well before the author thought of studying pandemic-related dream reports. She documented 99 dreams between March 1 and November 13, 2020, which is the day the author first contacted her for a second (unexpected) time. To avoid the risk of demand characteristics, he did not want any dream reports from after he contacted her that second time.

Nine of Zelda's dream reports mention COVID-19, which is 9.1% of the dream reports. Seven of them are notable for the passing way in which the virus is mentioned. On July 14, 4.5 months after fears of the virus were widespread, "a woman yells at her for walking past the row of homes she lives in," and the dreamer yells back, "there was a sign that allowed me to pass on this road because of COVID-19." On July 23, she is walking in a recreational area at "a respectful distance from the families who have paid to fish here (also perhaps in a COVID-19 distanced way)." On July 26, she is in a museum, but it is "emptier than usual: it must be shortly after COVID-19 has ended." On August 7, "the subway is crowded despite COVID-19" and people are wearing colorful leather jackets because they think "leather protects from COVID-19."

On August 8, she dreams that she has lost her job, which is very likely a personal concern to her at a time when layoffs were in the news, and she wonders if it is "because of COVID-19?" On August 23, she asks a store manager how things are going and the woman replies, "some things have had to change somewhat because of COVID-19, not too much." On August 29, in a dream in which she seems to portray a personal concern, she dreams she is in another recreational area: "we're all packed into these boats with the kids with a high risk of transmitting COVID-19, despite the masks." On September 18, she goes into a place that is closed, "either for COVID-19 or for construction," but she goes in anyway. On September 24, she wants to see an exhibit, but it is not clear if the exhibit is open, "due to COVID-19."

Allison and Jasmine: In Their 30s

Allison is in her mid-30s. She volunteered her dream series in the early spring of 2020. She has documented her dream reports for two different periods in her life, separated by a four-year gap, during which she married and had children. The author later asked her to update what she had provided earlier, a request that neither he nor Allison knew would occur. Allison lives in a northern state, which eventually had high infection rates. Her 79 dream reports between March 1 and October 31, 2020, which is the day the author contacted her after a hiatus of several months, revealed only one instance of a pandemic-related dream report (1.3%). On March 11, she had gone shopping for a school snack for one of her young sons. The first two cases of COVID-19 in her small town had been announced that day and "everyone was in a panic," and "all of the toilet paper and anything disinfectant was already gone." In the dream she had that night and documented the next morning, she "went shopping and everyone was in a hysteria. Everybody was out buying stuff. It was just crazy busy." Compared with many of her other dream reports, this is a very matter-of-fact dream.

Jasmine, also in her mid-30s, documented 58 dreams between February 29 and December 10, which is the date the author contacted her for the first time in a year or more. She lives in a warm southern state that had high levels of infection and did little to prevent them. She has been keeping a voice-recorded journal of her dreams on and off since her early adolescence. Only two of her 58 dream reports within the relevant time period, which is 3.4% of the total, had a mention of COVID-19. On May 31, three months after the pandemic began, she was trying to delete something she had just posted. She had decided she didn't want a certain person to see it. However, it was "slowly transforming into something more benign or something more COVID-19 and I'm realizing she probably won't know what it is anyway." On September 10, she was taking a Scantron test, which was supposed to determine if she had COVID-19. In a dream report from October 10, she mentions that she is afraid to fly because of "all the crazy crap in the world." If that dream is also counted, then 5.2% of her dream reports are pandemic related. (A detailed analysis of Jasmine's earlier dreams, from her early adolescence to her mid-20s, is presented in chapter 7.)

Overall, these five young women contributed 382 dream reports. Eighteen included a mentioned of COVID-19, which is 4.7% of the total. If the two dream reports that perhaps include an indirect mention of COVID-19 are added, then 5.2% have a mention of COVID-19. Few of the pandemic-related dream reports from these five participants expressed strong personal concerns. For 18-year-old Rebecca, neither of her two dream reports expressed a personal concern about COVID-19, although she was glad to be back sitting with her classmates in one of the dreams. For Penelope, her long-standing concerns about her sister's compromised immune system very likely led to her concern that her sister might succumb to the virus. For Zelda, there was concern about losing her job in one dream and of being on a crowded boat with "kids," ages unspecified, in another. For Allison, her one dream about the pandemic related to a personal concern she experienced that day: a buying panic while venturing out to purchase a snack for one of her school-aged sons. One of the two directly pandemic-related dream reports voice-recorded by Jasmine showed personal concern, in that she was being tested for COVID-19 by means of a Scantron test.

Beverly: A Frequent Pandemic Dreamer

Beverly, who is in her mid-60s, is a whole different story. She has been keeping a written record of her dreams for over 35 years. She showed high levels of consistency in an earlier analysis of three different time periods, based on generic word strings (Bulkeley, 2018). She has a lifelong interest in politics and current events. She dreamed frequently of former president Donald Trump from 2015 to 2020, whom she feared and disliked. She defines herself as a liberal Democrat and always has been regularly involved in helping others. She has an M.S. degree in clinical youth counseling. She documented 250 dreams in eight months between March 1 and the first few days of November. The reports are relatively brief, but they are more than adequate for the narrow purposes of this analysis. She also keeps a daytime journal of her thoughts and activities, which adds a further dimension to a study of her pandemic-related dream reports.

Beverly lives in a southwestern state, which had high rates of infection early in the pandemic. She first mentions COVID-19 in her daytime journal on February 22 and then discusses it at more length in an entry on February 28. She calls the virus a "threat," says its main impact so far has been on the stock market, and then adds: "So I worry. If things get worse, lots of people could die." On March 1, she documents her concerns about flu-like symptoms and a temperature of 99.9. On March 3, she awakened "with her head spinning," and says she blew her nose most of the day. She made further daytime comments on her symptoms on March 8.

On March 10, after digitally entering 11 dream reports between March 1 and March 9, she had her first pandemic-related dream. She was "wearing big mittens for picking weeds," and also dreams that the mittens "help stop the spread" of the coronavirus. On March 11, she writes in her daytime journal about COVID-19 being "really scary," and thinks the president "acts insane." She worries that she might "get the bad virus if I don't already have it." While continuing to enter daytime comments into her journal, she digitizes 11 more dream reports before she has another pandemic-related dream on March 18. In her dream she is helping the police at a "crime scene," which is very empty, but "we worry they [the many boxes strewn about] 'might be infected.'" They have to send the boxes to a lab "to be tested."

The dream about helping the police exemplifies what proved to be an ongoing pattern in her dream reports related to the pandemic, which less often express her concerns about herself. Instead she is helping gather data, helping scientists to analyze data, or finding new ways to provide scientific information to the general public. In a dream on March 31, she is helping to rewrite the script for a soap opera so that more people can be reached about the need to wear masks and take other precautions. In a dream in April, she is making a website for updating pandemic-related information. On May 2, she is "involved in a scientific investigation into the virus—we want to find out what it is and how it spread." She is also helping other people in other dream reports, such as people in a hospital. This emphasis on helping others, rather than enacting her worries and fears for herself, is revealed by the percentage of her pandemic-related dream reports in which she is in some way helping others. From March through August, when her concerns about the coronavirus were highest, 79.8% of her 47 dreams mentioning COVID-19 involved helping others in some way.

By April, Beverly's most frequent way of helping others became the sewing of masks for free distribution. She makes no mentions of masks in either her daytime journal or her dream journal during March, but this new personal interest—indeed, quest—is mentioned in both her daytime journal and her dream journal after her new sewing machine arrived in early April. Over the next six months she frequently mentions in her daytime journal that she is sewing masks, giving masks to family and friends, giving masks to nonprofits to distribute to the homeless, and mailing packets of masks to places as far away as Hawaii. She mentions this general topic four times in April, 13 times in May, nine times in June, and 16 times between July 1 and September 30.

However, she mentions masks only once in early October and then again on October 30, when she thinks back on the past several months. She noted how her "priorities had changed." In the early summer she had worked on another project related to a museum exhibit, but now she "hardly ever works on it." She further says, "I was making 100 masks a week or more." But in the past month, she "had only made 100." On election day, November 3, she wrote that it had been a busy day, including sewing masks, but she had "thought about when to quit. I have two boxes of sheets and a lot of elastic plastic bags, and thread." Then she writes, "maybe it's time to hit pause. Everyone has a mask."

Beverly's dreams dramatize this waking personal concern. They are not nearly as frequent as her daytime journal entries, but their rise and fall in frequency parallels the rise and fall of this personal concern in waking life. For example, one of her 32 dream entries in April is about sewing masks and six other dream reports are pandemic related, but the remaining 25 focus on other personal concerns, which is 78.0% of the dream reports for that month. Thirty-six of her 159 dream reports from May through September involve COVID-19 in one way or another, which is 22.6% of the total. Fourteen of those 36 (39.9%) relate to her concern about making and distributing masks. In October and early November, when the series ends for purposes of this research, she documented 27 dreams, only four of which (14.8%) related to the pandemic in any way, and even fewer, two, related to making and distributing masks.

When asked to comment on these findings, she said that "making masks helped ameliorate my fears about the pandemic," and added that "at times I was very concerned and frightened." Due to her passionate concern with helping others and for doing all she could to promote progressive social change, masks became a strong personal concern for her. This personal concern was enacted in her dreams. Her dreams about the pandemic thereby provide excellent evidence for the concept of continuity, which is a keystone of the neurocognitive theory of dreaming.

Overall, 21.2% of Beverly's dream reports between March 1 and November 7 concern the COVID-19 pandemic. When her 250 dream reports are added to the overall sample, they account for 39.6% of the 632 dream reports and 74.6% of the 71 dream reports mentioning the pandemic in any way. She is 3.2 times more likely to dream about pandemic-related topics than the other five dreamers combined. This comparison demonstrates she has a far greater personal concern with the general impact of COVID-19 than they do. Her frequent inclusion of events related to the pandemic is consistent with her early symptoms, and perhaps even more so with her ongoing interest in current events, including politics, as well as with her lifelong focus on helping others. Even with this very high level of concern, though, 73.2% of the 250 dreams she documented during this frightful time period were about her usual concerns with family, friends, and two or three long-standing personal avocations.

These overall results demonstrate that current events, even those as potentially life-changing as the COVID-19 pandemic, do not readily become personal concerns for most people. As revealed by the first five dream series, the pandemic had only a small impact on the dreams of people who were not suffering from COVID-19 or did not become highly concerned or traumatized by events related to the pandemic, or develop a sleep disorder due to it.

Methodologically, the results of this study demonstrate once again the importance of establishing a normative baseline for studying dream content under any circumstances. The results further show that questionnaires, including online questionnaires, are of limited value in dream research and of no value in developing a neurocognitive theory of dreaming.

Conclusions and Implications

The findings in this chapter support and extend the conclusions in chapter 3 with regard to the everyday nature of most dream content. More importantly, the results from the studies of these widely differing dream series demonstrate there is consistency over time in what people dream about. This discovery, and many others reported in this chapter, might not have been made with representative samples of dream reports. These findings also reinforce the findings in chapter 3, based on cross-sectional group data, which showed that dream reports change very little over the decades from young adulthood to old age. The very different dream series provided by Jason and Kenneth provide excellent evidence for this point. However, many other dream series make this point as well, sometimes in unique ways, such as in the case of the widower.

The findings also substantiate a conclusion drawn in chapter 3. There is continuity between what people dream about and their waking personal concerns. This continuity is perhaps most striking for the entomologist, due to his atypical activities and interests in waking life. Then, too, the Beverly series, in the space of eight pandemic-filled months, provided unexpected new evidence on continuity from a different angle, due to her daytime documentation of the rise and fall of her personal concern with contributing free masks to the efforts to control the spread of COVID-19.

However, the Barb Sanders series supports this point in the greatest detail by showing how her relationships with different characters in her dream reports provide an accurate picture of her conceptions and concerns about them in waking life. In addition, the rigorous quantitative network analyses of both her dreaming and waking social networks provide new and original evidence for continuity. There are also some understandable changes in *continuity* for some individuals, which leads to a further generalization about *consistency* in dream reports. There is consistency in people's dream lives to the extent there is continuity in their conceptions of people and activities in their waking personal concerns. In this regard, the consistency and continuity in the 4,254 dream reports from Barb Sanders are impressive, and perhaps definitive. In addition, the changes in some of Beverly's waking personal concerns at the height of the pandemic, which were followed by increases and decreases in her enactment of these concerns during dreaming, adds close-in evidence of this relationship.

Although it can be predicted that Barb Sanders will dream about one or more of the 13 most important people in her life in 30 to 35 of every 100 dreams, it cannot be predicted what she or anyone else will dream about on any one night. In that sense, dreams are based on a random spin of each person's cognitive rolodex, located in their semantic memory banks. Nor is it possible to know the full range of topics stored in a person's memory bank. Six individually tailored word strings relating to Kenneth's primary personal concerns, ranging from family and friends to driving trucks and playing sports, can account for at least some aspects of three-fourths of his 2,022 dream reports, but that still leaves out one-fourth of his dream reports. This figure is similar to an estimate in chapter 3, based on a study of "familiar" and "unfamiliar" activities in six different sets of dream reports, four from the United States and two from Germany: 70–75% of the dream reports contained "familiar" content. These and other findings make clear that consistency and continuity are very important, but they do not encompass all dreams.

The individual differences revealed in this chapter can be combined with the findings on the many cross-national and cross-cultural similarities in the large samples of dream reports overviewed in chapter 3. Based on these two different types of evidence, it seems plausible to suggest that dreams are first of all characterized by pan-human dimensions on the one side and by large individual differences on the other. The few gender, cross-national, and cross-cultural differences, although real, and often replicated in the case of gender, pale in importance to the pan-human similarities and the wide individual differences in dream content. Then, too, the character networks in dream series, which were discovered by mathematical psychologists with expertise on waking memory, provide indications that dreaming in general may be rooted in the same principles governing waking memory to a greater extent than is usually recognized.

In terms of assessing the theories of dreaming discussed in this book, the most relevant issue that emerges from this chapter concerns the extent to which readers find the methods and results in it to be convincing. If these methods and results are judged to be soundly based and useful in understanding dream content, then the conclusions in this chapter could be helpful to them in assessing the two comprehensive theories of dreaming examined in chapter 9, and the adaptive theories of dreaming discussed in chapter 10, as well as the neurocognitive theory of dreaming.

5 The Search for Symbolism in Dreams

Theorizing about symbolism in dream research began with the work of Sigmund Freud, who hesitantly added the age-old idea of symbolic interpretations into the second edition of his *The Interpretation of Dreams* (Freud, 1900/1909). He originally avoided symbolic interpretations because of their association with the popular code books of his and all previous eras. These code books date back to the second century AD, when a physician, Artemidorus, compiled a five-volume book on dreams. These volumes included many possible symbolic interpretations of dreams, which still appear in the popular guidebooks concerning dreams that are sometimes seen on the racks near the checkout counter in grocery stores (see Weiss, 1944, p. 5, and 1959 for an abridged version). Moreover, some of Artemidorus's interpretations were "derived in part from the large collection found on cuneiform tablets from Mesopotamia," which carries the history of symbolism back to about 4,520 years ago (Bulkeley, 2008, pp. 112–122; Noegel, 2001).

When Freud addressed the topic of symbolism nine years after the first edition, he was careful to differentiate his approach from the code-book tradition. At the same time, he also mentioned a major issue that makes systematic examinations of putative symbolic elements in dream reports extremely difficult: "Often enough a symbol has to be interpreted in its proper meaning and not symbolically"; in the same sentence, he then added another complicating factor: "on other occasions a dreamer may derive from his private memories the power to employ as sexual symbols all kinds of things which are ordinarily not employed as such" (Freud, 1900/1909, p. 352). Based on this one sentence alone, it is clear that potentially symbolical elements in dream reports are not easily studied. However, these initial concerns were more or less set aside in his most comprehensive statement on symbols, which appeared in a chapter in his *Introductory Lectures on Psychoanalysis* (Freud, 1916). It included references to word etymologies, jokes, fairy tales, and myths as evidence for the validity of Freudian symbolic interpretations of such objects and activities in dreams as hats, coats, tools, jewels, climbing stairs, and dancing.

Contrary to the strong expectations created by age-old cultural claims, as supplemented by the clinical tradition created by Freud and his followers, systematic empirical studies of dream reports suggest that the frequency of seemingly symbolic elements in dreams is very low. Moreover, most of this meager evidence appears within the context of studies of the content of a small number of "typical" dreams, such as flying under one's own power, appearing in public in little or no clothing, or losing one's teeth. Most studies of putative "typical" dreams involve yes-no responses to anonymous questionnaires, administered to large samples of students in university settings. They find that from one-third to two-thirds of those surveyed report having experienced one or more of the typical dreams (Griffith, Miyago, & Tago, 1958; Nielsen et al., 2003). However, studies of both group samples and dream diaries suggest that such dreams are less than 1–2% of all dreams, as discussed further later in the chapter (D. Barrett, 1991; Domhoff, 1996, p. 198; Mathes, Schredl, & Goritz, 2014).

In addition to presenting the evidence documenting the infrequency of symbolism in dreams, this chapter also suggests a neurocognitive explanation for this finding. This explanation is based on waking neuroimaging studies of participants who were asked to comprehend standard metaphors or to generate novel metaphors (Beaty & Silvia, 2013; Beaty, Silvia, & Benedek, 2017; Benedek et al., 2014; Holyoak & Stamenkovic, 2018; Rapp, Mutschler, & Erb, 2012; Vartanian, 2012). If the neural substrate that supports both metaphor comprehension and metaphor production is used as a starting point, then the many differences between it and the neural substrate active during dreaming may account for any differences in the nature and frequency of figurative thinking in waking thought and dreams. Put more exactly, the relatively rare appearance of metaphoric expressions in dream reports may be an example of a cognitive insufficiency during dreaming.

Experimental Studies of Symbolic Elements in Dreams

The first nonclinical studies of symbolism in dreams, inspired primarily by Freudian claims, were carried out as early as 1911 through the use of hypnosis to instruct participants to have a dream of a certain kind and/ or to interpret a dream (see Moss, 1967, for a detailed review of the classic studies and a reprinting of several related articles from the 1950s and early 1960s). However, later Freudian-oriented researchers criticized the "frag- mentary" and "unfinished" analyses in the original investigations of hyp- notic dreams and ended up questioning "the assumption that the hypnotic dream is a psychic production, which duplicates either in function or struc- ture, the spontaneous night dream" (Brenman, 1949/1967, p. 135). Despite the renewed interest in dream research in the 1950s, due to the discovery of REM sleep, there has been only one further study of dreams based on the use of hypnosis. It received very little attention, despite its methodological rigor (D. Barrett, 1979). It is likely that the social-psychological analysis of hypnosis in terms of role theory, which had become predominant within academic psychology, accounts for this disinterest (see Sarbin, 1993, for the development of this approach).

In light of later neuroimaging studies of hypnosis, its potential useful- ness in understanding dreams came into further doubt because the brain regions active during hypnosis are very different from those that are active during dreaming, as discussed in chapter 1. In fact, neuroimaging studies of hypnosis found some of the areas in the frontoparietal cortex, which sup- port focus and close attention, become even more active during hypnosis than during attentive waking thought (Abraham, 2016, p. 4206; Landry et al., 2017). These studies suggest that the neural substrate underlying the hypnotic relationship is different from the neural substrate that supports dreaming, so hypnosis is no longer considered to be a useful method for studying symbolism in dreams.

Freudian-oriented researchers also pioneered work on subliminal stimu- lation, in which they exposed participants to brief presentations of visual or auditory stimuli below the threshold of conscious awareness, just before bedtime. Participants were asked to report upon and/or to draw key aspects of their visual dream imagery the next morning. The evidence for incor- porations in these studies was based on seeming physical resemblances in the drawings and on symbolic interpretations. In a study by psychoanalyst Charles Fisher (1954), for example, it was assumed that a vague drawing by the participant after awakening had some connection to a picture shown to her for less than a second, below the threshold of awareness, before going to sleep. As a result of using symbolic interpretations to study symbols, the method used for interpreting the results is not independent of the theory being tested.

The most telling early criticisms of the subliminal stimulation studies, which were carried out sporadically between 1917 and the 1950s, were provided by psychoanalysts in the United States, beginning in the 1980s (e.g., Shevrin, 1986, 1996). One of them concluded "suitable controls were often lacking, and the various clinical interpretations could be seen as equivocal and, at times, arbitrary" (Shevrin, 1996, p. 96). These same investigators claimed that their own studies introduced the necessary controls but their studies relied primarily, if not entirely, on *waking* mental images produced *after* laboratory awakenings from either REM or non-REM sleep rather than on dream reports. Such studies therefore have nothing to do with dreaming. They are not based on dream reports, and more brain regions are active after awakening than during dreaming.

Nor are any of these claims likely in light of several carefully controlled experiments, which led most academic research psychologists to conclude subliminal stimuli are limited to small priming effects for one or two words, with no ability to influence concepts (Avneon & Lamy, 2018; Biderman & Mudrik, 2108; Greenwald, 1992; Greenwald, Draine, & Abrams, 1996). There are also neuroimaging studies that throw further light on this issue. They reveal that subliminal stimuli can register in the visual cortex but they cannot influence thinking unless they are "further processed by cognitive control networks underlying working memory" (LeDoux, 2019, p. 272).

The most striking Freudian-inspired study of dream symbolism was carried out in a sleep-dream laboratory. These results were labeled as "preliminary results." They were said to show, based on 57 REM dream reports from six of the men who participated, that the symbolic content could be used to predict the rise or decline of penile erections during REM sleep. The researchers also claimed bland dream content predicted long periods of REM sleep without a penile erection (C. Fisher, 1966, pp. 538, 541, 543, 550). However, this preliminary study was never extended to include an analysis of the full dataset.

In a more general study, with 12 adult men serving as participants, the pattern of penile erections during REM periods could not be influenced by presleep sexual arousal (Ware, Hirshkowitz, Salis, & Karacan, 1997). Erections were also found to occur outside of REM sleep (Hursch, Karacan, & Williams, 1972). The topic of penile erections during sleep was overviewed in great detail in the third edition of *Principles and Practices of Sleep Medicine* (M.

Schmidt, 2005), but with no mention of dreams. The topic of penile erections was abandoned as a separate chapter in future editions of that compendium.

Based on the overall findings in this section, there are no studies by Freudian-oriented researchers, whether based on hypnosis, subliminal stimulation, or the rise and fall of penile erections, that support the claims for the presence of symbolism in dream reports.

Systematic Empirical Studies of Symbolism Based on Content Analysis

A framework for empirical studies of dream symbolism, based on the content analysis of dream reports, first appeared in an article articulating a cognitive theory of dream symbols (Hall, 1953). According to this theory, symbolism may appear in dreams because it can express some concepts better and more succinctly than do literal enactments. It thereby disagreed with the Freudian emphasis on disguise. Different sexual symbols, the article continued, may express different conceptions of sexuality, which is just about the opposite of a disguise function (Hall, 1953).

To provide an empirical starting point, Hall (1953) began with the hundreds of slang terms he extracted from Eric Partridge's (1937/1984) *A Dictionary of Slang and Unconventional English from the Fifteenth Century to the Present Day*. He compared these terms with a list of putative sexual symbols that he compiled from the Freudian literature and found a considerable overlap (see Hall, 1953, p. 173, and 1964 for the complete results). He then suggested figurative thinking in both dreaming and waking thought is based on the laws of association, which were foundational to learning theory in the early decades of psychology. The *law of resemblance* is based on similarities in size, shape, function, and much else. The *law of contiguity* claims that events or objects that appear close to each other in time or space become associated. The *law of association of a part with a whole* assumes that a salient part of an entity or event can be used to refer to the totality, and the *law of contrast* claims that the sight or thought of one or another object or event brings about thoughts of something opposite in nature. In terms of dream symbolism, these laws potentially could support such symbolic interpretations as the equation of pen and penis (similar shape), dancing and sexual intercourse (similar actions), jewelry and female genitals (high value), king and father (high status), chocolate and feces (similar color), and basement and the unconscious mind (similar physical position). The 14 specific laws said to be based on resemblance, contiguity, the relation of parts to wholes, and contrast, are listed in table 5.1. Due to

Table 5.1
Hall's examples of how the "laws of association" relate to Freudian studies of symbolism, as adapted from Hall (1953)

Law of association	Example
1. By resemblance in shape	All circular objects and containers = vagina. All oblong objects = penis.
2. By resemblance in function	All objects that are capable of extruding something (e.g., gun, fountain pen, bottle) = penis.
3. By resemblance in action	Any act that separates a part from a whole (e.g., beheading, losing a tooth, an arm or a leg, having a wheel come off an automobile) = castration. Dancing, climbing stairs, riding horseback, going up and down in an elevator = coitus.
4. By resemblance in color	Chocolate = feces. Yellow = urine. Milky substance = semen.
5. By resemblance in value	Gold = jewelry = female genitals.
6. By resemblance in number	Three = penis and testicles.
7. By resemblance in sound	Sound of a trumpet/bugle or wind instrument = flatulence.
8. By resemblance in quality	Wild animal = sexual passion. Horse = virility.
9. By resemblance in personal quality	Policeman or army officer = father. Nurse = mother.
10. By resemblance in physical position.	Basement = the unconscious mind.
11. By resemblance in status.	King = father. Queen = mother.
In addition to association by resemblance, there are several other ways in which two items may become paired as symbol and referent:	
12. By contiguity	Church = virtue. Night club = sensuality. Bathtub = cleanliness.
13. Of part with whole	A specific accident = difficulties of life. A school test = a test of fitness for life.
14. By contrast	Crowd = being alone. Clothed = naked. To die = to live.

the general nature of these laws, they can be used to generate nearly endless numbers of symbols in any story or dream report, so the issue once again becomes if, when, and how to understand a term or phrase in a literal or figurative sense.

As a next step toward carrying out empirical studies of symbolism, Hall (1953, pp. 173–174) first proposed symbolism in dreams could be studied by making use of the four principal figures of speech: metaphor, irony, metonymy, and synecdoche, each of which, he claimed, makes use of one or more of the laws of association listed in table 5.1. Moreover, metaphor, irony, metonymy, and synecdoche are now considered by many cognitive linguists to be a form of figurative thinking, which is as basic as literal thinking. Figurative thinking is processed and understood by the human mind as fast as literal speech, which suggests thinking may be more poetic than past theories assumed (Gibbs, 1994, 1999). The figurative emphasis provides a context for the idea that symbols are a way to express conceptions, including the many different conceptions people have of sexuality or of significant people in their lives.

Metaphor, which comes from a Greek word for "transferring" or "carrying over," involves a comparison between people, events, or objects, which are not related in any literal or realistic way (Gibbs, 1994, chap. 4). Metaphors include such well-known phrases as a person being "fit as a fiddle," a "stick of dynamite" or "a stick in the mud." Many slang terms make use of metaphors, including obscenities (Gibbs, 1994, pp. 136–138; Tururen, 2016). In terms of the laws of association, the basis for a metaphor can be found in one of the laws of resemblance.

Irony, on the other hand, is a way of expressing what is meant by saying the opposite. Within the context of the laws of association, it would be considered to be an association by contrast. Metonymy uses an attribute to make a point about something related, such as "the pen (the press) is mightier than the sword (the army)" (Gibbs, 1994, chap. 7). Synecdoche involves the use of a part to stand for the whole ("we need a better glove at third base") or a whole to stand for a part ("Wall Street is having a panic attack"). In Hall's (1953) use of the laws of association, metonymy relates to the law of association by contrast and synecdoche to the law of association of a part with a whole.

Based on his theorizing about the possibilities of building on the laws of association, Hall (1966, p. 40) later created four nominal categories for locating potential symbolic elements in the dream reports in the large samples of laboratory and nonlaboratory dream reports he collected from 11 young male participants in his study in Miami, as overviewed in previous chapters. This "unusual elements" scale consists of four categories, which are demonstrated through examples from dream reports: (1) unusual activities, such as taking a shower in a pantry or sailing in a boat over dry land; (2) unusual occurrences, some of which are magical, such as an animal talking; (3) distorted objects or a distorted arrangement of objects, such as an object or person appearing misshapen or suddenly disappearing, or a mountain made of watermelons; (4) any metamorphosis of a person, animal, or object into another person, animal, or object, which includes people who become animals and objects that become animate. Based on these four categories, only 10% of the dream reports in either sample had one or more of the four unusual elements (Hall, 1966, p. 41).

Even if all of these unusual elements are assumed to be symbolic for the sake of argument, with no random cognitive glitches or other types of false positives, the results suggest symbols are not frequent in representative samples of lab and nonlab dream reports. Nonetheless, this low figure provides a starting point for comparisons with the frequency of potentially symbolic elements in later studies, which are discussed below. It is also useful to keep this figure in mind because it is at least somewhat similar to an estimate that 10–20% of phrases and sentences in various forms of waking discourse (conversations, fiction, news, and academic discourse) are figurative, with varying rates for different genres (Steen et al., 2010).

Three Similar Empirical Studies Using Other Categories for Unusualness
In a fine-grained laboratory study focused exclusively on the degree of "novelty" in dreams, a team of investigators first adapted several Hall/Van de Castle (HVdC) coding categories into six different scales to assess the degree of novelty in settings, objects, characters, and social interactions (Dorus, Dorus, & Rechtschaffen, 1971). Their conclusion stressed the rarity of novelty. The figures for the most improbable category were 1.3% of all characters, 4.9% of all physical surroundings, and 6.8% of all activities and social interactions. Similarly, major distortions occurred in only 6.2% of characters, 7.8% of all physical surroundings, and 16.7% of all the activities and social interactions (Dorus et al., 1971, p. 367). Based on a global rating, the researchers concluded only 8.9% of the dream reports were highly

improbable by waking standards and 25.8% showed large but plausible differences from previous waking experiences (Dorus et al., 1971, p. 367). Once again, even assuming that all of these novel elements are symbolic in some way, these percentages are relatively small in comparison to claims by Freud and other clinical theorists.

A study of 104 REM dream reports with 50 or more words, which were obtained by researchers at the Harvard Medical School from a University of Cincinnati sleep-dream lab, reported the same low figures for similar categories of unusualness. The sample was drawn from dream reports collected from 20 college men between the ages of 20 and 25 who were awakened from REM periods over a period of 20 consecutive nights. The sudden appearance or disappearance of a character occurred in 2.9% of the dream reports and abrupt shifts in time in 1%. There were no sudden appearances or disappearances of any objects. Violations of physical laws, such as dream reports of flying under one's own power, occurred in 2% of the dream reports and unusual environmental combinations in 17.3% (McCarley & Hoffman, 1981, pp. 904, 908, and figs. 4 and 6). Similar findings also emerged in a study of 217 dream reports written in dream diaries kept by 32 young adult women. The study examined 7,616 elements (objects, actions, people, and settings), which comprised fully two-thirds of all the elements in the dream reports, for any sign of unusualness (labeled as "bizarreness" in this study). It did so on the basis of a very minimal standard, namely, any deviation from real-world expectations. Despite the minimal standard, which was meant to capture as many instances as possible, only 15–20% of the four common elements were classified as bizarre (Revonsuo & Salmivalli, 1995).

Overall, then, results from both lab and nonlab studies demonstrate novel or unusual elements are not frequent in large representative samples of dream reports from women and men. Even if all of these unusual elements are assumed to be symbolic in some way, the results suggest symbolism in dream reports is rare.

Studies of Unusual Elements in Lengthy Dream Series

The findings on symbolism in group samples of dream reports are meager. However, dream series might provide a better context for detecting possible symbolism in dreaming. Hall (1953, p. 180) claimed the analysis of a dream series might provide "the best evidence for the validity of symbol translation." Such series may include repeated elements, especially unusual ones, which can be examined for possible metaphoric meaning (see Hall & Lind, 1970, for examples in a short dream series kept by Franz Kafka). Further, the straightforward enactments in a dream series can provide useful information for studying highly unusual elements, which may be symbolic. For example, a striking and seemingly unusual enactment in a short dream series of less than 20 dream reports concerned a search for a wedding gown. The dreamer last wore it for a wedding a few years earlier (Hall, 1953). In the dream she had been surprised to learn, through a call from her husband, that they were going to be married again on their first wedding anniversary. However, she was very disappointed when she found the gown: it was dirty and torn. She had the gown under her arm when she arrived at the church. To her further surprise, her husband asked her why she had brought the gown. In the dream she was "confused and bewildered and felt strange and alone" (Hall, 1953, p. 179).

Looking at the dream from a figurative point of view, the state of her shredded dress might express her conception of her marriage on the basis of a synecdoche, in which a part stands for a whole. To explore this possibility, a search was made for similar (seemingly more straightforward) dream reports in the series, which might suggest the marriage was in difficulty. The search led to the following plausible instances: (1) the stone from her engagement ring is missing; (2) her husband has tuberculosis; (3) one of her women friends is going through a divorce; and (4) a friend who is about to be married receives useless bric-a-brac for a wedding present. If the HVdC coding system had been available when this analysis was carried out, perhaps the case could have been improved by comparing the dreamer's Aggression/Character Index with her husband to the same index with other adult men. If it was higher with her husband than with other adult men, and if there was a lower rate of friendly interactions with him as well, then the metaphoric hypothesis would have been supported by means of a nonmetaphoric content analysis.

In an attempt to build upon this example, two longer dream series, which had already been quantified with HVdC coding categories, were examined. The first came from Barb Sanders and the second from the widower with the pseudonym Ed. As readers may recall, both of these dream series are discussed at length in chapter 4.

Potential Symbolism in the Barb Sanders Series

The Barb Sanders series is notable for including several instances of composite characters. The younger man she had a crush on when she was in her 40s appears as a composite character in 8 of the 43 dream reports in which he appeared. Most strikingly, he appears three times as a composite with her high-school boyfriend. Either he or her high-school boyfriend also were blended with two different movie stars, a favorite uncle, a former boyfriend, and "a man" (Domhoff, 2003, pp. 131–132 and fig. 5.4). Since they were the two men she loved the most but never had sexual intercourse with, the fact that they were blended together, or blended with other men she liked, suggests that these composites may make sense as instances of a conceptual category that could be labeled "likable men."

In addition, her entire dream series was studied for character metamorphoses, which are extremely rare in dreams. They occur only 14 times in the women's 500 normative dream reports, which is 1.8% of all characters (Hall & Van de Castle, 1966, p. 168). The also only occur a few times in Barb Sanders's representative sample of 250 dream reports. In an effort to find as many metamorphoses as possible within the entire series, the terms "changes into," "turns into," "becomes," and "is now" were entered into the search engine on DreamBank.net.

It is highly unlikely that these four terms capture all the potential metamorphoses in the dream series. But they do provide a large sample of dream reports, which can be examined on an instance-by-instance basis after false positives have been eliminated. Once the false positives were eliminated, the initial yield of 132 dream reports boiled down to 50 instances in 49 dream reports. This is a very small percentage (1.2%) of the total number of dream reports, even if a dream-by-dream reading of the entire series might uncover more instances. Thirty of the 50 metamorphoses included a human or animal character at the beginning or the end of the transformation. The other 20 involved object-to-object transformation.

At the risk of inducing tedium, the detailed breakdowns of the findings in this and the next paragraph demonstrate that there are no clear patterns involving character or object transformations. Thirteen of the 30 character transformations are human-to-human transformations. In seven dream reports, people change into animals, creatures, or objects. In another seven, animals or objects turn into people. There are also two occasions in which one animal turns into another and one in which a male puppet turns into a female puppet. Aside from a lack of any pattern, there were no obvious symbolic meanings for any of them. This conclusion is based on three types of information unrelated to figurative thought: the HVdC content codings for the representative sample, the biographical material gathered through interviewing the dreamer and four of her close women friends, and the nature of each particular metamorphosis.

The remaining 20 metamorphoses involved object-to-object changes. Fourteen of the 20 object metamorphoses could be placed in a category for travel or movement. Five of the remaining six object-to-object metamorphoses concerned household objects or wearing apparel. All of them involved small changes, and they did not immediately suggest any relation to metaphors. However, a banana, the one object in this sample involved in a dramatic transformation, has been used in slang expressions relating to sexuality (Partridge, 1937/1984). The following dream reads like a textbook example of sexual symbolism in dreams: "A pouch with a banana is handed to me. I get sexually excited, aggressively, powerfully so. I grin and pull two legs out of the pouch and the banana turns into a penis. I want it badly. I kick the little girl out of the way. I place the penis in me and I can feel it, real. I hold the man tightly and the closer I get to orgasm, the fiercer I am."

This dream report is the 142nd dream report in the first subset of Barb Sanders's two-part dream series. There are eight subsequent dream reports in which there are one or more bananas, a total of about .03% of the dream reports. However, bananas in these eight dream reports appear in the context of eating or in conjunction with other food items. In none of these other eight dream reports does a banana seem to have any symbolic meaning. One in nine seems to be a very small percentage, even within the context of Freud's admonition that potential symbols have to be taken literally "often enough" (1900/1909, p. 352).

The very small number of possible symbolic elements in a dream series about which so much is known leads to three conclusions. First, several of the examples seem highly plausible, but most unusual elements do not make any figurative sense. This leads to a second conclusion. Perhaps some unusual elements are metaphoric and others are due to cognitive glitches of some unknown and unstudied kind. These glitches may or may not be understandable as types of cognitive insufficiencies. Third, the nine dream reports with bananas in them, only one of which seems to be symbolic, suggest potentially symbolic elements should be taken literally much more often than not.

Potential Symbolism in Ed's Dream Reports

Ed's 143 dream reports were studied in detail for all types of potentially symbolic elements. This analysis used the unusual elements scale as a starting point. As in the case of the Barb Sanders series, the goal was to use the HVdC results to provide a context for understanding the 104 unusual elements identified in the Ed series. When examined independently by two researchers, one-third to one-half of the unusual elements did not make any obvious sense in terms of potential symbolism, nor did any of these potential instances of symbolism relate very often to the quantitative findings or to the available biographical information.

The unusual elements showing the most promise as possible symbolism occur in 10 dream reports in which Ed's deceased wife seemed to be alive. They are the most dramatic dream reports in the series. Ed is shocked and stunned to see her alive, or alternatively filled with joy. He then experiences disappointment when he realizes, within the dream itself, that she is deceased and therefore could not possibly be alive. Eight of these back-to-life dream reports seem to portray a sense of puzzlement or confusion due to seeing her alive. However, these dream reports appear to be straightforward cognitive appraisals of his confusion and surprise, which dampen his joy. Collectively, they may be the embodiment of his conflicted waking thoughts, since he wishes that she were still alive and at the same time realizes that she is deceased.

However, two of these dream reports do appear to include symbolic expressions of his confusion and doubt. In the first of these dream reports, "her face changes from a most beautiful complexion (as it had been before her death) into a waxen, artificial, dead face—as one sees after the mortician has dressed a body." In another back-to-life dream, in which his thinking processes show great depth, he seems to grasp its symbolic nature during the dream. It begins with him seeing Mary sitting in a car on the other side of the road from where he is standing. She is trying to talk to him. Even though she is talking, he realizes she is dead. Then he further realizes he cannot cross the road that divides them because "I know that she is dead, and that the road between us is the dividing line between life and death" (Domhoff, 2015, p. 21). Because crossing roads, bridges, or other boundary lines is very often used to express a life change or a different state of mind, this example seems to be highly plausible. However, it could also be understood as a literal enactment of what he knows. She is dead and there is no way to talk to her or to be in touch with her.

There are also two dream reports in which his second wife, Bonnie, changes into his deceased wife, Mary. These metamorphoses might be a symbolic expression of wishing Mary were alive and still his wife. In addition, there are several dream reports in which there are age changes in characters. For example, in one dream he and Mary are frantically searching for their 10-year-old son but he suddenly appears as the young adult he actually was at the time. This dream may imply a magical solution, which might be symbolic in nature. By way of contrast, in the dream reports in which Ed and Mary are playing with their children or are involved in a past activity, neither of them is anxious or upset. The settings in these dream reports suggest they are simply straightforward enactments of memories of the past.

Ed's unusual pattern of social interactions with his wife provided a context in which a search could be made for plausible symbolic expressions of the conception that he views himself as incompetent in his interactions with her. For example, in one dream report he spills the tea he is bringing her. In another, he sinks in quicksand while trying to keep up with her as they are walking along. In still another, he can't find their children when she asks him to do so. It is plausible that each of these dream reports enacts Ed's conception of himself as incompetent in his interactions with Mary, which leads to her annoyance with him. It also may be that these dreams enact Ed's belief that Mary conceives of him as incompetent.

Three of Mary's relatively few friendly gestures toward Ed are of interest because she gives him reassurances in dream reports in which she appears to have come back to life. Such dreams may possibly be a symbolic way of portraying his desire to be free of his feelings of sadness and guilt. In the first of these dream reports, shortly after her death, she tells him she wants him to be happy. In the second, she expresses her approval of his forthcoming marriage to Bonnie. In the third, she assures him she is fine and happy. These dream reports gain plausibility as symbolic portrayals because reassurance dreams were relatively frequent in the large samples of dreams about deceased loved ones, which were collected in psychology classes via dream diaries in the late 1980s at the University of North Carolina, Chapel Hill (D. Barrett, 1992).

However, as with the Barb Sanders series, it is difficult to make a case for symbolism in this series beyond the handful of plausible anecdotal examples, which have been enumerated here to provide a sense of the nature of unusual elements. This enumeration also demonstrates the difficulties of classifying or contextualizing unusual elements as a possible starting point for discovering instances of possible symbolism. To the degree that any of these instances can be construed as symbolic, they do not seem to disguise any potential hidden meaning meant to keep the dreamer from awakening with anxiety or fear, as Freudians theorize. Instead, they seem to make use of figurative expressions frequently used and fully understood in waking life by everyone. But these conceivable instances are relatively few in number and sometimes stretch the point.

Cognitive Linguistics, Figurative Thinking, and Dreams

Although the content analyses of a wide range of dream reports did not yield very many promising results, the search for symbolic meaning in dreams did receive a boost in the 1990s due to the development of a new theory of figurative thought by linguist George Lakoff and his collaborators (Lakoff, 1987; Lakoff & Johnson, 1980). The theory was applied to a wide range of topics, including poetry (Lakoff & Turner, 1989), and to dreams as well (Lakoff, 1993, 1997). According to this theory, figurative thinking is not based on simple laws of association but on general, experientially based cognitive categories. These categories are said to be organized around ideal prototypes that relate at least in part to events that happen regularly in many people's lives. These repeated basic experiences usually begin in early childhood, such as seeing, walking, and going on a journey. According to this theory, metaphors involve the mapping of abstract and difficult issues (the "target domain") onto experiential categories based on everyday experiences (the "source domain"). For example, love (the target domain) can be characterized as a journey (the source domain). Examples of metaphoric expressions that draw on this general category include the "rocky road" of a relationship and the "dead end" a relationship has reached. In a parallel fashion, a virtuous life can be characterized as a journey as well. Virtue and guilt are talked about in terms of walking carefully on the proper "path" and not straying from it. Examples include "stay on the straight and narrow" and "don't stumble in your quest for a virtuous life."

The pervasiveness of metaphoric thinking in waking life is demonstrated with the commonplace "I see what you mean." It is based on the experiential metaphor of "knowing is seeing." Since listeners do not literally "see" what speakers mean when their utterances are understood by the listeners, it is a metaphoric statement. This metaphor "works," and is taken for granted because some of the first concepts people understand from childhood onward involve seeing some person, object, or event repeatedly (Matlock, 1988). Indeed, the commonplace use of this metaphoric expression is so second nature in the English language that it is rarely thought about as a metaphor.

Many conceptual metaphors can be understood as a more abstract form of the concept of "embodied simulation," as discussed in chapter 1. According to this view, embodied concepts "contain representations of bodily states that customarily occur during interactions with concept-relevant stimuli" (Landau, Meier, & Keefer, 2010, p. 1054). In an embodied conceptual metaphor, to take one instance, the idea that two people can have a "warm friendship" is based on the "physical warmth" people experience as infants and young children, while being "safely cradled in a caregiver's arms" (Landau et al., 2010, p. 1054). More generally, metaphoric expressions of friendship are "rich with representations of bodily states, including temperature-related sensations (e.g., 'warm' embraces), that regularly occur during friendly interpersonal encounters" (Landau et al., 2010, p. 1054).

"Typical" Dreams May Fit the Cognitive Linguistics Theory

At least some "typical" dreams seem to have the potential to be understood in terms of the analysis of metaphor by cognitive linguists. The contents of such dream reports, which are readily definable as unusual elements, are either humanly impossible, such as flying under one's own power, or culturally implausible, such as appearing in public wearing only skimpy underwear or being completely without clothing. Two different waking metaphors may express the feelings present in these two examples. Flying in dreams could be based on the general conceptual metaphor, "happy is up," through which people express their feelings of happiness by saying they are "up," "high," "flying," "high as a kite," "walking on air," or "floating on air" (Lakoff, 1987). In the case of partial or complete nudity in dreams, it could be based on the conceptual metaphor, "embarrassment is exposure," which is used to express social embarrassment through metaphors such as "caught naked," "caught with your pants down," or "caught with egg on your face" (Holland & Kipnis, 1994).

Questionnaire studies find from one-third to two-thirds of those surveyed recall having one or more of the dreams they are asked about (Griffith et al., 1958; Nielsen et al., 2003). However, as mentioned at the outset of this chapter, studies of large samples of dream reports collected in lab and nonlab settings suggest typical dreams are extremely infrequent. One of the largest content analyses ever undertaken was based on 635 dream reports. They were collected in two different sleep-dream laboratories from 250 participants in the course of a variety of studies throughout the 1960s. The study included almost equal numbers of young adult women and young adult men. Only two of the 635 dream reports included flying by an individual, which is 0.31% of the total number of dreams (Snyder, 1970). Similarly, a study based on an even larger sample, 1,910 dream reports from students at the University of North Carolina, Chapel Hill, reported a low figure for flying dreams: 17 participants reported 22 flying dreams, which is 1.2% of the total sample (D. Barrett, 1991). Another large-scale dream-diary study, which included 1,612 dream reports from 425 university students in Germany, was coded for 144 "dream themes" (Mathes et al., 2014). Most of the themes are somewhat general and could not be considered as potentially "symbolic" within the context of the unusual elements categories. Focusing here on the narrow list of typical dreams used in past studies, 1.9% of the dream reports in this study contained "flying/soaring," 1.4% included nudity, 1.3% included finding money, and 0.6% included teeth falling out (Mathes et al., 2014, pp. 62–63, table 2).

The frequency of dream reports in which the dreamer is flying under her or his own power was even lower in a word string study of 3,309 dream reports from children, teenagers, and young adults, which are on Dream-Bank.net. The study used terms for flying, gliding, or floating. After dream reports were eliminated because they described realistic flying events, such as flying in an airplane, only 18 of the dream reports had human characters who were flying under their own power, which is 0.5% of the total reports. This miniscule percentage also was found when the search was restricted to the HVdC normative dream reports for men and women (Domhoff & Schneider, 2008b, p. 1243).

Finally, a study of 983 dream reports from two-week dream diaries, volunteered by 126 students at the University of California, Santa Cruz, as part of a psychology class, discovered virtually none of 10 typical dreams occurred more than a few times. For example, there were only five flying dreams in the sample, which is 0.5% of the total number of dreams. The figures for several other typical dreams were even lower. Two people dreamed of finding money, two became lost, two were taking an examination, and one lost his teeth (Domhoff, 1996, p. 198).

The Limits of Metaphoric Analyses

The findings on the low percentages of seemingly "typical" dreams in an individual's dream life means conceptual metaphor theory would have to account for a much higher percentage of dream reports to be useful. However, the theory's most promising metaphoric category in regard to dreaming, "the generic is specific," was met with skepticism in relation to waking applications by several psychologists and linguists, including a cognitive linguist (e.g., Grady, 1999, a cognitive linguist; McClone, 2001, a research psychologist who is a critic of cognitive linguistics). Moreover, as shown in the following paragraphs, subsequent changes in how the category is explained lead to resemblance analyses, which are very similar to those provided by Hall (1953).

According to the "generic is specific" concept, one instance (a dream) is related to an instance in the dreamer's life (such as general biographical information or a specific personal concern). This abstract statement of the concept becomes more understandable through the analysis of a dream report. In this dream report, the dreamer, an adult woman, is back in college. She is taking a course from a professor she describes in the dream report as her favorite professor. This professor says she is likely to fail the exam because she is not studying hard enough (Lakoff, 1993, p. 94). Based on knowing a small amount about her waking life, Lakoff concludes that the dream report shows she is afraid that her recent marriage to a professor may fail. She does worry in waking life about a failed marriage because she is not working in a job outside the home. Thus, a failed exam in a dream report is similar to a concern with a failed marriage in waking life. What they have in common is they are both "tests" or "trials," which can be passed or failed, and they both involve a professor. This analysis ends up as what some people might think of as an analogy (a comparison of two things for their similarities) or an allegory (a story with a moral to it).

There are two problems with this approach for research purposes. First, as Lakoff stresses, it is necessary to have information about the dreamer to figure out the plausible similarities between the dream and waking thoughts: "I claim that deep and extensive knowledge of the dreamer's life is essential to pinpointing the meanings of dreams" (1993, p. 88). He narrows this assumption slightly by writing that, under some conditions, "it might be possible to narrow the range of possible interpretations for a given dream without knowledge of the dreamer" (Lakoff, 1993, p. 88).

This is the same problem faced by Freud and any other therapists who make use of dream interpretation. They all need to have deep and extensive autobiographical information to provide context for a possible interpretation. Psychoanalysts try to discover this information through obtaining free associations to each element in a dream. (Free association consists of letting one's mind generate associations to a dream or an element of the dream, without censoring thoughts or starting with any assumptions.) But obtaining personal information from the dreamer, whether through free associations or interviews, may bias the dream interpreter, whether she or he realizes it or not (Foulkes, 1978, 1996a). There are demand characteristics, along with the potential for persuasion, in a therapeutic relationship, as discussed in chapter 9.

Second, and more important, most research psychologists and a few cognitive linguists doubt that the "generic is specific" concept fits with the original idea of metaphors being based on "target domain/source domain" type of constructions. Instead, some cognitive linguists argue that there are two kinds of metaphors—target/source metaphors, which "correlate" experiences, and a new type in terms of cognitive linguistics, known as resemblance metaphors (Grady, 1999; Sullivan & Sweetser, 2010). Resemblance metaphors are based on similarities and associations. They therefore seem to be at least somewhat similar to other claims about the basis for symbolism. The two examples the various discussants focused on in the 1990s were "the warrior is a lion" and "the surgeon is a butcher." Most people understand what those statements mean but neither has a source domain and a target domain, and most people have not experienced warriors, lions, or surgeons. However, people nonetheless know what qualities warriors and lions possess and what qualities surgeons and butchers possess. They know about those qualities due to their general knowledge, not through their direct experience. Based on general knowledge, people can figure out what qualities warriors and lions share (they are "courageous beings") and what qualities surgeons and butchers share (they are people who cut into bodies to make a living). They therefore can understand that "the warrior is a lion" refers to an extremely courageous warrior and the "the surgeon is a butcher" refers to an incompetent surgeon.

In a general theoretical analysis, which directly asks in its title, "Is 'Generic Is Specific' a Metaphor?," two cognitive linguists provide a more encompassing analysis. They note that this type of construction may be a metaphor "to a greater or lesser degree"; more generally, they "map the family resemblances characterizing a category prototype to other category members" (Sullivan & Sweetser, 2010, p. 309). They further emphasize that "extensive research on category structure has shown that natural categories do not themselves have properties—category members have properties" (Sullivan & Sweetser, 2010, p. 311). In doing so, they make reference to the original work by Eleanor Rosch (Rosch 1973; Rosch & Mervis, 1975). (Her work is discussed in chapter 2 of this book as having provided a new starting point for how cognitive psychologists think about categories.)

One example they use, a good academic journal is a "gem," is understood in academic circles because some journals share the quality, as with gems, of being valuable. The analysis on which this conclusion is based is a complex one, rooted in the discipline of cognitive linguistics. Assuming for the sake of argument the analysis is a sound one, the important point in terms of attempting to find symbolism in dreams is that it can incorporate the law of resemblance by value, which is one of the laws listed in table 5.1. Put another way, the "law of resemblance by value" may be part of a family of resemblances, based on value. It would thereby include one of the symbolic equations in table 5.1: "jewelry" is a symbol for something else that is highly valued, the female genitals.

With the distinction between target/source metaphors and correlational/ resemblance metaphors in mind, along with the general theoretical analysis provided by Sullivan and Sweetser (2010), it seems plausible that there are several resemblances in Lakoff's (1993) prototypical example of a dream report. In the woman's dream, recall she is likely to fail the test in the professor's class. In waking life she worries that she is failing in her marriage to a professor because she is not working outside the home. The resemblances between the dream and her waking-life personal concerns can be seen in the common functions the two men share (both are professors) and in the value she places on both of them (they are very highly valued). In addition, she is in the same situation in her dream report and in waking life in that they both involve a "test," which always includes the possibility of "failing" as well as "passing." Thus, this resemblance analysis maps "relations, scales, and temporal and causal sequences" (Sullivan & Sweetser, 2010, p. 310).

Although this analysis of the woman's dream report is complex and seems plausible, it still leaves the study of symbolism in dreams about where it started in Hall's (1953) early analyses. Lacking information other than the dream report, there are no solid criteria for discerning whether a term is being used literally or which of several possible figurative interpretations might be the "right" one. Within a therapeutic context, on the other hand, there is the possibility that the processes involved in the social psychology of persuasion are leading the therapist and the patient to accept one or another metaphoric interpretation. This process is also based on various social expectations, which are part of psychodynamically based therapeutic traditions. If it is assumed that there is symbolism in dreams, then symbolism is likely to be found.

In everyday waking life, on the other hand, people have a context for understanding the many figurative utterances they hear and read. This context includes a shared culture, the frequency with which a particular construction is used, their relationship with the person speaking, and the immediate context and situation in which they hear the utterance or read the statement. This multifaceted context makes it possible for the listener or reader to instantaneously sort out which contextual aids and conceptual metaphors can be used. But, to repeat, none of these possibilities is present in the context of a dream report or even in a dream series, as shown by the studies overviewed in the previous subsection.

The efforts to study symbolism in dream reports in the context of cognitive linguistics leads to two conclusions. First, the theory has its greatest potential through the mapping of repeated experiences to abstract constructs (e.g., "embarrassment is exposure"), which seem to be present in at least some typical dreams. However, the frequency of those dreams is extremely low. In addition, no empirical studies have been carried out to find out if flying dreams are more pleasurable than dream reports in general, as would be expected if "happiness is up." Nor is there any evidence that dreams of being inappropriately dressed contain more feelings of embarrassment than other dream reports, as would be expected if the "embarrassment is exposure" metaphor is being used. Second, the interpretations of most dreams by means of families of resemblances do not differ very much from the laws of resemblance listed in table 5.1. This conclusion holds true even though it is also true that the laws of resemblance are far better understood within the framework of cognitive linguistics (Kövecses, 2017; Sullivan & Sweetser, 2010).

Neuroimaging Studies of Waking Metaphoric Thinking

Given the limited evidence for symbolism in dreaming, it becomes of interest to examine what neuroimaging studies have to say about the production and comprehension of metaphors during waking. As suggested in the introduction to this chapter, if the neural substrate that supports waking metaphoric thinking differs from the neural substrate active during dreaming, then the neural network that supports dreaming may be cognitively insufficient in terms of its ability to support metaphoric productions. Similar reasoning suggests that other forms of figurative thought could not be supported either.

Several studies provide a good understanding of the neural network that supports metaphoric comprehension (e.g., Holyoak & Stamenkovic, 2018; Rapp, Mutschler, & Erb, 2012; Vartanian, 2012). Other studies have focused on the generation of metaphors (Beaty & Silvia, 2013; Beaty, Chen, Qiu, Silvia, & Schacter, 2018; Benedek et al., 2014). In both cases, functional connectivity analyses reveal a network that includes areas in the frontoparietal control network as well as in the default network. More specifically, there are direct functional connections between core regions in the frontoparietal, salience/ventral, and default networks. In terms of the cognitive networks active during metaphor production, they show "increased cooperation among brain regions involved in mental simulation, executive control, and semantic integration" (Beaty et al., 2018, p. 170). Recalling the several different analyses of neuroimaging studies of REM sleep and a meta-analysis based on such studies, the degree of cognitive control, cooperation, and semantic integration involved in metaphor generation is very likely beyond the capabilities of the neural substrate that supports dreaming. This neural substrate does not include any parts of the frontoparietal control network or the salience/ventral network. It is therefore unlikely that the active portions of the default network receive enough support to generate metaphors.

This conclusion is consistent with a meta-analysis of studies focused on metaphor comprehension. It found the comprehension of metaphoric thinking is supported by the inferior frontal gyrus and "sometimes the dorsolateral prefrontal cortex," as well as "broad regions of the temporal cortex" (Holyoak & Stamenkovic, 2018, p. 653). Since the inferior frontal gyrus and the dorsolateral prefrontal cortex are deactivated during all stages of sleep, this finding suggests that there is little or no complex metaphoric thinking during dreaming. Then, too, the meta-analytic studies of analogical reasoning and categorization, which are sometimes hypothesized to provide support for metaphoric thinking, "have established that complex analogical reasoning involves broad regions of the frontal and parietal cortices that form a frontoparietal network" (Holyoak & Stamenkovic, 2018, p. 653). Since the frontal regions of the frontoparietal control network are deactivated during sleep, it can be concluded that analogical reasoning could not be the basis for metaphoric thinking during dreaming.

Conclusions and Implications

Dreams are enactments, and they are often vivid and complex enough to invite comparisons with theatrical plays. They frequently dramatize personal concerns, which have been shown to be at least somewhat similar across a wide range of cultures. Dreams also dramatize revealing individual differences. At the same time, they sometimes contain unusual constructions, which remain to be understood. But it does not follow that dreams therefore also must be symbolic, as has been assumed throughout recorded Western history. Instead, the cognitive insufficiencies during dreaming support the conclusion that there is little or no symbolism in dream reports.

The empirical studies of potential symbolism that are reviewed and critiqued in this chapter do not provide systematic evidence for symbolism as an important feature in dreams, nor do the studies of dream series produce more than a few tantalizing and highly plausible instances. Typical dreams seem to fit a cognitive linguistic perspective, but typical dreams collectively add up to no more than a few percent of most people's extensive dream lives. Even if all of the plausible instances of symbolic elements in the various studies are combined, the best conclusion for now is that there are relatively few symbolic elements in dreams.

Most of the brain areas active during the generation and comprehension of metaphor, and other forms of figurative thought, are relatively deactivated during dreaming. It therefore seems likely that the paucity of symbolism in dream reports can be attributed to cognitive insufficiency. The few plausible instances of symbolism discussed in this chapter, such as those in the dream series provided by Ed, suggest that any future attempts to revive the claim that there is symbolism in dreams might begin with neuroimaging studies and the collection of REM dream reports in sleep-dream labs at the time of morning awakening. The most promising participants in such studies would be people who have provided instances of potential symbolic dream elements by means of already documented reports within the context of an ongoing dream series.

6   The Development of Dreaming in Children

This chapter adds a developmental dimension to the neurocognitive theory of dreaming. It does so at all three levels of the theory—neural, cognitive, and behavioral. It begins with neuroimaging research on the default network from preschool to early adulthood. This network is very immature in the preschool years and does not begin to approach adultlike levels until ages 9–13. Maturation involves changes in the connectedness within the default network itself and in its relationship with the frontoparietal control network.

The chapter then turns to a consideration of research on the development of dreaming, as documented in both longitudinal and cross-sectional laboratory studies. Strikingly, dreaming, like the default network, does not begin to reach adultlike cognitive complexity until ages 9–11 and only reaches adultlike content between the ages of 12 and 15. However, because the main focus of the chapter concerns neural substrates and cognitive processes, most of the substantive findings on dream content are presented in the next chapter, especially in the case of older children (ages 9–12) and adolescents (ages 13–18).

The findings on dreaming are then compared with the results from studies of the development of waking cognitive capacities. These studies were carried out by both laboratory dream researchers and a wide range of developmental psychologists using experimental designs. Taken together, these three separate literatures reveal dreaming is a gradual cognitive achievement.

The Maturation of the Default Network

Several studies find the default network matures gradually from infancy to adolescence, with much of that maturation occurring after ages 4–7. The basic outlines of the default network can be detected in infants (Bulgarelli et al., 2020). However, as noted above, the default network is immature until late in childhood and early adolescence and is not close to adult levels until the middle of the teenage years. In one of the foundational cross-sectional studies of the development of the default network, which spanned the ages from 7–9 to young adulthood, the default network was only minimally connected at ages 7–9 (Fair et al., 2008, p. 4029). One of the main hubs in the adult default network, the medial prefrontal cortex, is on the periphery of this generally sparse network at ages 7–9, and it is only sparsely and indirectly connected to the parietal areas, which contain important structures within the default network (Fair et al., 2009, pp. 6–7, 11).

The immaturity of the default network at ages 7–9 in part reflects the localized structure of most brain networks at that age, but other factors play a role as well: "synapse formation, the tuning of synaptic weights, synaptic pruning, and myelination all have unique developmental time courses that extend further into development" (Fair et al., 2009, p. 8). In this study, the frontoparietal control network had greater functional connectivity by ages 7–9 than did the default network, which had very few "short-range functional connections" (Fair et al., 2009, p. 2). It was not until two or three years later that the default network's main frontal structures had more than a few connections to each other, even though they are "fairly close in [anatomical] space" (Fair et al., 2009, p. 2).

The cross-sectional neuroimaging study of the default network in 7–9-year-olds also included 54 participants ages 10–15 and 91 participants who were 19–31 (Fair et al., 2009, p. 10). Studies of them confirmed ongoing changes in terms of increased within-network connections. These changes in connectivity and integration can be graphed for visualization purposes in a way that pictures the relationships within the default network as spatially closer as it becomes more functionally integrated. Based on this fictive anatomical configuration, the differences within the default network between ages 8 and adulthood are displayed in figure 6.1. This figure is a truncated and stylized version of a more complex graphic. It originally included three other brain networks, which interact with the default network to varying degrees (Fair et al., 2009, p. 5, fig. 2 and video S1). As figure 6.1 demonstrates, the default network becomes more integrated by age 11 but is not quite adultlike until age 15.

The findings in this foundational study were replicated in an even larger cross-sectional study of 447 Brazilian participants ages 7–15. They came from a wider socioeconomic spectrum and had greater genetic diversity

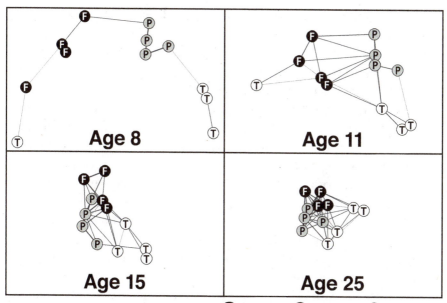

Lobe in which the nodes are located: **F** Frontal **P** Parietal **T** Temporal

Figure 6.1
The maturation of connectivity in the default network.
Note: The two bottom panels are "smaller" in a fictive anatomical configuration in order to depict the increasing functional connectivity within the default network. Adapted from Fair et al., 2009.

than the American participants (Sato et al., 2014). Once again, the centrality of the major hub regions in the default network increased with age, and the connectivity of the anterior and posterior regions of the network increased (Sato et al., 2014, p. 93). In addition, "the age-related changes in the central executive network [i.e., frontoparietal control network] were less widespread compared with the default network," which is consistent with the greater integration of the frontoparietal control network at an earlier age in the first cross-sectional study (Sato et al., 2014, p. 92). Similarly, a study of social cognition between ages 6 and 12 concluded the default network matures slowly during this age period (Moraczewski, Nketia, & Redcay, 2020).

These results were replicated and extended for the preschool years through adolescence in a cross-sectional study in Australia of 10 preschool children (ages 4–6), 14 school-age children (ages 7–12), and 24 adults (He et al., 2019). In a combined cross-sectional and longitudinal resting-state fMRI study of the default network in China, based on 305 children between the ages of 6 and 12, each of whom was studied twice within the space of a year, the results were close to those from a control group of 61 young adults ages 18–28 by age 12 (Fan, Liao, Lei, Tao, & He, 2021). In particular, the functional connectivity between anterior and posterior regions in the default network increased over this time period.

The results from these several studies are supported and extended in a longitudinal study of both the default network and the frontoparietal control network at ages 10 and 13, based on 45 participants (24 girls, 21 boys) (Sherman et al., 2014, p. 149). The investigators first concluded that "by age 10, the basic functional architecture of the default mode network is in place" (Sherman et al., 2014, pp. 151, 154). As reported in the cross-sectional studies, the integration between the medial prefrontal cortex and posterior cingulate cortex also increased between ages 10 and 13. The frontoparietal control network also became almost fully integrated by age 13. Since the connections found in this study at age 10 were more robust than those with 7–9-year-olds in the cross-sectional studies, the investigators suggest "it is possible that significant maturation between ages 7–9 and age 10 account for differences in our findings" (Sherman et al., 2014, p. 151).

This hypothesis is supported by two studies of the maturation of the cingulum bundle, a white-matter tract that in adults provides the major communication link between the medial prefrontal cortex and the posterior cingulate cortex. The first study, which compared 23 boys ages 7–9 with 22 young adult males, found only one difference between the child and adult groups on their degree of connectivity. It concerned the relationship between the medial prefrontal cortex and the posterior cingulate cortex. The investigators suggest that the relationship between the medial prefrontal cortex and the posterior cingulate cortex is "the most immature link" in the default network (Supekar et al., 2010, p. 290). Consistent with this finding, there is no structural or functional relationship in 7–9-year-old children's default network involving the cingulum (Supekar et al., 2010, p. 297). According to this team of researchers, this relationship develops during preadolescence or early adolescence.

A cross-sectional study, focused directly on the cingulum, confirmed there is a major change in the cingulum's role within the default network between ages 9–11 and 11–13. The study is based on data from nine girls and nine boys (Gordon et al., 2011). In the younger group, the correlation between functional connectivity within the default network and the integrity (completeness) of the cingulum was relatively small and statistically nonsignificant. However, the correlation was large (.73) and statistically significant in the older group. Very importantly in terms of the neurocognitive theory of dreaming, the researchers conclude that "this relationship emerges after age 9, but is more reliable as children enter the preadolescent years" (Gordon et al., 2011, p. 747). This "increased maturation of the cingulum bundle is related to more efficient communication between the medial prefrontal cortex and the posterior cingulate cortex" than is available through other connections at an earlier age (Gordon et al., 2011, p. 747).

In adults, the cingulum bundle "promotes far-reaching antero-posterior connectivity from the retrosplenial and precuneal cortices to the medial frontal lobe and other, partly diffuse, frontal areas of the brain" (Horn et al., 2013, p. 6). The retrosplenial cortex is a component of the default network, as are at least portions of the precuneus, which is also a part of several networks that subserve episodic memory, visual-spatial imagery, and self-reflection (Fransson & Marrelec, 2008; Kaboodvand, Bäckman, Nyberg, & Salami, 2018).

At the same time, there is additional evidence for significant changes occurring between ages 10 and 13 in the frontoparietal control network and in its relationship with the default network. For example, the frontoparietal control network is more integrated than the default network at age 8 years, 6 months, and only functionally connected to a few nodes in the default network. By age 13 years, 2 months, however, the frontoparietal control network is more functionally connected to nodes within the default network (Fair et al., 2009, p. 5, fig. 2). At this point, the default network and the frontoparietal control network become increasingly segregated, with the frontoparietal control network able to modulate and control the default network during waking. This finding is consistent with the findings in the longitudinal neuroimaging study of children at ages 10 and 13: the frontoparietal control network is very similar at age 10 to that of adults but the segregation between the default network and the frontoparietal control network is a work in progress between ages 10 and 13 (Sherman et al., 2014, pp. 151, 155).

The findings on increased segregation and integration are consistent with a major 10-year longitudinal study of changes during sleep in children between the ages of 9 and 18, with 3,500 all-night EEG recordings. Neuronal density declines during this time span, with the largest decline (by 65%) occurring between ages 11–12 and 17. This pruning shows that the brain goes through a dramatic reorganization during this time period, which may make complex thinking more possible (Campbell, Grimm, de Bie, & Feinberg, 2012; Feinberg & Campbell, 2010). Neuroanatomical analyses, based on fMRI studies, demonstrate cortical thinning in bilateral frontal, parietal, and occipital regions in children ages 6–10, which relates to the functional development of the frontoparietal, network, the default network, and sensorimotor and auditory networks (Zhong, Rifkin-Graboi, Ta, Yap, & Chuang, 2014).

Based on these several different studies, adultlike dreaming may not be possible until the default network matures into a more adultlike form between ages 9–11, or more broadly, 9–13 or even 13–17, when more pruning and reorganization have occurred. A more detailed age specification awaits results from future studies. In addition, the default network's relationship with the frontoparietal network becomes better articulated between the ages of 9 and 13, and perhaps a few years beyond. Thus, the expectation from a neurocognitive point of view would be that dreaming becomes more frequent and complex during those four years. This expectation is examined on the basis of laboratory studies of dream recall, dream complexity, and dream content in children's studies, which were carried out a decade or more before the studies of the maturation of the default network began.

Laboratory Studies of the Development of Dreaming

Just as the default network is immature until roughly ages 9–13 and not adultlike until age 15, so, too, dreaming is not fully developed cognitively or in the substance of dream content until the same age period. This conclusion is first of all based on two concurrent longitudinal studies that examined the years from age 3 to age 15 (Foulkes, 1982). It is also based on a cross-sectional study of children ages 5–8, which was designed as a replication and extension of the key findings in the first of the two concurrent studies (Foulkes, Hollifield, Sullivan, Bradley, & Terry, 1990). Finally, this conclusion is based on the findings in a five-year study for ages 9–15, which included 24 participants, 12 girls and 12 boys (Strauch, 2004, 2005; Strauch & Lederbogen, 1999).

In a largely successful attempt to minimize the loss of participants through moves to other cities, the main investigator in the concurrent longitudinal studies began by advertising for participants in the local newspaper in the city where he taught and researched: Laramie, Wyoming, in which the University of Wyoming is located (population 23,000 in 1970). He asked that only parents who intended to remain permanent residents of the city for the next five years should volunteer their children. The potential participants were interviewed and screened and were told they would be paid after each year and a larger amount if they completed the whole project.

The younger subset included 14 children, who were studied in the sleep-dream lab at ages 3–5, 5–7, and 7–9. All of them took part during all five years. They also were involved in two-week "nursery schools" in the first three summers, during which they were observed in playground settings and given a wide variety of personality and cognitive tests (Foulkes, 1982, pp. 14–18). They also took part in two home-based studies of the frequency and recall of dream reports, which determined that the low percentages of dream recall in the lab were not due to the laboratory setting (Foulkes, 1982, pp. 18–19). The 14 members of this younger cohort were supplemented when they were ages 7–9 by seven new female participants. They were added in order to control for possible practice effects resulting from the original cohorts' two previous experiences in the lab. This control condition showed there were no practice effects (Foulkes, 1982, p. 22).

Initially, there were 16 participants at the outset of the second subset of children, who were studied at ages 9–11, 11–13, and 13–15; 12 of them completed the whole study. They too responded to a variety of tests, but they did not take part in a summer program. This cohort was supplemented by six boys at ages 11–13 for control purposes, and once again there were no differences from those who had been in the study from the start (Foulkes, 1982, p. 22). Two nonlaboratory studies with this older group demonstrated that any content differences between lab and nonlab dream reports were due to methodological issues and selective recall of dreams remembered outside the lab (Foulkes, 1979). Altogether, 26 of the 43 participants in the two concurrent studies took part during all five years, eight for at least three years, and nine for at least one year (Foulkes, 1982, p. 22).

A major effort was made to ensure the participants were comfortable. The parents of the younger children were invited to help with the bedtime preparation (Foulkes, 1982, p. 27). After the children were prepared for the study, including the placement of electrodes, they often read, played, or explored the rooms in the laboratory. To help ensure their comfort and avoid introducing any confounds by using several different assistants to do the awakenings, the main investigator personally carried out all of the

2,711 lab awakenings (Foulkes, 1982, pp. 27, 33–34). The children's degree of anxiety and stress was monitored during the prebedtime period and through their sleep patterns. At ages 3–5, they were rated as calm or relaxed on 79% of the nights and as extremely anxious on only 5.4% of the nights. Their median time to fall asleep was 20 minutes. The same children took only 12 minutes when they returned two years later. The median number of spontaneous night awakenings was one or zero, and the median time awake during the night was eight minutes or less (Foulkes, 1982, pp. 33–34). The results on anxiety indicators were very similar to those for the older cohort (Foulkes, 1982, pp. 33–34).

The cross-sectional study of children ages 5, 6, 7, and 8 was carried out in an effort to replicate the striking cognitive changes that occurred at those ages in the first of the concurrent longitudinal studies. This time the study included more waking cognitive tests because the age changes in the first study did not correlate with the many personality tests. The recruitment began with ads in a major newspaper in the metropolitan area of Atlanta because the main investigator was now located at Emory University. The 80 participants who were chosen, half girls and half boys, were within one month of their fifth, sixth, seventh, or eighth birthdays when they were tested. Each participant was awakened a total of 10 times over two nonconsecutive nights. The efforts to reduce and monitor anxiety levels were similar to those in the longitudinal study. Once again, the parents often helped with the bedtime preparation for the youngest participants but the parents of the 8- and 9-year-old participants usually dropped their children off and then drove away. The participants scored low on the 5-point anxiety scale administered on each visit, and they fell asleep 12.5–14.5 minutes after the lights were turned out (Foulkes et al., 1990, p. 454). The main investigator made all 800 awakenings. Overall, the concurrent longitudinal studies and the cross-sectional study involved a total of 124 participants, who were awakened 3,511 times.

Dreaming Is a Gradual Cognitive Achievement

Both the longitudinal and the cross-sectional studies revealed dreaming to be a gradual cognitive achievement in terms of frequency, complexity, and content (Foulkes, 1982; Foulkes et al., 1990). Preschool children in the longitudinal study reported very brief dreams, with a range of 2 to 50 words, after a mean of 27% recall from REM awakenings. This mean recall figure was very likely inflated by two parent-pressured and eager-to-please participants. This hypothesis was supported by the fact that both of them, unlike the other children in the study, reported fewer dreams two years later, at ages 5–7. As a result, the median recall rate of 15% provided a more accurate picture at ages 3–5 (Foulkes, 1982, pp. 44–45, 51–53). The few characters in the reports were most frequently animals, and there were few social interactions. No dreamer was involved in those few interactions (Foulkes, 1982, p. 48).

The median rate of recall doubled to 31% at ages 5–7 and increased to 43% at 7–9. The median report length also increased between ages 7 and 9, to 70 words for girls and 48 for boys (Foulkes, 1982, pp. 113–120, 341). This recall first involved the last REM period of the night, which has parallels with the finding that the rates of recall are higher in later REM periods for adults as well. There were also changes in dream content, starting with more characters and more dreamer involvement in the few social interactions at ages 5–7. These social interactions were 2.7 times more "prosocial" than "hostile" for both girls and boys, and these trends continued at ages 7–9 (Foulkes, 1982, p. 48). Between ages 7 and 9, the normative Male/Female Percent began to appear (Hall, 1984, p. 115). The Animal Percent began to decline but there were no large-scale changes in dream content (Foulkes, 1982, pp. 113, 115, 234, 341). The most important correlates of the changing patterns of recall and complexity in young children from ages 5–9 were in visual-spatial skills, as measured primarily by the block design test and the embedded figures test (Foulkes, 1982, pp. 72–73).

Similar findings emerged in the cross-sectional replication and extension study involving 80 children at ages 5, 6, 7, and 8. The median rate of reporting from REM awakenings was only 20% for all age groups. This finding supported a conclusion in the longitudinal study—the frequency of dreaming does not change dramatically in this age period. However, the imagery in the dream content became more dynamic by age 7, and the dreamer was more frequently a participant in social interactions by age 8. The narrative quality of the dream reports improved considerably at age 8 (Foulkes et al., 1990, pp. 447, 456, table 1).

Several Hall/Van de Castle (HVdC) coding categories were used in a further analysis of these dream reports. The Male/Female Percent was present in all four age groups, and the Animal Percent was down to 12% in the boys and 9% in the girls at age 8 (Domhoff, 1996, p. 89, table 5.6; Hall, 1984,

p. 115). Still other HVdC content indicators led to results similar to those in the original analysis, which used slightly different scales. Generally speaking, there were few social interactions or negative events. The Aggression/ Character Index was .11 at ages 7 and 8, and the Friendliness/Character Index was .13 in girls and .05 in boys in those two age groups. There were very few (minor) misfortunes and no failures involving efforts to achieve a goal (Domhoff, 1996, p. 94, table 5.7).

The similar results for the Male/Female Percent and the Animal Percent in both the longitudinal and cross-sectional studies of younger children also provide evidence for the continuity of dreaming and waking thought. This evidence is based on the findings in a HVdC analysis of the characters in a large sample of brief fictional stories that were thought up and verbally reported at the request of investigators, 180 girls and 190 boys who ranged in age from almost age 3 to almost age 6 (Pitcher & Prelinger, 1963). The results for the Male/Female Percent in the children's stories were very similar to those in dream reports, although the gender differences on the Male/ Female Percent were even more extreme in the younger children, and the Animal Percent was higher in the stories than in dream reports (Domhoff, 1996, pp. 89–90).

Turning to the findings with the second cohort in the concurrent longitudinal study, the pivotal years in the development of increased frequency of dreaming proved to be in the first phase of the study (ages 9–11). For those years, the sample consisted of eight girls and eight boys. Collectively, they reported dreams after a mean of 65.8% of the 231 REM awakenings, which is about the same as they reported at later age levels (Foulkes, 1982, pp. 149–150, 178, 238, table B). The median recall percentage was even higher, 79% (Foulkes, 1982, p. 149). Moreover, the dream reports had the same cognitive complexity as the reports later provided by these same participants during their teenage years (Foulkes, 1982, p. 149). However, the rates of recall were lower than those in studies of young adults, and there were differences in content compared to later adolescence and compared to general findings with adults as well (Foulkes, 1982, pp. 74, 217). As with the younger cohort, visual-spatial skills remained "a strong predictor of both REM recall and new achievements in the constructional competence of dreams" (Foulkes, 1999, p. 114).

Although these are major changes in the general findings by ages 9–11, there are complexities within these overall findings. These complexities suggest considerable gender and individual variation. To begin with, the girls recalled a mean of 71.8% from REM awakenings and the boys only 60.0%, even though the median age of the girls was 9.8 years and the median age of the boys was 10.4 years. But there were also large individual differences. Two of the girls had mean recall rates of only 27% and 20% and two of the boys had rates of 27% and 25%, even though all four of these children were of average intelligence (Foulkes, 1982, pp. 150–151). On the other hand, the other 12 participants had an overall median report length of 86%, "a value in an expected range for adult research" (Foulkes, 1982, p. 151).

The surprising finding that dreaming does not begin to resemble adult dreaming until ages 9–11 was independently replicated in the first phase of the five-year longitudinal study of 12 girls and 12 boys in Zurich (Strauch, 2004, 2005). Based on a total of 171 REM reports collected for this age period, girls reported a higher mean percentage of dreams (77%) than the boys (58%). The girls remained at the same high levels in the last two sets of dream reports (74% and 82%) but below an adult control group of Swiss women (91%). The boys did catch up to some extent on recall in the second and third phases of the study, with mean recall percentages of 63% and 74%. However, they too remained below the figure of the adult control group (87%) (Strauch, 2005, pp. 157–158). In addition, the findings on wide individual differences in the Laramie study of children ages 9–15 were replicated. Three girls had 100% recall and another had 90.9% in the Zurich study, but two other girls were at 54.5% and 45.5%. The boys had an even wider range. Two participants recalled 100% and another recalled 83.4%, but four boys had very low recall rates of 36.4%, 33.3%, 25.0%, and 12.5% (Strauch, 2004, p. 223, Anhang 2). This longitudinal replication and extension study includes many new findings on dream content, using both new scales and HVdC categories. They are overviewed in chapter 7 as part of the discussion of dream content during preadolescence and adolescence.

For now, it is important to emphasize these three different large-scale studies took place in three very different cities (Laramie, Atlanta, and Zurich) in three different decades. This strengthens claims about the need for cognitive prerequisites for dreaming. These results show why it can be said with a considerable degree of confidence that dreaming is a gradual cognitive achievement. This conclusion is supported in the next subsection by findings on the gradual development of waking cognitive abilities.

Parallels between the Development of Waking Cognition and Dreaming

Drawing on findings generated by developmental cognitive psychologists for their own purposes, as well as on some of the results from cognitive testing carried out as part of the longitudinal and cross-sectional dream studies, it seems likely that there are five cognitive processes necessary for dreaming—the ability to form concepts, the capacity to generate mental imagery, the ability to make use of narrative skills, the capacity to simulate and imagine, and the acquisition of an autobiographical self.

The ability to form concepts is well established at the latest by age 2, and perhaps as early as late infancy (Mandler, 2004, 2012; Nelson, 2004, 2011). However, the other four cognitive capacities are underdeveloped or lacking altogether in preschool children. To start with, the ability to produce mental imagery seems to be lacking during waking in preschool children, a conclusion based on numerous different types of detailed studies of visual mental imagery (Frick, Hansen, & Newcombe, 2013, pp. 386–387, 395; Frick, Möhring, & Newcombe, 2014, pp. 536–538; Kosslyn, Margolis, Barrett, Goldknopfan, & Daly, 1990, p. 995). This conclusion is consistent with findings on visual-spatial skills as the best predictor of the frequency of dream reporting in children ages 5–8 in the longitudinal dream study (Foulkes, 1982, pp. 72–73). Similarly, the mental imagery tests used in conjunction with the cross-sectional dream study of children ages 5–8 did not detect sufficient capacity to create mental imagery at age 5. The investigators therefore concluded "the possibility of kinematic imaging emerges somewhere between 5 and 8 years of age, rather than being generally well-developed in 5-year-olds" (Foulkes, Sullivan, Hollifield, & Bradley, 1989, p. 450).

The findings on the relative absence of mental imagery in the children under age 7–8 in the cross-sectional dream study are also supported by the absence of visual imagery in people who are born blind or lose their sight before age 5. By way of dramatic contrast, people who become blind after age 7 continue to have visual imagery in both their waking lives and dreaming. They "continue to be able while awake to conjure up [visual] mental images of persons, objects, and events, and they continue to dream in [visual] imagery" (Foulkes, 1999, p. 15). This analysis is supported by a study of dream reports from 15 congenitally and adventitiously blind participants that used the word strings for sensory mental imagery for searches of the dream reports on DreamBank.net. The 10 women and men, who ranged in age from 24 to 73, each contributed at least six dream reports that were voice-recorded over a two-month period. Overall they provided 372 dream reports, 236 from women and 136 from men (see Hurovitz et al., 1999, p. 185, table 1, for the nature and degree of the participants' blindness, and the number of dream reports each person contributed).

All forms of the words "see," "saw," "watch," "look," and "notice" were used to provide a starting point for detecting visual imagery. All forms of "hear" and "listen" were used to locate possible auditory imagery. All forms of "taste," "smell," "aroma," "scent," "feel," "felt," and "touch" were used as a starting point for the other three senses. The sensory references were divided into three categories: visual, auditory, and taste/smell/touch, and the percentage of each type was determined. Once the terms were located through the word string searches, the investigators studied the context in which these terms appeared in order to eliminate possible instances of the figurative use of sensory references. The search for figurative usages was especially important with visual terms because of the frequent use of the conceptual metaphor, "knowing is seeing," as discussed in chapter 5, in which understanding something is expressed as "seeing" (Matlock, 1988). In addition, there are many other metaphoric expressions that build on other forms of sensory imagery (e.g., "he's deaf to new ideas," that idea "smells fishy to me," and "that comment left a bitter taste in my mouth").

The high frequencies of visual imagery in the dream reports of those who became blind after age 7 are similar to those for sighted adults in large-scale studies (Snyder, 1970; Zadra, Nielsen, & Donderi, 1998). For two women who lost their sight after age 8, only one of 36 sensory references could be classified in the taste/smell/touch category; five of their sensory references were auditory and the rest (86%) were visual. On the other hand, there were few or no visual references in the dream reports of those who were blind before age 4. The seeming visual exceptions in the dream reports of two such blind participants involved the use of the "knowing is seeing" conceptual metaphor in speaking. As expected, the taste/smell/touch percent was very high in the dream reports of those who were born completely blind, ranging from 48% to 67% in four participants (Hurovitz et al., 1999, p. 188, table 2). This figure is far higher than the 2–3% of gustatory, olfactory, or tactual imagery that is found in the dream reports of sighted people (Snyder, 1970; Zadra et al., 1998).

Taken together, the developmental findings and the results of word string searches in the dream reports of adult blind participants suggest visual mental imagery develops somewhere between ages 5 and 7 (Foulkes, 1999; Frick et al., 2014; Hurovitz et al., 1999; Kerr, 1993; Kosslyn et al., 1990). It therefore seems likely that individuals who become blind after that age period have retained a developmentally acquired system of visual imagery that is independent of their visual-perceptual capabilities, which makes it possible for them to create visual dream images of people they met after they became blind (Kerr, 1993, pp. 30–35). Although there are no developmental studies of the ability to generate auditory, tactual, and olfactory mental imagery, the best conjecture, based on adult studies of these other types of sensory mental imagery, is that these imagery abilities develop in roughly the same time period (Lacey & Lawson, 2013).

As for narrative skills, developmental studies of waking children find only half of young children's statements about an event are narratives by age 3, and they are limited in length. By age 5 or 6, many children can tell a story containing a beginning, middle, and end (Reese, 2013, pp. 197–198; Taylor, 2013, p. 803). In a waking cross-sectional study of children ages 7 and 11, the 7-year-olds included only three of the eight basic elements considered to be part of a well-developed narrative, whereas children's stories included six of the eight by age 11 (Bauer, Burch, Scholin, & Güler, 2007). Paralleling these results, the children in the cross-sectional laboratory dream study were only able to produce simple narrative scenes without chronology or sequence at ages 5–7. By age 8, they were able to generate a narrative with continuity in two temporal units, along with evidence of causality (Foulkes et al., 1990, pp. 456, 461).

In terms of simulation and imaginative abilities, children's ability to engage in "pretend dramatic play is typically delayed until age 4 or 5 in preschool environments" (Nelson, 2007, p. 170). They only become proficient at imagining and in acting out complex play scenarios after age 5 (Nelson, 2007, pp. 170–171). Consistent with these findings, the children ages 5–7 in the cross-sectional study did not recall dreams very frequently and showed only limited ability to produce complex imaginative narratives in response to story prompts. At age 8, however, the correlation between recall frequency and the total score on the imaginative narrative test was .45 (Foulkes et al., 1990, p. 458).

Preschool children seem to lack the ability to simulate versions of past and future events (Atance & Metcalf, 2013; Taylor, 2013, p. 791). Consistent with these earlier conclusions, a cross-sectional study of 93 Turkish preschool children showed statistically significant improvements at ages 4 and 5 on a series of cognitive tests that assessed their ability to think about the past and the future (Ünal & Hohenberger, 2017, pp. 248–250). In a larger study of American children at ages 5–7, 7–9, and 9–11, which included an adult control group ($n = 157$ overall), the participants provided narratives of past and future events. The children had difficulties in simulating past episodic memories through age 7 and continuing difficulties in simulating future episodic memories through age 11, as compared to the adult control group (Coughlin, 2016, p. 22; Coughlin, Robins, & Ghetti, 2019).

The American children and adults who simulated episodic memories also were asked to provide narratives of a make-believe event. The results for this task are of even greater interest for the construction of a neurocognitive theory of dreaming. Children at age 5 had difficulties in creating make-believe narratives but not thereafter, and improved considerably between ages 7–9 (Coughlin, 2016, pp. 22, 45, fig. 3). These results align very well with the findings on how rarely preschool children dream. On the other hand, children do almost as well by age 9 as older children and adults in developing imaginative narratives, which is also consistent with the conclusions showing the default network can support complex dreaming at ages 9–11.

Turning to the final necessary cognitive capacity, an autobiographical self, personal (autobiographical) memories develop only gradually during the latter half of the preschool years (Eisbach, 2013b; Gopnik, 2009, chap. 5; Nelson, 2005; Tulving, 2005). More specifically, episodic memories for specific events, which develop a year or two earlier, only slowly evolve into more personal types of memories. These personal memories have a sense of subjectively experiencing and reexperiencing. At this point, the memories are "infused with a sense of personal involvement" (Bauer, 2013, p. 521). One developmental psychologist concluded that children do not have "the basics of autobiographical memory," along with an inner mental life and an "autobiographical self," until they are around the age of 6 (Gopnik, 2009, p. 156).

Once again paralleling the waking findings, the longitudinal laboratory dream study discovered there were no signs of "self-involvement" until ages 7 and older, and the regular appearance of self-involvement was not yet fully developed even at ages 7–8. Instead, the "simple dream actions" were usually carried out by other characters (see Foulkes, 2017, p. 4, for a further analysis of his original findings).

Further Evidence: Preschool Children Do Not Understand the Origins of Dreams

There is one final piece of evidence concerning the rarity of dreaming in preschool children. This evidence unexpectedly arises in studies concerning children's understanding of the origins of dreaming. Children understand mental states and pretense by age 3, and they can distinguish between the real and the imaginary in stories (Woolley & Nissel, 2020). However, they have difficulties in understanding and answering questions about the origin of dreams. Based on the waking development findings presented in the previous subsection, their inability to figure out the origins of dreams may be due to their lack of personal experience with dreams.

Work on children's understanding of dreaming suggests they are "quite confused about their origins" at ages 3–4. Some of the 3-year-olds "appeared to conceive of dreams as shared fantasies, claiming that dream content is shared between sleeping individuals" (Woolley, 1995, pp. 189, 195). In a further study designed to determine at what age preschool children understand dreaming as a mental event internal to the person (e.g., "in the mind," "from inside you," "in your head"), as opposed to a nonmental event external to the person (e.g., "from your pillow," "TV," "the sky," "God"), the "majority of 3-year-olds cited non-mental explanations (76%)." The 4-year-olds "gave mental and non-mental explanations equally often, and 5-year-olds offered significantly more mental than non-mental explanations (83%)" (Woolley & Boerger, 2002, p. 26).

Additional results on preschool children's limited and gradual understanding of dreaming come from an interview study of 32 children, ages 3–7, at a school for faculty and staff children at a midwestern university. The study adds further cognitive specificity in terms of what children do and do not understand. Rather strikingly, only 1 of 11 younger children (9.1%), but 7 of 21 in the older group (33.3%), understood dreams take place in their heads (S. Meyer & Shore, 2001, pp. 186–187). Moreover, less than half of 11 children ages 3.5 to 5 (45.4%) understood that dreams are imaginary events, as compared to 17 of 21 at ages 5–7(80.9%). In addition, the order in which 25 of the 32 children understood specific dimensions of dreaming—their imaginary nature, the fact they are private to the dreamer, and their internal origins—is consistent with the age sequence in understanding similar issues in waking thought. Children who grasp the distinction between appearance and reality, and who know that someone can have a different perspective on an object than they do, did not always understand that dreams have no material reality and are private to the dreamer (S. Meyer & Shore, 2001, pp. 186, 190–191). These findings are consistent with two further observations by Gopnik (2009, p. 152): preschool children "don't understand that your thoughts can be internally generated" and they "don't understand that thoughts can simply follow the logic of your internal experience instead of being triggered from the outside."

Preschool children not only lack an understanding of thoughts as internally generated; they do not "experience their lives as a single timeline stretching back into the past and forward into the future" or "feel immersed in a constant stream of changing thoughts and feelings" (Gopnik, 2009, p. 153). In another study, only 44% of children understood what mind-wandering is at ages 6–7, compared to 86% by ages 10–11 (Eisbach, 2013a, table 2). Based on these findings, children very likely need to have an autobiographical self in order to experience dreaming.

If the gradual independent development of visual mental imagery, narrative skills, imagination, and an autobiographical self in young children are considered in combination, they explain why preschool children seldom dream, if ever. This necessary combination of cognitive abilities also may explain why the dream reports of children ages 5–8 often lack a sense of sequence, complexity, visual mental imagery, and a central role for the dreamer. More generally, both dreaming and a waking autobiographical self are very likely the gradually emerging results of both neural and cognitive development between ages 5 and 8. Children do not understand the origins of dreams because they rarely dream and almost never include themselves as a character in the few dreams they do report.

Conclusions and Implications

The results presented in this chapter strongly suggest there is a developmental dimension to the capacity to dream. The findings on the frequency, length, and complexity of dream reports in the cross-sectional and longitudinal laboratory studies of children between the ages of 3 and 15 are consistent with, and fully supported by, the findings on the gradual maturation of the default network. They are also consistent with the gradual development of waking cognitive capacities. The results seem all the stronger because the two longitudinal dream studies involving older children and adolescents were carried out by two different researchers in two different countries in two different decades (Foulkes, 1982; Strauch, 2004, 2005). Then, too, the findings on the Male/Female Percent and the Animal Percent in the dream reports of preschool children, and in the stories they tell, add new evidence on dreaming-waking continuity at a young age (Domhoff, 1996, pp, 89–94; Hall, 1984, p. 115).

Similarly, the studies of the maturation of the default network, and of the development of cognitive capacities, were carried out by still other independent investigators in different countries. Every empirical generalization for the years between ages 3 and 15 has been replicated at least once, with the exception of the work on dreaming with preschool children. Even in the case of the one exception, the fact that the same 14 children were studied at ages 3–5, 5–7, and 7–9, and showed an overall progression in the frequency, length, and complexity of their dream reports, provides good evidence for the robustness of the results with preschoolers.

The findings in this chapter concerning the maturation of the default network, along with the gradual development of cognitive capacities, make it possible to add the two remaining conditions to the list of six conditions necessary for dreaming to occur. First, the relevant neural substrates of the brain not only have to be intact, they also have to be mature. Second, the waking cognitive capacities for generating mental imagery, narration, imagination, and an autobiographical self have to be fully developed for adult-like dreaming to occur. In that regard, the differences in the frequency and complexity of dreaming between children and young adolescents therefore can be understood as another type of cognitive insufficiency.

Finally, to the degree that readers agree there is a need for a developmental dimension in order to understand dreaming and dream content, they may be more inclined to consider the claims made for the neurocognitive theory of dreaming in chapters 8–11. Before turning to those chapters, however, it is necessary to add a content dimension to the findings on the development of dreaming.

7 The Dream Reports of Preadolescents and Adolescents

The results of both the laboratory dream studies and the studies of waking cognitive development explain why there are few studies of dream content in early childhood, whether inside or outside a laboratory setting. Young children do not dream frequently or with any complexity, and do not seem to understand the origins of their dreams. The first systematic attempt to collect dream reports from children ages 4–7, carried out 100 years ago, concluded that very young children have "great difficulty" in "separating the dreaming from the waking element" (Kimmins, 1920/1937, p. 53). More generally, these findings may explain why the results of the few non-lab interview studies of young children's dreams often differ in their results and have been of limited value.

Studies of dream content based upon reports from late childhood and adolescence, which is also the time period in which the default network begins to become more adultlike, have shown more promise. As demonstrated in both of the longitudinal studies of older children and adolescents, participants between the ages of 9 and 15, and especially girls, are almost adultlike dreamers and are able to provide lengthy accounts of their dreams (Foulkes, 1982; Strauch, 2005; Strauch & Lederbogen, 1999). However, studies necessitating several visits to a sleep-dream laboratory are expensive and difficult to carry out. There are also problems in collecting dream reports from older children and adolescents in nonlaboratory studies, such as obtaining parental and school consent, gaining the cooperation of participants, maintaining high levels of participant motivation, and finding ways to encourage participants to voice record or write complete reports. This chapter therefore makes greater use of two underutilized non-lab methods of obtaining large and thorough samples of dream reports.

These studies provide several substantive findings consistent with the neurocognitive theory of dreaming.

Although the findings on dream content from the cohort ages 9–15 in the Laramie longitudinal study are noted at one or two points in this chapter, it begins with a primary focus on the lab and nonlab reports collected in the Zurich study for several reasons (Strauch, 2005; Strauch & Lederbogen, 1999). It then relies on Hall/Van de Castle (HVdC) findings from dream reports collected by means of the Most Recent Dream (MRD) method. This method, which has not been mentioned up to this point in the book, involves asking willing anonymous participants in a group setting to write down and date the most recent dream they can remember. It was originally created for use with adults (Domhoff, 1996, p. 67; Hartmann, Elkin, & Garg, 1991). It has been employed to collect samples of dream reports from college students in Hong Kong, Iran, South Africa, and Wales, as well as from animal rights activists in the United States and the elderly in Switzerland (Lewis, 2008; Malcolm-Smith & Solms, 2004; Malcolm-Smith, Solms, Turnbull, & Tredoux, 2008b; Mazandarani, Aguilar-Vafaie, & Domhoff, 2013; Strauch, 2003; Yu, 2008).

More importantly in terms of this chapter, the MRD method has proven to have many advantages in studying preadolescent and adolescent students. It overcomes the difficulties of collecting one-week or two-week diaries. Even adult participants, who initially agree to write down dream reports for a one-week or two-week period, very frequently turn in far fewer, which are often hasty and incomplete as well. In the case of teenagers, this point also has been demonstrated in two dissertations that focused on the dream reports of high-school students (Buckley, 1970; Howard, 1978).

As shown in studies in Greece, Italy, Spain, and the United States, it is far better to ask large groups of students in late childhood and adolescence to write down the most recent dream they can remember (Avila-White, Schneider, & Domhoff, 1999; Crugnola, Maggiolini, Caprin, Martini, & Giudici, 2008; Karagianni, Papadopoulou, Kallini, Dadatsi, & Abatzoglou, 2013; Oberst, Charles, & Chamarro, 2005; Saline, 1999). The collection of MRDs often does not require permissions beyond a classroom teacher because the dream reports are voluntary and anonymous, and the students need only 20 to 30 minutes to write their reports. Moreover, the MRD method gives the young participants greater autonomy and privacy than may be possible when keeping a dream diary at home, where parents may influence the participants. There also may be a lack of privacy for writing down a dream report during free time and lunch breaks at school (Saline, 1999, p. 173). As in the case of the laboratory studies, the results with this method lead to the conclusion that girls provide more useful samples than do boys in terms of the frequency of their participation and the length of their dream reports.

In addition, the usefulness of dream series in studying this age group also is demonstrated in this chapter. It does so through analyses of four dream series, which were kept for varying amounts of time for varying reasons, beginning in adolescence. Two of them are very extensive because they began during middle school and extend into young adulthood. The findings with these four dream series are based on HVdC codings, generic word strings, and individually tailored word strings. They establish that both consistency and continuity are present in individual dream series by age 15 at the latest, shortly after the default network and most cognitive abilities are almost fully adultlike.

Finally, and in spite of the general criticisms of two-week dream diaries, this chapter nonetheless makes use of an excellent set of two-week dream diaries that were collected from highly motivated girls in the fifth through eighth grades at a private American school in the 1990s.

Lab and Nonlab Findings in the Zurich Longitudinal Study

Developmental psychologist and dream researcher Inge Strauch's (2004, 2005; Strauch & Lederbogen, 1999) longitudinal study of 24 Swiss children at the University of Zurich, based on 12 girls and 12 boys at ages 9–11, 11–13, and 13–15, included both lab and nonlab conditions. It provides solid substantive results of theoretical importance, as well as a good baseline for discussing findings with nonlab dream reports. As also mentioned in chapter 6, girls in this study recalled dreams from 77% of the REM awakenings at ages 9–11, whereas boys reported dreams from only 58% of the awakenings. The girls remained at this high level of recall in the second and third stages of the study but did not reach the 90% level of the young Swiss adults serving as a comparison group. The boys reached the second-year level of the girls' recall in the final year of the study (Strauch, 2005, p. 157, fig. 1). Due to their higher recall frequency, the girls contributed 303 of the 551 REM reports and the boys contributed 248. These findings on gender differences in the frequency of reporting replicate similar findings in both cohorts in the Laramie longitudinal study (Foulkes, 1982). The dream reports on which this study is based can be found on DreamBank .net, separated into lab and nonlab dreams, under the name "Swiss children." They are in the original German that the children spoke and wrote.

The content findings from the Zurich and Laramie longitudinal studies of participants ages 9–15 are generally similar in terms of increasing numbers of social interactions and in increasingly sophisticated dream scenarios, and both studies show the girls were generally ahead of the boys in the complexity of their dream reports (Strauch, 2005, pp. 163, 166). However, detailed comparisons cannot be made in several instances because the Laramie study used rating scales created especially for that study. The Zurich study therefore has greater usefulness on dream-content issues for the purposes of this chapter. In addition to larger sample sizes, its basic content analyses utilized HVdC coding categories, which makes it possible to draw comparisons with the several studies in this chapter that also deploy HVdC categories. In addition, as a replication and extension study, the Zurich study makes use of several additional scales, which add new understandings on a variety of topics. These scales provide strong indications that the dream contents of children by ages 11–13 are becoming similar to those of young adults.

Using a scale for determining the primary activity and setting in each dream report, the study found that being around the house, going from place to place, and playing sports (albeit different sports for the girls and boys) contributed equally in accounting for 60% of the dream reports. Another 10% occurred in a school setting, usually involving a recreational or other nonclassroom setting. This finding is in line with earlier findings with Swiss adults: dream reports more frequently involve social and recreational settings and activities than they do focused work settings (Strauch & Meier, 1996). Still another 10% of the reports in the longitudinal study of preadolescents and adolescents involved highly unusual settings that were part of what the researcher described as "adventurous" dreams (Strauch, 2005, p. 161). The final 20% of the dream reports could not be classified in terms of a major setting and activity.

Based on a scale for categorizing dream reports as "realistic" if they "approximated experiences that happen in waking life," "inventive" if they "combined familiar waking experience" in a creative manner, and "unrealistic" if they "lacked any relation to the waking world," the researcher found that about 40% were realistic at each age period. However, inventive dream reports increased from 29% at ages 9–11 to 44% at ages 13–15. Conversely, unusual reports declined from 31% to 15% from the first to the third stages of the study (Strauch, 2005, pp. 160–162). The researcher also examined self-involvement in the dream reports at each of the age periods by determining the percentage of speech acts in which the dreamer was a speaker. The figure rose from 17% at ages 9–11 to 39% at ages 13–15, which was still below the adult figure of 47% (Strauch, 2005, p. 164, fig. 4). Both sets of comparisons demonstrate developmental trends concerning the nature of dream content.

In terms of character categories, both girls and boys showed consistency at all three age periods for most character categories, including the usual gender differences on the Male/Female Percent. In addition, the Animal Percent for boys declined to the low levels found in girls and in adults. The percentage of dream reports with at least one aggression increased from 18% to 31% for the girls and fluctuated between 24% and 28% for the boys. However, the percentage of dream reports with at least one friendliness fluctuated slightly, from 28% to 20% to 23% for the girls and increased gradually for the boys from 16% to 18% to 24% (Strauch, 2005, p. 159). Both girls and boys were more likely to be victims than aggressors in aggressive interactions (Strauch, 2005, p. 160). Some of these findings are similar to those with adults, but the findings on aggressions and friendliness are not yet entirely adultlike.

The Zurich study added to past findings on young dreamers by using HVdC categories to compare social interactions between the dreamer and adults with those involving peers. As a result, it discovered the difference between girls and boys on the Male Female Percent was even more pronounced when only individual peers and groups of peers are considered. The researchers concluded the girls clearly preferred same-sex peers at ages 9–13, although they became a little more inclined to include boys in their dreams at ages 13–15. In a similar fashion, the boys interacted almost exclusively with their male peers. Both girls and boys were more likely to be criticized or punished by adults, especially men, but they were just as likely to be aggressors as victims in interacting with their peers. They were also more likely to be befriended by adult men and women than to initiate a friendly interaction, but they were as likely to initiate or receive friendship with peers (Strauch, 2005, p. 160). These findings parallel those in the Wyoming longitudinal study for this age group using a different coding system (Foulkes, 1982, chaps. 6–8). As shown in both studies, these differences very likely portray the way in which older children and adolescents conceive of and interact with the people in their waking lives and thereby demonstrate continuity between dreaming and waking thought for this age group.

The Zurich study also included a very large collection of nonlab dream reports from the same participants. They were given five tape cassettes at the end of each of the three years of laboratory participation and then instructed to mail the dream reports they voice-recorded to the lab. The 299 nonlab dream reports, 164 from girls and 135 from boys, were 91% of what the researchers expected from the girls but only 57% of what they expected from the boys. At ages 13–15, two girls reported only two dreams apiece and two boys did not send in any dream reports (Strauch & Lederbogen, 1999, p. 155). The researcher concluded that lack of motivation and a reported increase in feelings of fatigue were the main reasons for the especially low rates of reporting in the fifth and final year. She further concluded that nonlab studies of teenagers "should be restricted to highly motivated and cooperative participants with high rates of dream recall" (Strauch, 2005, p. 168). The suboptimal response to the request for brief dream diaries from outside the lab setting, even though the participants were being paid, is the best-controlled evidence in relation to the problems that usually arise in asking either adolescents or adults to keep a one-week or two-week dream diary.

The lab and nonlab dream reports in the Zurich study were very similar in content for most HVdC categories. Both samples of dream reports showed the same age changes and gender differences. However, the nonlab dream reports "showed higher frequencies for some social interaction categories"; this generalization includes higher percentages and rates for aggression indicators (Strauch, 2005, pp. 167–168). These results replicate the findings on lab and nonlab dream reports in adult studies that use the same participants in both conditions (Domhoff & Schneider, 1999; Weisz & Foulkes, 1970) and as discussed at the outset of chapter 3.

The content findings in the Laramie and Zurich longitudinal studies serve three important purposes in terms of understanding the development of dreaming and in building a neurocognitive theory of dreaming. Theoretically, they support the idea that the maturation of the default network between ages 9 and 11 is very likely a key turning point in the development of dreaming, but also that further maturation of the default network and further cognitive development is occurring in adolescence. Methodologically, they provide an anchor point or baseline for comparisons with results from exclusively nonlab studies of dream reports. However, the differences between lab and nonlab dream reports in the Zurich study during this age period also serve as a reminder on the likely bias in nonlab studies toward dream reports with greater salience, in this case related to the frequency of aggression in dream reports. Finally, the Zurich study provides new evidence for continuity in what people dream about and what they think about in waking life by the middle teenage years.

A Dream-Diary Study of Preadolescents and Adolescents

As the lab and nonlab comparison in the Zurich longitudinal study makes clear once again, nonlab dream reports are difficult to collect from preadolescents and adolescents, so only highly motivated and cooperative participants should be considered for dream-diary studies. In terms of those criteria, 45 volunteer participants who attended a private school for high-achieving girls in the San Francisco Bay Area provided a good sample of two-week dream diaries (Latta, 1998). They were not a representative sample of girls their age in terms of demographic and socioeconomic variables, but they demonstrate what is possible when children have the environmental support to be successful in their schoolwork.

The participants used in this analysis, who were in the fifth through the eighth grades, first wrote down as much of the recalled dream as they could remember and then explained who the characters were and any thoughts or feelings they had in relation to the dream report. They were asked to record as many dreams as they could remember, but at the same time they were reassured that no set number of dream reports was necessary. These instructions and reassurances led to varying numbers of dream reports from each participant, which suggests there were once again individual differences in dream recall (Latta, 1998, p. 91). The original study also included six girls in the fourth grade who provided only 36 brief reports as well as 10 girls in the ninth grade who were asked to contribute only four dream reports each. Neither the fourth graders nor the ninth graders are included in this analysis because they provided too few dream reports. All of these dream reports, including those of the fourth and ninth graders, can be found on Dream-Bank.net under the pseudonym Bay Area Girls.

In all, the sample consists of 312 dream reports from the 45 participants. The year-by-year sample sizes varied from 120 reports contributed by 14 of the fifth graders, to 55 reports from 11 of the eighth graders. However, they are sufficient at each grade level to detect any general age trends, as displayed in table 7.1. As the table also conveys, the findings for the fifth graders were significantly higher than the HVdC norms for women on the at least one aggression percent, the at least one friendliness percent, and the Physical Aggression Percent. The at least one aggression percent remains higher than the norms for the sixth graders, and the Physical Aggression Percent remained higher for the sixth and seventh graders. The Male/Female Percent is lower for the fifth graders and the eighth graders, but in general the eighth graders are similar to the women's norms. Methodologically, these results demonstrate that representative samples of dream content can be collected under the right conditions from girls in middle schools. Empirically, the results add to the evidence that dream content is close to adultlike at ages 13–14.

Table 7.1
Content of the Bay Area Girls' dream reports at grades 5, 6, 7, and 8, compared to the Hall/Van de Castle female norms

	Grade				
	5	6	7	8	Female norms
Characters					
Male/Female %	38*	48	46	37*	48
Friends %	34	38	35	44	37
Social interactions					
Aggression/Friendliness %	49	47	52	58	51
Befriender %	39	48	50	40	47
Aggressor %	38	43	25	27	33
Physical Aggression %	57**	67**	56**	47	34
Percentage of dreams with at least one:					
Aggression	33*	31*	37	36	44
Friendliness	31*	37	39	35	42

* $p<0.05$; ** $p<0.01$.

MRD Findings with Children and Adolescents

The possible usefulness of MRDs with children in late childhood was examined in a study of third (ages 8–9), fourth (ages 9–10), and fifth (ages 10–11) graders in two private elementary schools on the Berkeley–Oakland side of the Bay Area in California (Saline, 1999, p. 175). The study is based on 30 girls and 32 boys. Their parents were highly educated (e.g., 16% of their mothers had doctoral degrees, another 32% had an M.A., and 40% had a B.A. as their highest degree), so the sample is not representative of American children. The study revealed the reports from the third and fourth graders were brief, vaguely dated, or were said to have occurred "long ago," all of which suggested it is highly unlikely they were recent dreams (Saline, 1999, p. 175). By the fifth grade, however, the MRD sample showed more promise, especially in the case of girls. This finding led the researcher to the conclusion that studies of fifth graders are feasible, "especially for girls" (Saline, 1999, p. 178). This conclusion is supported by the findings with fifth graders in the dream-diary study of the Bay Area girls.

In terms of the recency of the reports, all but one of 19 reports from girls in the fifth grade were from within two or three weeks of the date of collection and many were from the past few days. Similarly, their reports were twice as long as the reports from girls in the fourth grade. They had a median word length of 106, which compares very favorably with the findings in the Laramie longitudinal study for girls in that age group (Foulkes, 1982, p. 328). Both of these results are also consistent with the results in the two-week diary study of the Bay Area Girls. More generally, the fact that dream reports can be collected at this age by means of either dream diaries or the MRD method fits with Foulkes's (1982, pp. 74, 217) conclusion that children ages 9–11 have the same dream recall frequency and median report lengths as they do at ages 11–13 and 13–15.

The feasibility of collecting MRDs with fifth graders was demonstrated for a second time in schools in Milan a few years later with 182 children, 102 girls and 80 boys. They provided MRDs twice toward the end of the school year, April and June, when they ranged in age from 10 to 11.5. The results were "quite stable" from April to June, which suggests the reliability of the method (Crugnola et al., 2008, pp. 212–213). Dream reports from the two samples with 50 words or more were used for the content analysis, leading to a combined sample of 187 dream reports from girls, which is 82% of the reports provided by the girls, and 128 from boys, which is only

62% of the reports provided by the boys (Crugnola et al., 2008, p. 205). The authors conclude that their work "confirms the possibility of using the Most Recent Dream Method" with children ages 10–11, "particularly for girls" (Crugnola et al., 2008, p. 214).

As might be expected from the promising results with fifth graders, excellent MRD reports can be obtained from children in the sixth grade and higher. For example, in a large-scale study of seventh graders in 16 classrooms in a suburban middle school in Santa Cruz County in California, it was possible to collect MRDs from 83% of the girls and 60% of the boys. This level of return led to a sample of 162 reports from girls and 110 from boys. Based on careful observation of when the participants finished their reports, most of them needed about 20–25 minutes to complete their dream reports. Several children turned in a blank sheet or wrote that they could not recall a dream, which is a good indication that the anonymity and instructions reduced the demand characteristics of the study (Avila-White et al., 1999).

Once again, there were gender differences on report lengths: 92.5% of the girls' reports included 50 or more words, with a median word length of 125. Only 74.3% of boys' reports were this length or greater, and the median word length of 89 for boys was 36 words less than the median for those provided by girls. These figures are roughly comparable to the mean of 100 for girls and 94.9 for boys at ages 11–13 in the dream diaries in the Zurich longitudinal study at the same age (Strauch & Lederbogen, 1999, p. 155). In addition, the results were similar to those in an unpublished pilot study in different schools in the same suburban area in Santa Cruz County. This was especially the case for the girls, except for somewhat less friendliness and slightly more aggression in the final study (see Avila-White et al., 1999, p. 166, table 1, for a comparison of the two studies, and Domhoff, 1996, pp. 95–96, for a one-paragraph summary of the pilot study with 80 girls and 64 boys).

Then, too, confidence in the published results is bolstered by their similarity to the nonlab findings in the Zurich study at ages 12–13 in terms of the Animal Percent, Male/Female Percent, and Aggressor Percent for both girls and boys (Strauch & Lederbogen, 1999). In a comparison with the HVdC norms, the Santa Cruz girls and boys differed in the same way women and men differ on several indicators, but both girls and boys had a higher Aggression/Character Index (A/C Index) and Physical Aggression

Percent. The boys differed even more from the men than the girls did from the women (Avila-White et al., 1999, p. 167, table 2).

A study employing the MRD method with 45 girls and 45 boys in several schools in Barcelona in late childhood and adolescence replicated the finding that the method is feasible with preadolescents ages 11–12. It also documented the method's usefulness at ages 14–15 and 17–18 for the first time (Oberst et al., 2005). Although the sample sizes at each age are small, the girls and boys at ages 11–12 were more often victims in aggressive interactions than the two older groups and the boys had a higher Male/Female Percent and more aggressive content. Both of these results are similar to those in the Santa Cruz County study. The results at ages 17–18 were close to those in the HVdC norms, although there were fairly large differences on Aggressor Percent and Physical Aggression Percent (Oberst et al., 2005, p. 175).

The results from the largest and most far-reaching MRD study of children, carried out in Thessaloniki with 756 children and adolescents ages 8–18, provide the most substantial findings to date. This study demonstrates the full potential of the method with adolescents when there are large sample sizes, thanks to the cooperation of the school system in the city and surrounding areas (Karagianni et al., 2013). The sample consisted of 273 children ages 8–12 (150 girls and 123 boys) and 483 adolescents ages 13–18 (278 girls and 205 boys). The sample is very likely representative of the children in that city, based on the diversity of the schools (center city, suburbs, and rural areas), along with the size of the sample and the breadth of the age range (Karagianni et al., 2013, p. 92). The girls were higher than the boys on familiar characters, family members, familiar settings, and indoor settings, which parallels the HVdC normative findings. The boys were higher on the Male/Female Percent, Aggression/Friendliness Percent, Physical Aggression Percent, and A/C Index, which also parallels the HVdC normative findings. In terms of age, the girls started out higher on familiar characters, family members, animals, and Physical Aggression Percent, and lower on dreams with at least one friendliness, and then moved toward the norms. The boys started out higher on aggression indicators and lower on friendliness, and then moved toward the normative figures for men. The full results for the girls in both age groups, along with HVdC norms for women, are presented in table 7.2.

As the Thessaloniki research team states on the basis of their detailed literature review, their results confirm similar findings in earlier MRD studies

Table 7.2
The most recent dreams of Greek girls ages 8–12 and 13–18, compared with the HVdC normative findings for women

	Girls 8–12	Girls 13–18	HVdC norms
Characters			
Male/Female %	45	52	48
Friends %	22	38	37
Social interactions			
Aggression/friendliness %	59	56	51
Befriender %	40	36	47
Aggressor %	16	13	33
Physical Aggression %	75	65	34
A/C Index	.25	.21	.24
F/C Index	.13	.13	.22
Percentage of dreams with at least one:			
Aggression	42	34	44
Friendliness	28	26	42

with nonadult participants. They do so on gender and age differences relating to the indicators for Animal Percent, Male/Female Percent, Familiarity Percent, Family Percent, and Friends Percent, as well as for the A/C Index, the Aggression/Friendliness Percent, the Physical Aggression Percent, and the Aggressor Percent (Karagianni et al., 2013, pp. 94–95). The researchers conclude that their findings also are similar to many of those in the Swiss longitudinal study (Karagianni et al., 2013, pp. 94–95). They further suggest that dream content "expresses the same interests, thoughts, and concerns children have in waking life." In addition, they add that "children in Europe and the United States seem to have similar concerns as well as some differences that may relate to cultural divergences" (Karagianni et al., 2013, p. 96). If they are right, then they have contributed impressive new findings and conclusions for children and adolescents from ages 8 to 18, which may be all the more valuable because their results are similar to the findings and conclusions concerning adult dream reports, as discussed in chapter 3.

The Thessaloniki researchers further conclude that dream contents and waking thoughts express the same personal concerns and interests. Their conclusion gains credibility on the basis of waking studies of gender similarities and differences. Although there are in fact relatively few waking gender differences, meta-analyses of children find boys are higher in activity level, self-assertion (including a striving for dominance), and direct aggression (physical and verbal). However, the effect sizes are often small or modest, except in the case of physical aggression (Hyde, 2014; Leaper, 2013; Leaper & Farkas, 2015). The developmental changes are also similar to those in cross-national studies of children and adolescents (Achenbach & Rescorla, 2007; Tremblay, Hartup, & Archer, 2007).

Based on the generally similar findings with children from four different countries, and the similarities with the results of studies using dream reports collected in laboratory settings and by means of a dream diary, the feasibility and usefulness of the MRD methodology receives further support. The results of these MRD studies also provide new evidence that dreaming has a developmental trajectory. In conjunction with the results from laboratory and dream diary studies, the MRD results suggest that studies that focus on girls may lead to larger and more complete samples, which may be more useful for extending and refining the scientific understanding of the development of dreaming from the preschool years to late adolescence.

Consistency and Continuity in Four Dream Series

Building on the evidence derived from the representative samples of dream content from the fifth grade through high school in both lab and nonlab settings, it is also possible to make use of the very few adolescent dream series available from what is almost certainly a miniscule pool of teenage dream journals kept for more than a month or two. The similarity of the results with just four teen dream series (three from girls and one from a boy) suggests that consistency and continuity are present by age 15 and may begin as young as age 13–14. Moreover, there is very good evidence that this consistency and continuity is stable into young adulthood.

Continuity in Bea's Dream Reports at Age 15

Bea wrote down every dream she recalled between ages 14 years, 9 months and 15 years, 11 months, for a total of 189 dream reports. The 139 continuous dream reports containing 50 or more words were coded with HVdC categories, which provided information for 19 content elements. The Wald-Wolfowitz runs test for autocorrelation indicated no statistically significant results at the .05 level for any of those indicators (Domhoff & Schneider, 2015a, pp. 73–74). One of her three family members (father, mother, and a younger brother) appeared in 39.6% of her dream reports and her teenage friends and acquaintances accounted for half of all the human characters in her dream reports. The few known adults, aside from her parents, were primarily her teachers (Domhoff, 2018a, pp. 146).

Bea was high on a character category that is usually minor in adult dreams, the "Dead and Imaginary Percent." She sometimes dreamed that she was interacting with fictional characters in her favorite television shows or that a prominent person, such as a movie star, had died. However, the main reason for this high figure was her frequent mention of—and sometimes direct involvement with—key characters in the Harry Potter books, which she and many other teenagers were reading passionately at the time she was in high school. These dream reports can be called up on DreamBank .net by marking the Bea series and searching for this personalized string: Potter | Weasley | Hermione | Voldemort | Quidditch | Gryffindor | Ravenclaw | Hufflepuff | Slytherin | Hogwarts | Dumbledore (Domhoff, 2018a, pp. 146–148).

Bea also was very high on at least one friendliness, and also high on the Friendliness/Character Index (F/C Index). These results are consistent with what she later reported about her interests and interactions in waking life during high school. She was totally preoccupied with both her female and male friends in high school, and especially her teammates on sports teams, which is consistent with the findings in the Zurich longitudinal study (Strauch, 2005). Bea's dream reports had more friendliness than aggression, which is atypical, but her percentage of dream reports with at least one misfortune was also very high at 59 ($h = .52$, $p = .01$), so her dream reports were not universally positive in their content (Domhoff, 2018a, pp. 145, 147, fig. 44). She said that she often felt upset by the dreams she remembered upon awakening, even though they contained many friends and a great many friendly interactions. She also said she worried about many things in waking life (see Bulkeley, 2012, for a study using generic word strings, which also included Bea's 87 high school and college dream reports, and Domhoff, 2018a, pp. 145–151 for full details on the HVdC findings on Bea's dream reports at ages 14–15).

Izzy and Jasmine: Consistency and Continuity from the Teens to Young Adulthood

The findings on continuity in the Bea series were replicated with two much longer dream series that extend from ages 12–13 into the dreamers' mid-20s. The young women who contributed these dream series were born and raised in different countries in the English-speaking world. They differ in age by six years and they do not know each other. In addition, both participants were willing to answer questions that were asked after the quantitative analyses were completed. At that point they were queried via email about the important people in their lives and their thoughts about these people, as well as about their personal interests, avocations, and preoccupations. The analyses of these two dream series are presented in considerable detail for both methodological and theoretical reasons, including the fact that they provide the best possible case for the claim of consistency and continuity by age 15 and perhaps by ages 13–14.

The study of these two dream series also best demonstrates the usefulness of individually tailored word strings with dream series at any age level (see Domhoff & Schneider, 2020, for full details on methods and findings, as well as evidence for the authenticity of both series). In addition, the study shows the HVdC normative sample for women can be used for some of the character categories constructed with the word strings. This combination of two methodologies contributes to the usefulness of personally tailored word strings. Further, the conclusions are strengthened by the fact that the two dreamers are very different from each other, and both of them often differ from the HVdC normative women's sample as well. Any vagueness about dates and places in this account is part of the effort to protect the anonymity and respect the privacy of the two participants. However, sequences and time intervals are always maintained. Their dream reports, as well as the individually tailored word strings, are available on DreamBank.net for further analyses. The personal word strings and more details on the findings can be found in the detailed published article on these findings (Domhoff & Schneider, 2020, pp. 146–147).

Izzy recorded 4,329 dream reports between the ages of 12 and 25. She was born in the early 1990s and she has a brother who is two years younger. Their parents never married but they lived together until Izzy was about 9 years old and her brother was 7. When their parents separated around the year 2000, Izzy and her brother had rooms in both parents' residences. By about 2008, when Izzy was 17, she and her brother were living exclusively at their mother's house, along with their mother's partner of the previous seven years. Around 2010 her brother moved in with their father after he was caught stealing money from their mother. In late spring 2013, Izzy moved to another English-speaking country for 14–15 months. When she returned home she lived with her mother and her mother's partner until mid-summer 2016. She then moved into a rented house with three male classmates from a local community college. She had lived there for 4.5 months when her dream series ended for purposes of this analysis.

Jasmine voice-recorded the 664 dream reports in her dream series between the ages of 14 and 25, usually shortly after she awakened on any morning on which she recalled a dream. The recordings were transcribed by an experienced transcriber who had done previous work for DreamBank .net. Jasmine's dream series consists of 39 dream reports from her middle-school years, 268 dream reports from her high-school years, 261 reports as a college student, and, after a four-year hiatus, 96 at ages 24 and 25, when she returned to college and earned an M.A. in education and a teaching credential. (Jasmine's later dream reports from her mid-30s were used in the study of dreams about the pandemic in chapter 4.)

Jasmine's dream series also offered an unusual opportunity because she was born with an atypical form of blindness: optic nerve hypoplasia. Her type of hypoplasia makes it possible for her to recognize people by means of their hairstyle, hair color, and the way they walk (and of course she can recognize their voices because her hearing is normal). She can read 26-point type if it is very close to her face, so she did not rely on Braille growing up. For reading regular-sized print she has a video magnifier, which takes a picture of the document and uses optics to enlarge the image on the screen. For working with her computer she uses text-to-speech and screen-magnifying programs.

Jasmine was born in the mid-1980s and she has lived as an only child with her parents in the same house for most of her life. However, she has two older half-sisters, who are nine and 10 years older, and a half-brother, who is seven years older, from a first marriage by her father. In her early years they lived in the same city as Jasmine but then moved several hours away. She also has an extended family that lives over a thousand miles from her. When Jasmine was younger, her mother took her to visit these relatives for long vacations when school was not in session.

Jasmine attended a regular public high school and graduated in about 2003. After a year of community college she moved to a nearby city and spent close to a year finishing an accelerated program at a technical school. She then lived at home again with her parents and earned a B.A. from a local university in about 2010. After graduation she took a job with a local school for the visually impaired. The school asked her to learn Braille as an essential skill in working with blind youngsters who were not prepared to attend public schools. In this context, she returned to college to earn an M.A. in education and a teaching credential.

The two dream series were initially studied in sets of 200 or 100 dream reports, respectively, as well as by school levels (middle school, high school, and college) and then by one-year age intervals. These analyses demonstrated considerable similarity in the general findings and the same trends with all three types of subset samples. Because age is an important variable in developmental psychology, and in this chapter as well, it therefore is useful for substantive psychological reasons to present the results by age to the degree it is possible.

Since Izzy recorded over 100 dreams for each year except the first, when she was age 12 and documented 52 dreams, little if anything was lost by analyzing her dream reports in sample subsets based on her actual age. However, a school-level approach had to be used in analyzing Jasmine's series to increase sample sizes. This necessity still made it possible to examine her dream reports in relation to the important developmental transitions indicated by the maturational and cognitive changes that mark the differences between middle school, high school, college, and full adulthood.

Izzy: Personally known characters Overall, the most frequent character in Izzy's dream series is her mother, who appears in 34.9% of the dream reports. She is the most frequent character in every year except one, which comes very close to being an example of relative consistency in a dream series. The appearances by Izzy's mother showed two different patterns of developmental consistencies. There was a gradual overall increase in her appearances from ages 12 to age 18, followed by a gradual developmental decline until age 22. There was a relatively small increase in the next year, and then a further decline. Izzy said that her mother is the person she thought about most frequently in waking life through age 22, so her mother's dream appearances provide evidence of both developmental consistency and continuity between the ages of 12 and 22.

Slightly condensing what Izzy said in reply to written questions about the people she thought about the most at different ages, she portrayed her relationship with her mother as "somewhat complicated [before seventh grade]" and as "mellowing out during the year I started writing down my dreams." She added, "Generally, I didn't really have any strong feelings towards her, though I did resent her 'uselessness' as she was often very late for things or didn't really seem to listen to my concerns." Izzy seldom thought about her mother after she rented a house with other students (see Domhoff & Schneider, 2020, p. 149, for further detail). The same patterns of consistency and continuity were found with her brother and father.

Izzy also dreamed consistently about other family members. When her immediate family is combined with the rest of her extended family, 56.3% of her dream reports contained at least one family member, with individual years showing the same two developmental consistencies as found for her immediate family—high points when Izzy was 18 and 19 and the lowest percentages at 24 and 25. Similarly, she dreamed about her friends and acquaintances (26.9% of the dream reports), along with a wide range of people known from school or work (28.3%). When friends, acquaintances, and people from school and work are merged into a list containing all family members, then at least one person she knew personally appeared in 75.9% of her dream reports. This figure is very close to the HVdC normative percentage for women dreamers (76.8%) (Hall & Van de Castle, 1966).

Izzy's dream series was very atypical in regard to the frequency with which she dreamed about "prominent characters," a category including "any character who is well known by general reputation, but not known personally to the dreamer" as well as "fictional, dramatic, imaginary, and supernatural figures" (Hall & Van de Castle, 1966, p. 59). In Izzy's case, this means actors in TV shows, movies, and video games, which leads to the appearance of at least one prominent person in 35.7% of her dream reports. This figure is far above the HVdC normative figure of 3.5% for women (Domhoff & Schneider, 2020, p. 121, table 1, for details).

Romantic infatuations, which Izzy calls "crushes," are also an important part of her dream life. (She uses the term to refer to the subject of the crush as well.) In fact, a crush on a celebrity actor at age 12 led her to write down a few dreams about him, and then a crush on another celebrity actor motivated her to write down all of the dreams she could remember each day (see Domhoff & Schneider, 2020, pp. 151–153, for details). In all, 18.1% of her dream reports contained celebrity crushes, with a high of 41.5% when she was 22. However, as important as these celebrity crushes were in Izzy's dreams, the 14 classmates she became infatuated with at various times between ages 12 and 19 appeared in more of her dream reports (19.6% vs. 18.1% for celebrity crushes). Her deepest and most lasting crush, between the ages of 16 and 19, was on "Eugene," a classmate she had never spoken to in waking life. He appeared in 11.8% of her dream reports overall, including 2.0% of her dream reports after the crush on him ended in waking life at age 19. However, the actor who appeared most frequently appeared in only 3.1% of the dream reports. Overall, crushes who were either movie stars or acquaintances, appeared in 35.0% of Izzy's dream reports. There was absolute consistency from ages 14–19 (23.4%–38.2%, a range of 14.8%) and then a decline to 25.5% and below at ages 23, 24, and 25. This decline fits with her claim of developing no new crushes after age 22. The findings on the frequency and consistency of selected characters in Izzy's dream reports are presented in figure 7.1.

Izzy: Dream content and waking avocations As might be expected on the basis of her crushes on a long succession of male actors, Izzy dreamed frequently about popular culture, as it is expressed in visual media. Taken together, the word string encompassing terms related to specific shows, movies, and video games revealed at least one of these closely related interests appeared in 45.7% of her dream reports. A related word string for horror and fantasy appeared in 19.5%. When these two word strings were combined into a word string for hobbies and interests, they appeared in 54.1% of her dream reports. For all three of these word strings the percentages are lower from ages 12–17 and from ages 23–25 than they are between ages 18 and 22. However, there is variation from year to year in the percentage of dream reports with at least one mention of a popular culture term, so there are no consistencies in relation to her main avocations.

Izzy's response to a general question about the nature of her dreams provides very good evidence for continuity with waking thoughts in terms of her avocations: "I dream a lot about pop culture, since I watch a lot of films and TV. Also, from that I dream a lot about zombies because I love horror." Her comment confirms the frequency of her main waking avocational interests in her dream series relates to the intensity of her concern with this interest in waking life, just as the intensity of her personal concerns with specific individuals relates to the frequency of their appearances in her dream series.

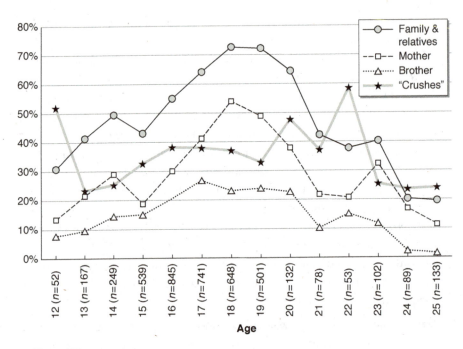

Figure 7.1
Percentage of Izzy's dream reports containing family members and "crushes," by age.

Izzy: Sexual interactions   Izzy had active sexual interactions in a few of her dream reports from the outset of her dream series. (The HVdC coding system defines a sexual interaction as anything from sexual thoughts to sexual intercourse, and this definition was used as far as was possible in studying her dream reports) (see Domhoff & Schneider, 2020, p. 154, for examples, which include instances of sexual intercourse involving herself or others). When asked if she ever had any sexual contact with any of her school crushes, she replied, "I never had any physical contact with anyone until age 25, so I never had any sexual contact with any of my crushes." Izzy's sexual activities in her dream reports from age 12 on provide a striking contrast to her shyness and avoidance of any type of romantic physical contact in waking life until she was in her mid-20s. However, her sexual involvements in her dream reports are continuous with her waking fantasies about her crushes.

Whether the issue is characters, avocations, or sexuality in Izzy's dreams, her waking thoughts are continuous with the concerns and interests she dramatizes in her dreams.

Jasmine: Personally known characters   As in the study of Izzy's dream series, Jasmine's mother is the most frequent character in her dream reports in all four subsets, appearing in 55.6% of the 664 dream reports. Her father appears in 42.5% of the dream reports and is the second-most-frequent character in all four subsets, which is an example of relative consistency in comparison to Jasmine's mother. Overall, at least one of her two parents appears in 66.7% of her dream reports. The dream frequencies for her mother and father were analyzed in terms of subsets, which in her case are based on school levels in order to have adequate sample sizes. Both parents also provide strong evidence for absolute consistency, with a range of 13.8 percentage points (48.7%–62.5%) for her mother and an even narrower range of 5.6 points (39.6%–45.2%) for her father. In addition, her response to how she perceived each of her parents confirmed their relative importance, and her attitudes toward them, in waking life (Domhoff & Schneider, 2020, pp. 154–155).

Overall, personally known people, including all family members, appear in 85.1% of Jasmine's 664 dream reports. This figure is almost 10 percentage points higher than the HVdC normative figure of 76.8% for at least one familiar character. Jasmine also dreamed about a wide range of prominent characters, both real and imaginary. They included both well-known popular singers and fictional characters who appear in movies, such as Snow White, The Little Mermaid, and Pocahontas, and also a few political figures. Many of these celebrity characters, real and imaginary, appeared in Jasmine's dream reports in relation to her strong interest in music, which is discussed shortly.

When all the findings on family, friends and acquaintances, familiar characters, prominent characters, and known characters for both Jasmine and Izzy are compared with the HVdC norms for women, the results reveal that they both have a higher percentage of dream reports with at least one familiar character than is found in the HVdC female norms. However, this shared deviation from the norms on known characters occurs for different reasons. While Izzy does dream of her family members more than the female normative sample, the most notable discrepancy in her case is the incredibly frequent appearance of celebrities, whom she only knows from the media. In contrast, Jasmine's high number of familiar characters is mostly driven by her family members. More generally, the results can be summarized by noting that Izzy and Jasmine differ significantly from the HVdC women's norms, *and* from each other, on virtually all the character indicators. The main exception to this generalization concerns the category of "friends & acquaintances," in which Izzy and Jasmine differ by only 3.4 percentage points. At the time they both are at least 16.7 percentage points below the women's norms. These findings on characters in both the Izzy and Jasmine series are compared with the HVdC women's norms in table 7.3.

Jasmine: Avocations   Aside from the very high percentage of known characters, the most striking feature of Jasmine's dream reports is her interest in her audio and electrical equipment, including her voice recorder, her visual magnifier, and her computer. Music is her other strong interest, including her keyboards and singing. These two general interests are somewhat intertwined in some instances, such as her use of her electrical keyboards and listening to recorded music. One or more words related to her audio and electrical equipment—along with the wires, cords, and other equipment that support them—appeared in 37.2% of her dream reports. In the case of music search terms, the overall total was 31.5%. When the two word strings are combined, one or both of these interests appeared in 55.1% of her dream reports, which is very similar to the

Table 7.3
Number of dreams containing various familiar character subclasses in the female norms and in the dreams of Izzy and Jasmine

	Female norms (n = 491)	Izzy (n = 4,329)		Jasmine (n = 664)	
	Percent	Percent	h	Percent	h
Family	33.6	56.3	+.46*	72.7	+.81*
Mother	12.6	34.9	+.54*	55.6	+.96*
Father	8.8	19.6	+.32*	42.3	+.82*
Parents and siblings	26.3	53.3	+.56*	70.8	+.92*
Relatives	9.2	15.9	+.20*	11.9	+.09
Friends and acquaintances	59.1	42.4	−.34*	39.0	−.40*
Celebrities	3.5	35.7	+.91*	12.7	+.35*
All familiar characters	78.2	86.5	+.22*	86.9	+.23*

Note: h values are from a comparison of each individual series to the female norms.
* p < 0.001.

54.1% of dream reports for Izzy's avocational interest in visual expressions of popular culture.

As with her character categories, the overall findings on Jasmine's avocational interests are continuous with her waking thoughts and interests. In waking life she is constantly putting together, repairing, working with, or worrying about her audio and electrical equipment and her computer. In terms of music, she and her parents listened to a wide range of music while she was growing up. She "enjoyed singing in the school chorus" and "felt drawn to the piano." She hesitated to take music lessons during high school but she bought "a cheap keyboard" while she was away from home at technical school. She learned to play it with some help from her music professor. She began to take lessons on the piano when she returned to live with her parents. She became proficient enough on the piano, and then on the organ, to play in nursing homes, hospitals, and churches (Domhoff & Schneider, 2020, p. 157).

Jasmine: The absence of sexual interactions Sexuality appears in fewer than 10% of HVdC women's normative dream reports, and it is even more infrequent in most of the relatively few available teenage dream samples. It is completely absent from Jasmine's dream reports. For reasons of ethics and respect for Jasmine's privacy on a potentially sensitive topic, sexuality was not a topic it seemed appropriate to ask her about. However, she later volunteered the information, at about age 30, that she had a "significant other" who is older than she is (Domhoff & Schneider, 2020, pp. 157–158). When then asked about her possible interest in romantic relationships during high school, she replied that she did not "focus too much on boys or dating when I was a teenager or in my 20s. No one really seemed interested in me, and I just never let it be an issue." Her reply suggests that the atypically low frequency of sexual elements in her dream reports reveals what is very likely an atypically low waking interest in romance and sexuality. In this instance, the relative lack of interest may be at least in part due to the fact that no one had shown any romantic interest in her, which might have sparked her own interest.

Izzy and Jasmine kept track of their dreams on a daily basis for very different reasons, and their interests in waking life are extremely different as well. Nevertheless, they are similar in that their dream series provide evidence of consistency and continuity for several characters and for their primary avocations. They are similar in waking life in that they never engaged in any sexual activities during high school or their college-age years. At the same time, their waking thoughts about sexuality were very different: Izzy often daydreamed about sex and Jasmine did not. In both instances there is continuity between dream content and waking thought. Both of these dreamers once again reveal the overriding importance of imagination, not waking behavior or reality, in dreaming.

Consistency in Phil's Dream Reports at Ages 14–15 and 29

Phil, a retired professor of humanities, taught for several decades at an eastern university. He offered his dream series after he had been retired in a western mountain state for two years. He first wrote down 29 dream reports in the two months before his 15th birthday and another 77 in the first several months after that for a total of 106 dream reports. He also wrote down his dreams regularly for a year or two when he was in his late 20s. No detailed personal information on Phil was ever obtained, which is why the study is primarily focused on the issue of consistency. His dream series was studied with both HVdC categories and two generic word strings. The dream reports can be found on DreamBank.net under the pseudonym Phil.

The first analysis, based on HVdC categories, focuses on 86 dream reports between 50 and 300 words for ages 14–15 and 86 reports of the same length range at age 29. There were only three substantive differences on 16 HVdC content indicators between the two age periods. His very high teenage F/C Index declined to a figure closer to the norms (.36 to .26, $h = .20$, $p < .05$) and his low teenage A/C Index increased, also moving closer to the norms (.16 to .23, $h = .19$, $p < .05$). The decline in friendliness and rise in aggression is also captured by the large change in his A/F%, from a very low 26% to 47% ($h = .45$, $p < .001$)—which is still well below the men's norm of 59%.

Phil's dream reports for those two ages were also compared using generic word strings for religion and sexual intercourse. These comparisons were based on all the dream reports in the two subsets, not just the dream reports that were coded for 16 HVdC content indicators. The comparisons were possible because the median word lengths did not differ significantly. There were 141 in the first subset and 153 in the second subset. There were differences between the adolescent and mature adult subsets on both of the word strings, which were consistent with his changing interests and preoccupations since his teen years. He was less interested in religion and more interested in sex. Based on this quick overview of the findings with the

Phil series, it provides solid evidence for consistency in what people dream about between their mid-teens and their late 20s. This evidence supports the findings based on the Izzy and Jasmine series.

Conclusions and Implications

Based on replicated laboratory studies with girls and boys ages 9–15, two-week dream diaries from a sample of girls at a private school, MRD studies of children in four countries, and four adolescent dream series, this chapter first shows that adolescent dream content is in general similar to adult dream content by ages 13–15, just as the default network is becoming close to the maturation level of young adults. However, dream content is not yet similar to the norms on some indicators. The findings from the four adolescent dream series add strong evidence to the overall case for consistency and continuity in dreaming, as discussed in chapter 4. In particular, Izzy and Jasmine are of special interest as outliers on many character categories when they are compared to the HVdC women's normative sample. Their specific avocations may be somewhat unusual as well. But the different ways in which they are atypical greatly strengthens the case for consistency and continuity in dreams. This is because they are atypical individuals while nonetheless showing consistency over time and continuity with waking personal concerns by ages 13–15. As with the cross-sectional studies of late childhood and early adolescence, this excellent longitudinal evidence is consistent with the maturation of the default network at about this time and with the full development of the cognitive capacities necessary for imagination to flourish.

This chapter first examines the empirical findings from the mid-1960s to 2020 on the most difficult and contentious issue in the study of dream content: the frequency and nature of emotions in dream reports. In the process, the chapter of necessity discusses relevant methodological issues. The studies serving as the basis for the chapter were carried out in both laboratory and nonlaboratory settings and include dream reports from children and adolescents as well as adults. Using the emotions categories in the Hall/Van de Castle (HVdC) coding system or very similar coding categories devised by other dream researchers, these studies led to an unexpected discovery. Only half of women's dream reports and one-third of men's dream reports contain an emotion, and as many as 17% of dream reports contain no emotions in situations that would have produced emotions in waking life. Moreover, even fewer dream reports collected from children and younger adolescents include an emotion. These studies also reveal that the majority of the emotions in samples of adult dream reports is negative, which is consistent with the predominance of other negative elements in dream reports, such as aggressions and misfortunes, as discussed in chapter 3. However, in children and adolescents' dream reports, the preponderance of emotions is positive, as is the case with other elements in their dream reports, as discussed in chapters 6 and 7.

These descriptive empirical findings have been called into question by some of the results in four later studies. These studies make use of the dreamers themselves to make ratings of their own dream reports on 4-point or 5-point rating scales. The ratings concern the presence or absence of emotions, the intensity of the emotions, and the degree to which the emotions in each dream are positive or negative. Self-ratings by participants lead to much higher percentages of dream reports containing at least one emotion and higher percentages of dream reports in which the emotions are predominantly positive, as opposed to predominantly negative. Independent raters of the same dream reports, who are given the same instructions as the dreamers, sometimes produce results different from those provided by the participants and also from the results in studies based on quantitative coding systems. These differences are discussed in a section of the chapter that compares coding systems and rating systems.

The final section begins with an overview of Darwinian-derived ideas on emotions. It next turns to waking neuroimaging studies of the neural substrates and associative networks underlying the emotions of fear, anger, sadness, disgust, and happiness. The waking neuroimaging results are then compared to the neuroimaging findings during dreaming, which reveals large differences between the neural substrates active during waking emotions and the more truncated neural substrate active during dreaming.

The neurocognitive theory of dreaming therefore suggests the relative lack of emotions in dream reports, as found in studies using coding categories, is very likely due to cognitive insufficiencies during dreaming. More specifically, only portions of one of the association networks involved in the experiencing of emotions, the default network, has a relatively high level of activation during dreaming. This analysis is supported by studies of people who frequently experience highly emotional and frightening nightmares. Unlike people who do not suffer from frequent nightmares, people with frequent nightmares have continuing high levels of brain activation during sleep, as revealed by both neuroimaging and EEG studies.

Quantitative Content Analyses of Emotions in Dream Reports

The most rigorous coding categories for studying emotions in dream reports were created as part of the detailed process that led to the comprehensive HVdC coding system, which was overviewed in chapter 3 (Hall & Van de Castle, 1966). In carrying out this general work, the emotions categories proved to be the most difficult to construct: "The problem of reducing the hundreds of words in the English language that represent affective states to a fairly small number of classes that seemed to be fairly comprehensive, yet discrete in coverage, was a formidable one" (Hall & Van de Castle, 1966, p. 110). In addition to the difficulties in classifying emotion terms, there were problems in establishing high intercoder reliability, even with a slightly more general set of categories than was initially considered. Since high reliability is an absolute necessity for any coding system, the two researchers had to collapse a few of the initial categories into the categories that comprise the final system (Hall & Van de Castle, 1966, p. 110).

In creating the category labeled as "Fear/Apprehension," the two researchers note that there are "recognizable" differences among the emotion states in this category but nonetheless conclude that they share a "common denominator." In each case, "a threat of some potential danger exists," whether physical or psychological (Hall & Van de Castle, 1966, p. 110). There is "the possibility of physical injury or punishment, or the possibility of social ridicule or rejection" (Hall & Van de Castle, 1966, p. 111). In theoretical terms, these are the same type of categories that cognitive psychologists and cognitive linguists now understand as natural categories. They are based on prototypes and families of resemblances (Rosch, 1973; Rosch & Mervis, 1975; Rosch, Mervis, Gray, Johnson, & Boyes-Braem, 1976; Sullivan & Sweetser, 2010).

Due to the fears people have about social ridicule or rejection, this category also includes guilt, embarrassment, and shame. Disgust is also included in this category because it seems to have its origins in a fear of contamination by specific immoral actions, often related to sex, and a similar fear of bodily excretions and specific types of food (Gifuni, Kendal, & Jollant, 2017, p. 1171). However, guilt, shame, embarrassment, and disgust are infrequent in the normative samples.

Hall and Van de Castle concluded four of their five categories are intuitively sound: Anger (AN), Fear/Apprehension (AP), Sadness (SD), and Happiness/Joy (HA). (Joy was relatively infrequent and difficult to reliably distinguish from happiness, so joy was made part of a Happiness/Joy category.) However, the two researchers debated back and forth on whether to include what they decided to call Confusion (CO) as an emotion. Confusion is indexed by "surprised, astonished, amazed, awestruck, mystified, puzzled, perplexed, strange, bewildered, doubtful, conflicted, undecided, and uncertain," as mentioned in chapter 3 (Hall & Van de Castle, 1966, p. 112). Although they thought "it may be debatable as to whether confusion is a condition possessing the same degree of autonomic involvement as the preceding emotions," they decided the "feeling state" that accompanies uncertainty may begin to shade toward "a type of free-floating anxiety."

They therefore decided it belonged "most appropriately in the classification of emotions" (Hall & Van de Castle, 1966, p. 112).

As explained in chapter 3, the introduction of the concept of cognitive appraisals into cognitive psychology has made Confusion a better fit as a cognitive category. Surprise or confusion may or may not lead to an ensuing emotion state (Dixon, Thiruchselvam, Todd, & Christoff, 2017; I. Roseman & Evdokas, 2004; I. Roseman & Smith, 2001) This rationale fits well with Hall and Van de Castle's (1966, p. 112) starting point as to the origins of confusion: "Confusion is generally produced either through confrontation with some unexpected event or else through inability to choose between available alternatives."

Coders are instructed to code explicit mentions of emotions, with one exception that happens infrequently in the normative ample: "If the dreamer describes definite autonomic activity accompanying an event, and it is clear from the combination of context and autonomic description that the dreamer was experiencing an emotion," then an emotion would be scored. Two examples are provided. If the dreamer says, "Tears began running down my face when I received word of my mother's death," that would be an instance of Sadness. If the dreamer begins to "sweat and tremble, and tries to cry out, but no words would come," that would be an instance of Fear/Apprehension (Hall & Van de Castle, 1966, pp. 112–113).

The normative findings for the four remaining emotions categories in the HVdC coding system—Fear/Apprehension, Anger, Sadness, and Happiness/Joy—are presented in table 8.1 for women, men, and women and men combined. In addition, this table includes figures for the percentage of dream reports in the normative samples that do not include any emotions, which leads to a "no emotions percent" or its inverse, an "at least one emotion percent." It is calculated by dividing the total number of dream reports with no emotions by the total number of dream reports with or without an emotion. There is also a percentage for at least one instance of one of the three negative emotions categories: Fear/Apprehension, Anger, or Sadness. The table also includes the normative findings for a "Fear/Anger/Sadness Percent" (FAS%), which is determined by dividing the total number of "negative" emotions by the total number of coded emotions in the four emotions categories combined. The FAS% replaces a former measure, the "Negative Emotions Percent," which included confusion as one of the negative emotions. Although this change to a new FAS% is conceptually

Table 8.1

Frequency of emotions in the Hall/Van de Castle norms for women and men

	Women's norms ($n=491$)	Men's norms ($n=500$)	Women and men combined ($n=991$)
Percentage of dreams with NO emotions	49.7	66.8	58.3
Percentage of dreams with at least one:			
Happiness/Joy (HA)	13.8	9.2	11.5
Fear/Apprehension (AP)	26.5	17.2	21.8
Anger (AN)	10.0	7.2	8.6
Sadness (SD)	10.2	4.8	7.5
Fear/Anger/Sadness (AP+AN+SD)	40.9	26.4	33.6
Any emotion	50.3	33.2	41.7
Fear/Anger/Sadness Percent (AP+AN+SD ÷ all emotions)	75.9	75.1	75.6
Negative emotions (F/A/S) compared to positive emotions, expressed as a ratio	3.0	2.9	2.9

necessary, it leads to a difference of only 4.6 percentage points for women and 5.4 percentage points for men.

The detailed results presented in table 8.1 provide a baseline for examining several of the studies that follow. The combined norms that are included in it, along with normative figures for women and men separately, are useful for comparisons with studies that combine women and men in presenting their results.

The normative HVdC normative findings concerning emotions in non-lab dream reports are replicated in detail in samples of men's dream reports from the methodological study of lab and nonlab dream reports in the Miami sleep-dream lab in the mid-1960s (Hall, 1966a). Since the study used the same men as participants in both the lab and nonlab conditions, the dream reports make it possible to compare nonlab findings with those in the HVdC normative study and to provide norms for lab findings on emotions as well. However, Hall (1966) did not code these dream reports for emotions. They were coded for emotions in 2014–2015 by two of the author's research assistants, who were not aware of their origins. The few disagreements were resolved by a highly experienced coder, Adam Schneider. In order to have equal numbers of dream reports from participants in both the lab and nonlab settings, the analysis is based on 167 extant dream reports from seven of the young adult male participants (see Domhoff & Schneider, 1999, for information on how these comparable samples were created).

The percentage of dream reports with no emotion in the Miami nonlab reports, 67.0%, is almost identical to the percentage in the HVdC norms, 66.8%. This percentage is higher in the lab dream reports, 76.0%. The FAS% for the nonlab reports, 82.9%, is a little higher than it is in the HVdC normative findings, 75.1%. The percentage for the lab dream reports, 81.5%, is very similar to the Miami nonlab result. These findings, along with a few others from this new study of older data, are displayed in table 8.2. These results can serve as another possible baseline for comparing HVdC results with those based on other coding systems.

Following the introduction of the HVdC coding categories and the normative findings, the frequency and nature of emotions in dream reports was next analyzed in a study based on 635 REM dream reports, which were collected in two different sleep-dream laboratories over the space of "250 subject nights." Although the exact number of participants is not stated, the participants were almost equally divided in terms of gender. Only one-third of the dream reports contained at least one emotion, which also can be expressed as 67.0% of the dream reports having no emotions. A subset of the overall sample had been collected from 20 young adult male students

Table 8.2

Emotions in dream reports collected in a sleep-dream lab vs. nonlab-collected dreams, compared to the male norms

	Miami lab ($n = 167$)	Miami nonlab ($n = 106$)	Male norms ($n = 500$)
Percentage of dreams with NO emotions	76.0	67.0	66.8
Percentage of dreams with at least one:			
Fear/Anger/Sadness (AP+AN+SD)	21.0	27.4	26.4
Happiness/Joy (HA)	4.8	7.5	9.2
Any emotion	24.0	33.0	33.2
Fear/Anger/Sadness Percent (AP+AN+SD ÷ all emotions)	81.5	82.9	75.1

for the explicit purpose of studying emotions. The percentage of dream reports with no emotions was only marginally lower, 65.0% (Snyder, 1970, p. 141). These figures are reasonably close to the findings in the combined HVdC normative sample, based on nonlab dream reports. Anxiety, fear, and anger were the most frequent emotions in the overall sample, and negative emotions outnumbered positive emotions by a 2-to-1 margin, which is close to the HVdC findings (Snyder, 1970, p. 141). (Readers may recall from earlier chapters that there were only a few "typical" dreams and unusual elements in this large REM-based sample.)

A study of 104 REM dream reports, obtained by the researchers from colleagues in a University of Cincinnati sleep-dream lab, reported similar low emotions figures. (This study, based on a sample from 20 college-aged men, was also briefly overviewed in chapter 5 in terms of the few unusual elements in it.) Using their own categories for anger, anxiety, guilt, sadness, shame, surprise, and joy, the researchers found anxiety in 14% of the reports, surprise in 9%, anger in 9%, joy in 8%, sadness in 4%, shame in 2%, and guilt in none at all. These categories are very similar to those in the HVdC coding system, except that the HVdC system includes shame and guilt in the Fear/ Apprehension category. The study does not provide the percentage of dream reports with at least one emotion, but the sum of the percentages for the six emotions listed in the previous sentence adds up to only 46%.

However, the researchers do say that 8.7% of the dream reports included "more than one affect," which means a maximum of 38% included at least one emotion. If surprise is removed from their emotions scale, due to the fact that it is now considered to be a sign of cognitive appraisal, then the percentage is very likely in the 30–35% range, which is similar to the HVdC normative finding for men, 33.2%, and can be expressed as 66.8% of the dream reports having no emotions (McCarley & Hoffman, 1981, pp. 908, 912). The three negative emotions appeared 3.7 times more often than joy, which occurred in only 8% of the dream reports (McCarley & Hoffman, 1981, p. 908). In the HVdC men's norms, negative emotions appear 3.0 times more often than Happiness/Joy, so the difference between the two is not large on the basis of this ratio comparison.

As one part of a study based primarily on rating scales, which is discussed below, two researchers at the Central Institute of Mental Health in Mannheim also coded 180 nonlab dream reports on the basis of a German translation of the HVdC coding categories for emotions. Their findings are similar to those in the normative study. Based on their combined findings for women and men, they first found that 39.4% of the dream reports contained at least one emotion, which is close to the HVdC combined normative finding of 41.7% (Schredl & Doll, 1998, p. 640, table 1). They also found that 24.4% of the dream reports contained predominantly negative emotions, whereas only 9.4% contained predominantly positive emotions (Schredl & Doll, 1998, p. 640, table 1). In the HVdC combined normative sample, 75.1% of the emotions were in the Fear/Anger/Sadness content indicator, which incorporates all the negative emotions in the updated norms, and only 24.9% in the Happiness/Joy category. The HVdC findings, based on the total number of emotions in all the dream reports, are not the same as the indicator based on the predominance of negative or positive emotions in each dream report in this study. However, if the two separate sets of findings are expressed in ratios, then the negative/positive ratio is 3.0 in the study carried at the Central Institute of Mental Health and 2.9 in the HVdC combined normative sample.

Generally speaking, then, the three studies by three different groups of independent researchers reviewed above provide very similar results to those in the HVdC normative study and the Miami lab/nonlab study. They all conclude that only 30–40% of the dream reports included at least one emotion, and negative emotions outnumbered positive emotions by at least a 2-to-1 margin.

The relatively high percentage of dream reports with no emotions in the studies overviewed in this section, which included a large gender difference in the HVdC normative sample (49.7% for women, 66.8% for men, $h = .35$, $p < .001$), also led to an unanticipated research question. From the outset of the studies of emotions in dream reports, Hall and Van de Castle (1966, p. 110) were "generally surprised at how few emotions are actually reported," and then added a question about why there would be a difference in emotions between dreams and waking life: "Situations that would undoubtedly be terrifying or depressing for the average individual may be reported in some detail, but a description of their emotional impact upon the dreamer is often curiously lacking."

This impressionistic observation was supported in a later study of the appropriateness of emotions in dream content. The study is based on REM awakenings of 17 young adult college students, nine women and eight men. Each participant was awakened five or six times over the course of two nonconsecutive nights in a sleep-dream lab, leading to 94 dream reports from 106 awakenings, a recall percentage of 88.7% (Foulkes, Sullivan, Kerr, & Brown, 1988, p. 32). The participants were asked in detail after each awakening about the presence or absence of emotions in the dreams they recalled. They were also asked about the appropriateness of the emotion, or any lack of emotion, in relation to the nature of the dream content. The researchers then asked independent raters, working separately, to make their own ratings based on copies of the dream reports. Both the participants and the independent raters agreed that emotions were absent in situations in 17% of the dream reports in which emotions would have been present in similar situations in waking life. On the other hand, 3.2% of the dream reports included emotions in situations in which there would not have been emotions in waking life. The presence or absence of feelings was appropriate to the dream situation in 60% of the dream reports (Foulkes et al., 1988, pp. 34–35).

The high percentage of negative emotions in all of these studies may come as a surprise to many readers, so it needs to be stressed that this finding is not a function of having fewer categories for positive emotions than for negative emotions, as is the case of all of the studies overviewed in this section. For example, the HVdC coding system for emotions has only one category for positive emotions, Happiness/Joy. However, all of the synonyms for happiness and joy are included in this HVdC category.

The possibility that having fewer categories for positive emotions leads to bias toward overcounting negative emotions has been refuted in a nonlab study. This study used 10 categories for "positive" emotions and 10 categories for "negative" emotions. The analysis was carried out by determining the percentage of dream reports with predominantly negative emotions and the percentage of dream reports with predominantly positive emotions. The percentage with predominantly negative emotions was higher than the percentage of reports with predominantly positive emotions by over a 2-to-1 margin, 28.1% as compared to 12.5% (Sikka, Feilhauer, Valli, & Revonsuo, 2017, p. 374, table 1). If these percentages are expressed as a ratio (2.2), then that ratio is similar to the ratio based on a comparison of the percentage of negative emotions to the percentages of positive emotions in the combined HVdC normative sample (2.9). However, it should be emphasized that this comparison provides only a general idea of the degree to which the two sets of findings are similar. The HVdC ratio is higher, and is based on the total number of emotion elements, not the percentage of dream reports with predominantly negative or predominantly positive emotions.

Studies of Emotions Using Rating Scales

Studies of emotions during the first 30 years of dream research relied on quantitative content analyses by independent coders. Since the 1990s, however, most of the studies of emotions have made use of rating scales. The participants, and often independent raters as well, are asked to indicate the presence or absence of emotions in each dream report and the degree to which the emotions in a dream report are negative or positive. The results from self-ratings by each participant in these studies are often very different from the HVdC normative findings. However, the results from ratings by independent raters are sometimes very similar to the HVdC norms on the percentage of dream reports with at least one emotion. When the findings concerning negative and positive emotions are expressed in terms of ratios, the findings by independent raters are often more similar than different in comparisons with the HVdC norms. Although the studies using rating systems also included findings on intensity, the focus in this chapter, as also mentioned in the introduction, is on the presence or absence of emotions and on the degree to which the emotions are reported to be negative or positive. (Generally speaking, however, the emotions in dream reports are not usually rated as intense.)

The first large-scale study of emotions in dream reports using ratings by both the participants and independent raters was carried out by the aforementioned dream researchers at the Central Institute of Mental Health in Mannheim (Schredl & Doll, 1998). The first author of the original study and a new coauthor then carried out a second study almost two decades later, which reported similar results (Röver & Schredl, 2017). In these studies, based on nonlab dream reports, the participants and independent raters were instructed to rate each dream report for the degree to which emotions were present on a 4-point rating scale: (0 = none, 1 = mild, 2 = moderate, and 3 = strong). Using a similar rating scale, participants and independent raters assessed the degree to which any emotions in a dream report were negative, and then, on a separate page, they rated the degree to which they were positive (Röver & Schredl, 2017, p. 66; Schredl & Doll, 1998, p. 638). Both the participants and the independent raters were asked to include any "implicit" emotions and moods in their ratings as well (Schredl & Doll, 1998, p. 638).

The 964 self-ratings by the 293 participants, 190 women and 73 men, led to the finding that 92.7% of the dream reports contained at least one emotion. Ratings by two independent raters, based on one dream report from each of 180 participants, led to the conclusion that 85.6% of the dream reports contained at least one emotion (Schredl & Doll, 1998, p. 640, table 1). The 180 dream reports rated by the two independent raters are the same 180 dream reports they coded with HVdC categories, the results of which were discussed above. Readers may recall that the two researchers' HVdC findings led to the conclusion that only 39.4% of these dream reports included an emotion, which is a difference of 46.2 percentage points between the codings and the ratings by the same two researchers. Readers also may recall that the two researchers' findings with the HVdC coding categories are very similar to the HVdC combined norms for nonlab dream reports, 41.7%.

Ignoring here the 8.3–11.9% of the dream reports in this study that had balanced emotions, the participants rated 42.6% of the dream reports as predominantly negative and 38.2% as predominantly positive. When these two percentages are translated into a predominantly negative/predominantly positive ratio, it leads to a ratio of 1.1. Following the same procedures for the independent raters, the predominantly negative/positive percentage is 56.7%/20.6%, a ratio of 2.8, which is more than double the ratio for ratings by participants. By way of further comparison, the same ratio was 2.5 for the two researchers' HVdC findings in their study, so the independent ratings and codings led to very similar negativity/positivity ratios (Schredl & Doll, 1998, p. 640, table 1). The differences between the ratings by the participants and the two independent raters concern two issues. The participants found a higher percentage of dream reports with emotions than did the independent raters, and they also made higher estimations of the percentage of predominantly positive emotions in the dream reports.

The researchers therefore concluded that there is an "inconsistency" between self-ratings and independent codings or rating by independent raters, which they think "is best explained by the underestimation of emotions by external raters in general but particularly by the strong underestimation of positive emotions and the low rate of explicitly mentioned positive emotions in a dream report" (Schredl & Doll, 1998, pp. 642–643). However, it may be that the wake-state bias, as heightened by instructions to include implicit emotions, could be the main factor. The tendency toward a wake-state bias also may be increased by the absence of emotions in many situations in dreams that would include emotions in waking life, as evidenced by the finding that emotions were absent in such situations in 17% of dream reports (Foulkes et al., 1988).

The second study, based on 413 participants, 350 women and 63 men, who contributed 1,207 diary dreams, presented similar results. According to the participants, 95.3% of their dream reports included an emotion, whereas the one independent rater in this study reported 82.5% had an emotion. For the participants, the predominantly negative/predominantly positive findings were 44.4% vs. 34.7%, a ratio of 1.7. The independent rater in this study saw things differently: 53.9% of the dream reports contained predominantly negative emotions, as compared to 18.3% containing predominantly positive emotions. This comparison can be expressed as a ratio of 3.8. Once again, an independent rater differed considerably from the participants, which led to an even higher negativity/positivity ratio than provided by the HVdC combined normative figure of 2.9. The researchers conclude that "external judges underestimate emotional intensity in general but especially for positive emotions" (Röver & Schredl, 2017, p. 68).

Very clearly, participants and independent raters agree in these two studies that far more dream reports contain emotions than was found in studies based on coding categories. However, participants and independent raters disagree in both the 1998 and 2017 studies on the degree to which negative emotions predominate. At the same time, the estimates of the degree of negative predominance by the independent raters are similar to those that result from HVdC normative findings based on coding categories. The researchers once again explain this difference in terms of underestimations of both the frequency and positivity of emotions by independent raters (Röver & Schredl, 2017, p. 68). The possibility that self-ratings may be problematic is not addressed. (There is a further discussion of self-ratings as compared to independent codings in the next section of this chapter.)

A series of three studies at the University of Turku in Finland, building on a database created by an academic research psychologist, Pilleriin Sikka, and her coworkers, examined the frequency and the negativity/positivity of emotions in lab and nonlab dream reports (see Sikka, 2020, p. 76, table 9, for a useful summary of all the results that are expressed in percentages; see also Sikka, Revonsuo, Sandman, Tuominen, & Valli, 2018; Sikka,

Valli, Virta, & Revonsuo, 2014; Sikka et al., 2017). Both the participants and the two independent raters assessed the dream reports for the presence or absence of any emotions on a 5-point scale, which ranged from (0) "I did not experience any of these feelings at all" to "I experienced these feelings" (1) a little bit; (2) moderately; (3) quite a bit; (4) extremely. The participants were asked to rate the dream they just had experienced, not the dream report itself. The researchers therefore stress that participants may be drawing upon both their memory of the experience, including mood states, as well as the report they had made. In the case of the independent raters, they were relying exclusively on the transcriptions of the spoken dream reports. The participants and independent raters were instructed to focus on explicitly expressed emotions in making their ratings. They were aided by a list of 10 positive and 10 negative emotions. Three examples were also provided for each emotion.

Due to the fact that the two independent raters worked from lists of 10 positive and 10 negative emotions, made frequency counts, met to resolve differences, and reported reliability scores on the basis of the percentage of perfect agreement method, they were essentially doing a coding study. As stated in a summary of this work, within the context of a full review of the literature on emotions in dreams: "In dream research, external ratings—or third-person ratings—have been the traditional method for measuring dream affect. With this method, narrative (written or oral) dream reports are collected and content analyzed by 'blind' external judges, using a particular scale" (Sikka, 2020, p. 40). Based on this explanation, the independent raters in these studies will be called "rater/coders" in the remainder of this chapter.

Generally speaking, there were dramatic differences between the results derived from the participants' self-ratings and the codings by the independent rater/coders. In addition, as shown below, the results for the two independent rater/coders are much closer to the HVdC normative findings than they are to the findings by the participants. For purposes of this comparison, the presence or absence of emotions in the three studies is expressed in the same terms as in the HVdC normative study, which means the percentage of dream reports with at least one emotion. In discussing the percentage of dream reports rated as having predominantly negative or predominantly positive emotions, the comparisons are made with the HVdC normative findings for the percentage of dream reports with at least one instance of fear, anger, or sadness, as well as the percentage for dream reports with at least one instance of Happiness/Joy. The FAS% is sometimes employed as well, which readers may recall is determined by dividing the total number of Fear/Anger/Sadness codings by the total number of emotions codings (fear, anger, sadness, and happiness). Finally, this analysis ignores the small percentage of dream reports rated as balanced on the positive and negative nature of the emotions they contained, as well as the even smaller percentage of dream reports for which a rating could not be made. Combined, these percentages vary between 5.0% and 9.8% for participants and independent rater/coders in all three studies (Sikka, 2020, p. 76, table 9).

The first study focused on 115 lab dream reports provided by 17 participants, 10 women and seven men (Sikka et al., 2014, pp. 54–55). The participants found that 100% of their dreams included an emotion, compared to 28.7% for the independent rater/coders. The comparable percentage for the presence of at least one emotion in the HVdC combined norms, 41.7%, is far closer to the findings by the independent rater/coders than it is to the findings provided by the participants. The participants rated only 9.6% of their dreams as containing predominantly negative emotions, compared to 79.1% with predominantly positive emotions. The two independent rater/coders concluded that 11.3% contained predominantly negative emotions, compared to 9.6% with predominantly positive emotions. This is a large disagreement in that the independent rater/coders found more negatively toned dream reports as well as far fewer dream reports with either predominantly negative or predominantly positive emotions (Sikka et al., 2014, pp. 56–57).

The second study focused on 552 nonlab dream reports provided by 44 participants, 28 women and 16 men. It reported that the participants and the two independent rater/coders found very different percentages for at least one emotion, 97.5% compared to 47.8%. The rater/coders' at least one emotion percentage is close to the HVdC normative percentage, 41.7%. In terms of the percentage of dream reports with predominantly negative emotions and predominantly positive emotions, the participants' parentages of 35.3%/55.8%, is the opposite of what the two rater/coders concluded, 28.1%/12.5%. The independent rater/coders' ratio of negative to positive, 2.2, is not much lower than the comparable HVdC ratio for the combined norms, 2.9 (Sikka et al., 2017, p. 367).

In the third study, which has many parallels with the Miami lab/nonlab study discussed above, the researchers focused on ratings by two independent rater/coders, who once again worked independently and then met to resolve most of their few disagreements (Sikka et al., 2018). This study is based on 151 nonlab dream reports and 120 lab reports from REM awakenings, which came from the same 11 women and seven men. Less than half of the dream reports from either setting contained one or more emotions: 29.3% in the lab dream reports, 45.4% in the nonlab dreams. In the Miami lab/nonlab study, which included only men as participants, the figure was only slightly lower in the lab dream reports, 24.0%. In comparison to the University of Turku nonlab dream reports, however, the Miami nonlab results were lower, at 33.0%. In the case of the combined HVdC norms for women and men, based on nonlab dream reports, the figure for dream reports with at least one emoton, 41.7%, is very close to the aforementioned percentage in the University of Turku study (45.4%).

Similarly, the University of Turku researchers found that predominantly negative dream reports were more frequent than predominantly positive dream reports: 36.7%/2.5% in the nonlab dream reports and 12.6%/8.4% in the lab dream reports. These two percentages lead to ratios of 14.7 and 1.5. In the Miami lab/nonlab study, 21.0% of the lab reports had at least one negative emotion and 4.8% had at least one positive emotion, a ratio of 4.4. The Miami nonlab reports had a negativity ratio of 3.7 and the ratio for the HVdC combined sample of nonlab dream reports was 2.9. Although those ratios vary, they uniformly show a negativity bias in the emotions in dream reports.

The three studies carried out at the University of Turku are the largest and most detailed studies of emotions in dream reports, collected both inside *and* outside of a laboratory setting from the same participants, since the HVdC normative study and the Miami lab/nonlab study in the 1960s. The University of Turku studies and the HVdC studies were undertaken five decades apart, took place in two different Western countries, and used different coding categories. Nevertheless, the similarities in their results are remarkable when the coding is carried out by independent rater/coders, who were using defined categories to study written reports. Based on the well-controlled nature of the University of Turku studies and the similarity of the results with coding studies, it is likely that the results are highly credible.

A Comparison of Coding Systems and Rating Systems for Emotions

As the results above show, coding systems and rating systems for emotions often lead to different results. These differences are in part due to the different levels of measurement on which they are based. However, there may be other issues involved as well, such as a wake-state bias, which may be even stronger in an analysis of emotions.

Coding systems are based on the categorical (nominal) level of measurement. They involve yes/no, present/absent, and other types of categorical judgments. Rating systems are based on rankings that involve judgments as to the degree to which a person, place, or object possesses certain qualities. Coding systems usually take longer to create and always take longer to learn and longer to apply to large datasets, but they usually have high reliability (e.g., Krippendorff, 2004; Charles Smith, 2000). Rating systems are readily created, although they often have to be fine-tuned. They are easy to learn and do not take as much time to use with large datasets. However, rating the different degrees of the intensity of emotions or of the positivity or negativity in the emotions in a dream report is a far more difficult and subjective judgment than may at first seem to be the case. Individual differences among raters on such judgments may lead to lower reliability scores, which may be part of the explanation for why the various rating systems created by dream researchers in the past were never used by researchers in other laboratories (see Winget & Kramer, 1979, for a discussion of the numerous rating scales from the first 25 years of systematic dream research, which were rarely used more than once or twice).

Coding systems make full use of the potential information available in a document, as seen in the case of the HVdC coding system, in which every emotion is categorized and tabulated, and used in a variety of indicators. Rating systems, by contrast, do not make detailed use of the information available, and the outcome variables are more limited in number. This is because the "unit of analysis" in the HVdC coding system is the elements in the dream content. In rating systems, the unit of analysis is the whole dream report (see Domhoff, 2003, pp. 58–60, 75–76; Hall, 1969a; and Van de Castle, 1969, for discussions and critiques of specific rating scales once used in dream research).

In terms of the coding and rating systems for emotions discussed in this chapter, the results from the use of coding systems are more similar than they are different on the frequency and negativity of emotions (McCarley & Hoffman, 1981; Snyder, 1970). This result may be especially impressive because the similar results include a study based on a German translation of the HVdC coding system, which was used with impressively high reliability rates, "ranging from 93.4% to 98.3" (Schredl & Doll, 1998, p. 40). Moreover, the findings based on the participants' self-ratings are extremely different from the results produced by the independent raters in the same studies (Röver & Schredl, 2017; Schredl & Doll, 1998; Sikka et al., 2014, 2017).

Nor do the very high percentages of dream reports said to contain positive emotions seem credible in the face of the findings by independent raters and by those using coding systems. The laboratory study reporting that 17% of dream reports do not include emotions in situations that would elicit emotions in waking life suggests that a wake-state bias may account for this large difference (Foulkes et al., 1988). Similarly, the high percentage of positive emotions claimed by self-raters does not fit with the findings by independent raters and coders. Nor does the preponderance of positive emotions fit with the fact that there are more aggressions, misfortunes, and failures in the HVdC normative sample than there are friendly acts, good fortunes, and successes. It is highly unlikely that the relatively fewer friendly acts, good fortunes, and successes would lead to positive emotions, but the larger number of aggressions, misfortunes, and failures would not lead to higher percentages of negative emotions.

For several reasons, then, it is very doubtful that self-ratings by participants can contribute to dream research. The evidence suggests that they provide vast overestimations on both the frequency and positivity of emotions in dream reports, not underestimations. Self-raters appear to be subject to all of the factors that can create confounds in systematic studies during waking, starting with people's self-conceptions and including social desirability factors and expectancy effects. In addition, self-ratings are very likely to be inflated by the instructions to include implicit emotions and by the expectation that there must have been emotions because the same situation would almost certainly include emotions during waking life. As shown when the chapter turns to neurocognitive studies of emotions, the current best conclusion is that the neural substrates that support dreaming are not sufficient to support the frequent appearances of emotion found by dream researchers who make use of self-ratings for emotions. It is also likely that the negativity bias is present during dreaming as well as in waking thought.

However, it should be noted that one group of neurocognitive researchers apparently disagrees with the work on the negativity bias in waking thought and with the findings on the relative lack of emotions in dream reports, which are based on codings by independent investigators using established coding categories. Instead, they conclude that most self-generated thought is positive or neutral and claim that "*at least* 70–75% of dreams from REM sleep contain emotion" (Fox, Andrews-Hanna, et al., 2018, p. 39; emphasis in the original). They make their claims about the frequency of emotions in dream reports in fair measure on the basis of the studies in which participants carry out self-ratings of their own dream reports and on the basis of the contested claim that the HVdC coding system underestimates the frequency of emotions in dream reports (Schredl, 2010; Schredl & Doll, 1998).

The Frequency of Emotions in Adult Dream Series

The conclusion that half or more of dream reports do not include any emotions and that the FAS% is usually around 75%, based on large group samples of lab and nonlab dream reports, can be examined from another angle by studies of individual differences in dream series, using HVdC coding categories. (As far as can be determined, there are no peer-reviewed studies of dream series that make use of self-ratings for emotions.) The most comprehensive study of emotions in a dream series is based on the sample of 250 dream reports drawn from the Barb Sanders series, as discussed in chapter 4: 44.8% of the dream reports did not include at least one emotion, which is not very different from that for women in the normative sample, 49.7%. The FAS% (84.1%) is also in the same range as earlier findings in lab and nonlab samples. The findings with the Barb Sanders series are supplemented by results from a sample of 100 dream reports, drawn from a dream series documented by "Emma" between the ages of 40 and 77. Fully 67.0% of the dream reports in the sample did not include at least one emotion, and there was a relatively low FAS% of 62.5% (see Domhoff, 2003, pp. 103–105, for findings from samples drawn from the 1,221 dream reports in this series).

Two of the dream series discussed in chapter 4, those from the Natural Scientist and Ed, also provide useful information on emotions in dream series. Only 20.9% of the Natural Scientist's dream reports contained at least one emotion, so 79.1% were without emotion. This finding fits with his description of his dream reports as "emotionless," as mentioned in chapter 4. On the other hand, emotions were present in all but 28.7% of Ed's 143 dream reports, which were highly emotional when he had a social

Table 8.3

Frequency of emotions in four individual dream series and the Hall/Van de Castle norms

	HVdC female norms	Barb Sanders (n=250)	Emma (n=100)	HVdC male norms	Natural Scientist (n=187)	Ed (n=143)
Percentage of dreams with NO emotions	49.7	44.8	67.0	66.8	79.1	28.7
Percentage of dreams with at least one:						
Positive emotion	13.8	11.2	15.0	9.2	6.4	32.2
Negative emotion (F/A/S)	40.9	49.6	20.0	26.4	16.0	51.0
Any emotion	50.3	55.2	33.0	33.2	20.9	71.3
Fear/Anger/Sadness Percent	75.9	84.1	62.5	75.1	72.7	65.1

interaction with his deceased wife. However, both men experienced a high percentage of negative emotions. The FAS% is 72.7% in the case of the Natural Scientist and a somewhat lower 65.1% in the case of Ed, who frequently experienced Happiness/Joy when he saw his wife in his dreams. The complete findings on emotions in the four dream series that are discussed in this and the previous paragraph appear in summary form in table 8.3, which also includes the HVdC normative figures for women and men so that readers can examine individual differences more readily. The main conclusion is that there are wide individual differences, which is an example of why it is necessary to have large sample sizes drawn from non-patient populations in order to develop a foundational theory of dreaming.

Emotions in Dream Reports from Children and Adolescents

The longitudinal studies of children and adolescents discussed in chapters 6 and 7 revealed dreaming to be a gradual cognitive achievement in terms of frequency, complexity, and content (Foulkes, 1982). Here it can be added that only a very small percentage of the dream reports from children contained an emotion, and the figures are well below the adult level in young adolescents' dream reports. Once again, as in the case of studies of dream series, no studies of emotions in children and adolescents' dream reports make use of self-ratings.

At ages 3–5, only 8% of the dream reports had any "feelings," the term the researcher used for discussing fear/apprehension, anger, sadness, and happiness, as well as feelings of hunger and fatigue. Most of the instances of feeling at this early age involve hunger or fatigue, along with a few mentions of happiness and sadness. There were no instances of apprehension or anger (Foulkes, 1982, p. 331, table B.1). At ages 5–7 there were still few emotions. The majority of them were positive, although two of the 64 reports had fear content (Foulkes, 1982, pp. 50, 86). Generally speaking, the percentage of dream reports with at least one emotion increases in each age period for both girls and boys, although the figures decline in the girls' dream reports from a high point of 50% at 9–11 to 42% at ages 11–13 and 40% at ages 13–15. At every age, the girls had a higher percentage of dream reports with at least one emotion than did the boys. The percentage of dream reports with one of the three negative emotions gradually increases for both genders, but the percentage of dream reports with happiness still remains higher (Foulkes, 1982, pp. 337, 345, tables B.2 and B.3). Looking at his findings as a whole, Foulkes concludes that "feeling itself is a cognitive achievement" (Foulkes, 1982, p. 55).

Even lower figures were found in the REM dream reports from lab awakenings in the Zurich replication and extension longitudinal study of 12 girls and 12 boys, which was based on the five-year period from ages 9–11 to 13–15. In the first year, only 8.8% of the girls' dream reports included emotions in any of the five original HVdC emotions categories, which at that time included the Confusion category. The figure for boys was 11.6%. After that first year in the lab, however, the girls had more dream reports with at least one emotion than did the boys (15.0% vs. 11.8% at 11–13, a difference of 3.2 percentage points, and 21.3% vs 14.1% at 13–15, a difference of 7.2 percentage points). The figures at ages 13–15 are well below the findings for the adult Swiss control group, 62.0% for women and 29.3% for men (Strauch, 2005, p. 161, table 4). The emotions findings in this study were not discussed in terms of specific emotions categories. However, in the researcher's book-length presentation of her findings, she reported in a table and in a figure that the emotions were more positive than negative in all three phases of the study (Strauch, 2004, p. 161, tabelle 9-5; p. 177, abbildung 10-4).

The findings on emotions using the two-week dream diaries provided by the Bay Area Girls are similar to those in the two laboratory studies in that only a small minority of dream reports (20%) contained at least one emotion. However, unlike the Zurich longitudinal study, which included both lab and nonlab dream reports, negative emotions predominated in this study. The FAS% was 80%. The finding on emotions at ages 14–15 in Bea's dream series supplement the picture. Based on the 139 dream reports in her series with 50 or more words, 56% of the dream reports contained at least one emotion. Her FAS% was 89%, which is higher than the 75.9% normative figure for women in the HVdC normative findings. As a very mature teenager, as discussed in chapter 7, she is both on the cusp of adulthood and a person who frequently experiences emotions.

To summarize, the frequency of emotions in dream reports is lower for children and adolescents than it is for adults. Once again, however, there are individual differences, as best shown in the case of Bea's dream series.

The Steps Leading to a Neurocognitive Theory of Emotions in Dream Reports

On the basis of the overall results concerning emotions in dream reports from childhood to adulthood, this section sets the stage for suggesting the neurocognitive theory of dreaming can explain the relative infrequency of emotions in dreams through the concept of cognitive insufficiency. Since this claim does not fit well with many traditional assumptions about emotions, it may be useful to briefly trace the scientific steps that led to the relatively recent neurocognitive explanation for waking emotions as cognitive categories. The first one-third of the section relies on psychologist Joseph LeDoux's neuroscience work on the "survival circuits" in mammalian brains, which he originally named a "threat-detection" system. The section makes even more use of his conclusions concerning the basis for emotions in humans (e.g., LeDoux, 2003, 2012, 2015, 2019, 2021). The middle third of the section makes use of the pathbreaking neuroimaging work on emotions by psychologist Lisa Feldman Barrett and her several coworkers (e.g., L. F. Barrett, 2017; Lindquist, Wager, Kober, Bliss-Moreau, & Barrett, 2012; E. Siegel, Sands, Chang, & Barrett, 2018; Touroutoglou, Lindquist, Dickerson, & Barrett, 2015). In combination, the work of these two pioneers in the neurocognitive study of emotions provides an ideal context for incorporating emotions into a neurocognitive theory of dreaming, which is the focus of the final third of this section.

From Darwin to the Prefrontal Cortex

Due to Charles Darwin's (1872/1998) highly influential emphasis on the similarities in the emotions of humans and most other mammals in the face of threats and other high-wariness situations, some early brain researchers attempted to develop what they saw as a Darwinian analysis of brain evolution. They organized their observations on brain differences among mammalian species as a "layering of new structures in the forebrain" on top of the earlier reptilian brain. They therefore focused on a seemingly new feature in mammalian brains, the amygdala, which was seen as the seat of emotions (LeDoux, 2019, pp. 183, 185). Since a larger forebrain is a key feature of later species of mammals, the evolution of the brain purportedly led inexorably to a "neocortex" on top of a "paleocortex" (see LeDoux, 2019, chap. 38, for a historical account).

This "ladder" view of the brain, which in fact does not fit well with Darwin's metaphor of a branching tree as a good way to view evolutionary change, was updated in the 1950s by a psychiatrist concerned with brain evolution in relation to emotions. He did his update on the basis of his work on high blood pressure, stomach ulcers, and other putative "psychosomatic" conditions (LeDoux, 2019, chap. 39). He proposed that each part of the "triune brain," as he renamed it, had a different function. The reptilian level is responsible for instinctive reactions, such as aggression and dominance, and the paleocortex (renamed the "limbic region") is responsible for the appearance of emotions in mammals and is the basis for "unconscious emotion" (LeDoux, 2019, p. 188). At the top level, the neocortex was said to be responsible for thinking, planning, and deciding, but it purportedly had relatively little control of the instinctive and emotional layers of the brain. Psychosomatic problems were claimed to be one result, with higher mental states "somehow disrupting" the "regulation of body physiology" (LeDoux, 2019, p. 188).

This model gained widespread acceptance, but it was directly challenged in the early 1960s by two cognitively oriented research psychologists (Schachter & Singer, 1962). As succinctly summarized by LeDoux, these two early cognitive theorists of emotions said that emotions are "constructed by the appraisal, interpretation, and labeling of biological, including neural, signals in light of the social and physical context of the particular experience," and other cognitively oriented researchers gradually provided increasing evidence for this analysis (see LeDoux, 2019, p. 351, for this quotation and for a list of 20 of the many cognitively oriented psychologists who helped develop the cognitive view of emotions). In particular, this cognitive orientation was strengthened by the new understanding of categorization introduced into psychology in the 1970s, as discussed in chapter 2 (Barsalou, 1982, 2003, 2009; Rosch, 1973; Rosch & Mervis, 1975; Rosch et al., 1976).

Moreover, all aspects of the triune brain theory had been proven to be wrong by the late 1970s. This refutation started with the discovery that the brains of reptiles and birds did in fact share homologous features with early mammals. In addition, reptiles have an amygdala, which leads to the conclusion that the differences among vertebrate animal brains are "mainly one of degree, rather than in kind" (LeDoux, 2019, p. 186). Then, too, areas in the "limbic system" also "contribute in significant ways to cognitive functions like memory and attention, and neocortical areas contribute extensively to felt emotional experiences" (LeDoux, 2019, p. 191). These studies discovered that the concept of a "limbic system," as it later came to be called, was not accurate. Nevertheless, the idea that the brain is "an onion with a tiny reptile inside," is still "widely shared in introductory psychology textbooks," even though this belief "has long been discredited among neurobiologists" (Cesario, Johnson, & Eisthen, 2020, p. 255).

In addition to rejecting the model of a triune brain, LeDoux (2019, p. 364) concluded there are four reasons for doubting that the activation of the amygdala is "the defining condition for the conscious experience of fear." To begin with, the activation of the amygdala cannot be solely equated with fear because many other circuits include at least portions of the amygdala, such as the search for food, drink, and sexual partners. And, as shown in one neuroimaging study, parts of the amygdala also are included in a perception network that detects and interprets social signals from other people. Other parts of the amygdala are active in the neural substrate underlying an affiliation network, which promotes prosocial behavior, empathy, and concern for others (Bickart, Dickerson, & Barrett, 2014, p. 238). Nor does fear invariably occur when the threat-detection system is activated. On the other hand, fear can be experienced when the amygdala is damaged, as shown in a study that induced experiences of fear through the inhalation of 35% CO_2 by three participants, all of whom had rare lateral amygdala lesions (Feinstein et al., 2013). Then, too, other survival circuits can generate fear experiences caused by dehydration, hypothermia, and fear of death from starvation.

Drawing on his experimental work in tracing the survival circuits, LeDoux (2019, p. 359) concluded that several areas in the prefrontal cortex, including the dorsolateral and ventrolateral prefrontal cortices, are necessary for the conscious experience of emotions. The focus on the prefrontal cortex, when combined with the evidence that the amygdala is not necessary or sufficient to experience fear, led LeDoux (1991, 2003, 2012) to abandon the concept of a limbic system. His emphasis on the primary importance of the prefrontal cortex is supported by an integrative analysis of emotions, based largely on neuroimaging studies (Dixon, Thiruchselvam, et al., 2017).

LeDoux also expressed skepticism about the assumption that subjective feelings can be inferred on the basis of similarities of behavior. Since the neural substrates underlying human consciousness differ from those found in other animals, caution is necessary in assuming the subjectively experienced phenomena labeled as "emotions" are present in other animals on the basis of generally similar behavioral responses. According to this reasoning, it is not possible to "use behavioral similarity to argue for similarity of conscious feelings functionally," if "the circuits that give rise to conscious representations are different in two species" (LeDoux, 2012, p. 20). Emotions are "cognitive interpretations" of the "situations in which we find ourselves," and they are the "products of the same general cortical cognitive circuits that generate other kinds of conscious experience" (LeDoux, 2019, pp. 350, 351, 359).

This conclusion is strengthened by two separate studies, both of which demonstrated it is inaccurate to assume that the activation of the amygdala leads to the subjective experience of emotion. In the first of these two studies, 11 epilepsy patients received electrical brain stimulation by means of electrodes implanted in the amygdala. It found that declarative memory for specific images of neutral objects was improved without eliciting a "subjective emotional response," which led the researchers to conclude their findings demonstrate "that the amygdala can initiate endogenous memory prioritization processes in the absence of emotional input" (Inman et al., 2018, p. 98). The second study was based on 15 young adult women and 16 young adult men who were shown hundreds of images of commonly feared animals, animals not feared, and human-constructed objects while undergoing functional magnetic resonance imaging (fMRI) scanning. The researchers found that "subjective fear and objective physiological responses" are correlated in general but that "the amygdala and insula appear to be primarily involved in the prediction of physiological reactivity, whereas some regions previously associated with metacognition and conscious perception, including some areas in the prefrontal cortex, appear to be primarily predictive of the subjective experience of fear" (Taschereau-Dumouchel, Kawato, & Lau, 2020, p. 2342). In particular, the researchers suggest "regions of the middle frontal gyrus, dorsomedial prefrontal cortex, and lateral orbital cortex were more closely related to the subjective reports of fear" (Taschereau-Dumouchel et al., 2020, p. 2349). They therefore conclude there is reason to doubt "a one-to-one mapping between subjective sufferings and their putative biosignals, despite the clear advantages in the latter's being objectively and continuously measurable in physiological terms" (Taschereau-Dumouchel et al., 2020, p. 2342).

Based on LeDoux's past findings and more recent studies, it is by now well established that the concept of a triune brain is outmoded and that the amygdala is neither necessary nor sufficient for consciously experiencing emotions. However, it is also known there is a limbic network with nodes in the prefrontal cortex, as well as in subcortical regions, which is often involved in emotions (e.g., Fox et al., 2018; Yeo et al., 2011, p. 1135, fig. 11). As noted in chapter 2, this fifth and final association network includes the orbitofrontal cortex, insula, medial prefrontal cortex, anterior cingulate cortex, and the temporal pole at the cortical level, as well as portions of the amygdala and other subcortical regions. As shown by neuroimaging studies discussed in the next subsection, the amygdala node in this network is therefore involved in emotion, as well as in supporting memory, reward, and empathy (Bickart et al., 2014; Inman et al., 2018).

Neuroimaging Studies of Emotions during Waking

Neuroimaging studies of waking emotion states by a group of cognitively oriented emotion researchers fit well with the emphasis on prefrontal cortical areas as necessary to the experiencing of emotions (L. F. Barrett, 2013, 2017; Bickart et al., 2014; Lindquist, Satpute, Wager, Weber, & Barrett, 2016; Lindquist et al., 2012; Polner-Clark, Wager, Satpute, & Barrett, 2016; Raz, Touroutoglou, Atzil, & Barrett, 2016; E. Siegel et al., 2018; Touroutoglou et al., 2015). Based on four separate meta-analyses, their studies focus on fear, anger, sadness, disgust, and happiness. They begin with the autonomic nervous system and move "up" to hedonic tone (emotional valence), activated brain regions, and the role of the five association networks, as well as other neural networks. This series of studies led to the conclusion that interactions among the five association networks are necessary for the conscious experience called "emotion."

Starting at a basic neurophysiological level, the autonomic nervous system, neuroimaging research consistently shows different emotions do not have separate neural substrates. Most of the autonomic nervous system variables are generally involved in all five of the emotion states that were studied. There is some variation in their roles within each emotion category as well, as shown in a meta-analysis of 202 relevant studies of induced emotions in representative samples of adults (E. Siegel et al., 2018). It was further discovered, based on a second meta-analytic study that included 397 neuroimaging studies, that there is no evidence of a neural substrate underlying a positive-negative (hedonic tone) dimension in the brain. The existence of such a dimension was assumed in the past on the basis of "hundreds of studies of semantics, self-reports of experience, and emotions perceptions in faces and vocal acoustics." Nor is there evidence for an alternative hypothesis on hedonic tone, which claims that positivity and negativity are independent dimensions (Lindquist et al., 2016; see Polner-Clark et al., 2016, p. 147, for a summary statement on their findings).

Instead, a wide range of brain regions are involved in shaping hedonic tone, which is also called "valence of expression" in these studies. This result suggests the same brain regions are involved in both positive and negative hedonic states. Some of these regions are not usually associated with valence or emotions. In particular, two neural substrates mentioned frequently throughout this book, the dorsal medial prefrontal cortex and the ventral medial prefrontal cortex, are involved in the valence of expression during the waking state. This new finding casts doubt on the idea that every mental state has a hedonic tone. In addition, it is consistent with the conclusion, based on neurological findings, that pleasure and pain are not emotions: "While pleasure and pain are often treated as emotions, they are actually different" (LeDoux, 2019, p. 218). Pleasure and pain "are tied to sensory receptors that detect particular kinds of stimuli," but there are "no sensory receptors for fear, anger, sadness, joy, or other emotions—the content is determined by the brain" (LeDoux, 2019, p. 217).

Still another meta-analytic study by the neuroimaging group of emotion researchers provided important new evidence on another longstanding issue. It showed that a variety of brain regions, primarily in the prefrontal cortex but also in regions in the amygdala, are involved in the experiencing of fear, anger, sadness, disgust, and happiness, and that the same brain regions are often involved in supporting nonemotional cognitions as well (Lindquist et al., 2012). However, the various brain regions are activated to varying degrees in the five different emotions that were induced (Lindquist et al., 2012, p. 132). These findings therefore provide an unexpected negative answer to what is "perhaps the question that has garnered the most interest in the science of emotion." Does each emotion category have a "specific, dedicated neural circuitry in the brain?" (Polner-Clark et al., 2016, p. 152).

Studies of the degree of involvement of the five association networks in the five emotions states provided the fourth and final step in this series of studies. It revealed that the frontoparietal, dorsal attention, salience/ventral, default, and limbic networks, along with the subcortical basal ganglia nuclei, are common to these emotion states. The researchers stress that emotions are therefore very similar to "other basic psychological functions" in this regard, such as attention, language, and memory (Touroutoglou et al., 2015, p. 1263). Very relevant in terms of the neurocognitive theory of dreaming, the salience portions of the salience/ventral network are involved in supporting all of the negative emotional states, and parts of the salience/ventral network are involved in supporting happiness as well (Touroutoglou et al., 2015, pp. 1257, 1261–1262). In the case of the default network, which is central to the neurocognitive theory of dreaming, it is similar to the salience/ventral network in that it is "routinely engaged across all categories of emotional experience, during typical and atypical instances of emotion, and in the representation of emotion concepts, as well as of concepts more generally" (Raz et al., 2016, p. 721).

Overall, these researchers conclude that their "emotion discovery maps reflected combinations of domain-general networks," which is "consistent with the hypothesis that different emotions arise from the interaction of domain-general systems within the brain" (Touroutoglou et al., 2015, p. 1262). Put another way, this study shows that emotion categories can be "distinguished by their profiles of relative activation across networks" (Polner-Clark et al., 2016, p. 155). The fear and anger categories, for example, show greater activation in the default, dorsal attention, and frontoparietal control networks, along with secondary visual areas. However, those areas were less activated when sadness, disgust, or happiness was induced.

Instead, sadness, disgust, and happiness share greater activation of the salience/ventral and sensorimotor networks (Polner-Clark et al., 2016, pp. 155–156).

In concluding this discussion, it is important to add that the conceptualization of emotion categories used by this research group is similar in its rationale to the one developed by cognitive psychologists (e.g., Barsalou, 2003; Rosch & Mervis, 1975). They conceive of emotion categories as a "statistical summary," which makes it possible to "identify the features of a prototypical chair, but no chair need have *all* those features to be a chair" (Polner-Clark et al., 2016, p. 155; emphasis in the original). Put another way, "each instance of the category has some set of features that are sufficient for category membership, but none of them are necessary for category membership" (Polner-Clark et al., 2016, p. 155).

Because this chapter attempts to link the recent neurocognitive findings on waking emotions to quantitative studies of dream content, it is also relevant to emphasize again that the four HVdC coding categories for emotions are based on a similar conceptualization of emotion categories in that they rest on a very general characterization of a category. To return to the example of fear/apprehension used earlier in the chapter, Hall and Van de Castle (1966, p. 111) state that the emotions in this category have "differences" that are "recognizable," but in all of them "the person feels apprehensive about the possibility of physical injury or punishment, or the possibility of social ridicule or rejection." Therefore, "the common denominator" is that "some potential danger exists," whether physical or psychological (Hall & Van de Castle, 1966, p. 111). It is also striking that all five emotions studied by the waking neurocognitive researchers—fear, anger, sadness, disgust, and happiness—are also included in the HVdC coding system for studying emotions in dream reports, although disgust is rare in dreams and is part of the Fear/Apprehension category.

Explaining the Infrequency of Emotions in Dreams through Neuroimaging Findings

Based on the waking neuroimaging findings, the task facing a neurocognitive theory of dreaming is to explain the quantitative findings on emotions in dream reports. Why are there no emotions in 59.3% of the 991 dream reports in the combined normative sample from 491 women and 500 men (as documented in table 8.1)? Why do 17% of dream reports in another study include no emotions, even though there very likely would be emotions in similar situations in waking life? The foregoing meta-analyses of waking neuroimaging studies provide a basis for plausible answers to these questions. Due to the relative deactivation of four of the five association networks, the neural substrate underlying dreaming appears to be very limited in its ability to support emotion states.

This point is perhaps best demonstrated in a study mentioned in chapter 2, which analyzed 20 separate datasets, both individually and collectively, using a "combination of techniques" (Uitermarkt et al., 2020, pp. 1–3). Consistent with a meta-analysis (Fox et al., 2013), as well as the findings in the neuroimaging study of high and low recallers (Eichenlaub et al., 2014), the "REM-activation networks" involved regions within the default network, secondary visual cortices, and sensorimotor regions. On the other hand, the "REM-deactivation network" included regions within the fronto-parietal control network as well as the frontal/right orbitofrontal cortices, the right anterior insula (which is a key substrate in suppressing the default network when need be during waking), and the right antero-ventral thalamus (Uitermarkt et al., 2020, p. 5 and supplementary table 3). (Readers may recall that the relative deactivation of these and other neural substrates during all stages of sleep is also presented in table 2.2 of this book.) These findings demonstrate that portions of the default network are activated at various times during sleep, but it is important to emphasize that it is insufficient for supporting the full range of waking cognitive capacities. More specifically for purposes of this discussion of the relative infrequency of emotions in dream reports, it is unlikely that the default network alone is able to support the full range of waking emotions.

The concept of cognitive insufficiency also may help account for the fact that young children and young adolescents dream even less often of emotions than do individuals over age 15. Several of their association networks, and in particular the default network, which is the only association network that has activated portions during dreaming, do not reaching adultlike-levels until ages 9–13. Young children do not have an autobiographical self until ages 5–7, and they do not reach full cognitive development in terms of imagination and narrative skills until the middle years of childhood and the early years of adolescence.

The neurocognitive theory of dreaming is also able to explain why the emotions in the dreams of adults are more negative than positive, as measured by the FAS%. It does so by drawing upon the concept of a "negativity bias" in thinking, as already discussed in chapter 3 as evidence for the continuity between the personal concerns expressed in both waking thought and dreaming (Baumeister, Bratslavsky, Finkenauer, & Vohs, 2001; Norris, 2021; Rozin & Royman, 2001). Moreover, the negativity bias in the HVdC normative samples is higher for emotions in dream reports than it is for other "negative" elements (aggressions, misfortunes, and failures) when they are compared to the combination of "positive" elements in dream reports (friendly acts, good fortunes, and successes). Women's negativity ratio for emotions in dream reports is 2.97, compared to only 1.40 for the ratio that makes use of aggressions, misfortunes, and failures. For men, the negativity ratio for emotions is almost has high as it is for women, 2.87, as compared to 1.65 for the ratio that involves aggression, misfortunes, and failures.

This analysis can be made more comparable to the self-rating studies discussed earlier in the chapter, which analyzed the percentage of dream reports with predominantly negative and predominantly positive emotions. For women in the HVdC normative sample, 40.9% of their dream reports had at least one negative emotion, compared to only 13.3% with at least one positive emotion, a negativity ratio of 2.96. For men, 26.4% of their dream reports had at least one negative emotion, compared to 9.2% with at least one positive emotion, a negativity ratio of 2.87, which is very close to the negativity ratio for women.

In terms of adding a developmental dimension that might explain the higher portion of positive emotions in the dream reports of children and young adolescents within the confines of the concept of a negativity bias, one plausible hypothesis, based on the dream findings, suggests that the negativity bias develops gradually and is not fully developed until middle adolescence. The waking negativity bias emerges at ages 5 to 7 in children, at about the same time as the autobiographical self develops, but it is not known when it becomes as strong as it is in adulthood (Haux, Engelmann, Herrmann, & Tomasello, 2017). The developmental trajectory of the negativity bias aside, there are two other plausible hypotheses that may explain the higher proportion of positive to negative emotions at younger ages. Perhaps children and young adolescents do not have as many enduring negative personal concerns as older adolescents and adults, or perhaps they do not have the cognitive capacities to imagine and articulate enduring personal concerns during dreaming. However, these possible explanations for the far lower FAS% in the dream reports of most individuals under the age 16–18 have yet to be tested.

The neurocognitive analysis of the infrequency of emotions during dreaming, whether in children or adults, may provide a plausible starting point for creating a translational model for understanding the strong and upsetting emotions experienced in nightmares, and in particular in the nightmares experienced by victims of PTSD. People who suffer from frightening nightmares, and especially those with PTSD, can be viewed as the victims of a cruel natural experiment in which sleep is flooded with greater activation in the attention and limbic networks. As briefly mentioned in chapter 2, the frightening dreams of those who report great suffering from their nightmares are correlated with the expansion of the neural substrates active during dreaming beyond their usual boundaries (see Levin & Nielsen, 2007, and Nielsen & Levin, 2007, for a somewhat similar neurocognitive view). Based on past EEG studies of nightmare sufferers and victims of PTSD, it is known that their brains are more activated during sleep (Germain, Jeffrey, Salvatore, Herringa, & Mammen, 2013; Marquis, Paquette, Blanchette-Carrière, Dumel, & Nielsen, 2017; Simor, Horváth, Ujma, Gombos, & Bodizs, 2013). This heightened activation in turn leads to atypical dream reports, with high levels of fear and anxiety (Robert & Zadra, 2014; Zadra & Donderi, 2000a; Zadra, Pilon, & Donderi, 2006).

As also briefly mentioned in chapter 2, a highly activated and vigilant brain in victims of PTSD, whether waking or sleeping, points directly to the salience/ventral network, which has been shown in the previous subsection to be active in the five emotion states that have been studied. This point is demonstrated in a waking fMRI study, which was based on 102 active-duty US Army soldiers, 50 who suffered from PTSD and 52 who did not. Each of them was studied during the resting state and in an experimental condition in which participant-specific script imagery was used to trigger the fear and anxiety that are hallmarks of PTSD. The salience/ventral networks of those who suffered from PTSD were more activated during the experimental condition than were those of the soldiers in the control group (Abdallah et al., 2019). The likely importance of an activated salience/ventral network as a source of nightmares during sleep is supported by EEG findings in an earlier study. Although this nighttime EEG study did not analyze the findings with regard to the salience/ventral network, it showed that the relevant brain regions are as highly activated in both REM and NREM sleep in nightmare sufferers as they are during waking (Marquis et al., 2017).

Conclusions and Implications

Due to cultural beliefs and earlier scientific understandings of what dreams are like, the empirical findings on the infrequency of emotions in dream reports do not fit with most people's expectations. Nor does it seem feasible within this cultural context that the dreaming brain may not be able to support the experiencing of emotions, except in unusual circumstances. These expectations also may be in part based on the Darwinian-derived assumptions about the similarities in the emotional experiences of most mammals, including humans, which are consistent with the intuitive sense that emotional feelings, as humans see and experience them, are deeply rooted in human nature and often are not controllable.

In addition, the clinical theories of dreaming, which came out of psychoanalytically oriented psychotherapy, also continue to influence cultural and scientific assumptions about emotions. Moreover, all too many people have suffered from one or more frightening nightmares, whether during a physical illness, the transition to a new medication, or when under great stress, to conceive of nightmares as atypical. Then, too, as shown in chapters 9 and 10, emotions figure prominently in both of the traditional comprehensive theories of dreaming and in two adaptation theories.

However, in contrast to the widespread emphasis on emotions as central to dreaming, the neurocognitive theory of dreaming suggests the neural substrates that enable dreaming do not have the capacity to support the cognitive abilities necessary to experience emotions as often and as fully as they are experienced in relevant situations in waking life. Of the five association networks involved in creating the conscious experience of fear, anger, sadness, disgust, and happiness during waking, only portions of the default network are activated at various times during sleep. The frequent absence of emotions during dreaming, as most dramatically seen in situations in dreams that would have led to emotional experiences in waking life, can therefore be explained within the context of the neurocognitive theory of dreaming by the concept of cognitive insufficiency.

Nor are nightmares a prototypical example of the importance of emotions in dreaming. They are the exception. A focus on clinical symptoms during times of illness, medication changes, or great stress once again leads away from the development of a foundational theory of dreaming. Instead, nightmares actually reveal the lack of sufficient support for emotions in everyday dreaming. Both EEG and fMRI studies of people who suffer from frequent nightmares demonstrate high levels of brain activation in brain areas that are not usually activated during sleep. Based on a waking neuro-imaging study of heightened activity in the salience/ventral network when experiencing PTSD symptoms, it is likely that the salience/ventral network remains highly activated in the sleep of PTSD victims as well (Abdallah et al., 2019). Within the context of a foundational theory of dreaming, such as the neurocognitive theory of dreaming attempts to provide, the findings on the brain activation levels of those who suffer from nightmares provide a good starting point for developing a translational model of this atypical syndrome, with the dorsal attention, salience/ventral, and limbic networks as the primary focus of attention.

9   An Assessment of Two Comprehensive Traditional Dream Theories

This chapter assesses the two most comprehensive, impactful, and widely known traditional theories of dreaming and dream content, the Freudian and Activation-Synthesis theories. Many dream researchers consider them passé and are critical of numerous aspects of both of them. However, some tenets of these theories, including their emphasis on unusual elements and emotions, continue to have an influence on the study of dreaming and dream content. These two theories, and especially Freudian theory, also continue to figure prominently in introductory psychology textbooks. Nor have the critics produced theories with the same scope. Instead, they are primarily focused on a variety of adaptive theories, which are discussed in detail in chapter 10.

The chapter begins with the oldest and most complete of the two theories, Freudian theory, which dates to the year 1900. It has been used and extended by psychoanalysts ever since, as well as by a few academic research psychologists into the 1990s. The chapter then turns to activation-synthesis theory, a product of the late 1970s, which has much to say about the nature of dreaming and about every aspect of dream content.

Freudian Dream Theory

Sigmund Freud's theory of dreams, as first presented in *The Interpretation of Dreams* (1900), proved to be the most impactful theory of dreams of the twentieth century, and it changed very little over that time span. Every theory in the first six decades of the century was an offshoot of it or a strong reaction against it. The functional aspect of the wide-ranging Freudian theory of dreaming begins with the assumption that dreams are infrequent and brief. They are said to be similar to "a firework that has been hours in the preparation, and then blazes up in a moment" (Freud, 1900, p. 377).

According to this theory, dreams occur because the sleeping brain is try-ing to cope with basic physiological urges by turning them into psychologi-cal wishes that temporarily and partially satisfy these urges by means of a hallucinatory image. This stratagem is said to be in the service of preserving a vital bodily necessity, namely, sleep. The theory thereby linked dreaming to survival itself by claiming that dreams "serve the purpose of continuing sleep instead of waking up" (Freud, 1900, p. 180). More metaphorically, *"the dream is the guardian of sleep, not its disturber"* (Freud, 1900, p. 180; emphasis in the original).

However, laboratory findings nearly 60 years later revealed that adult dreaming is frequent, appears regularly in REM sleep throughout the night, may continue for 20 minutes or more in the two or three hours before awakening, and increasingly occurs in NREM 2 sleep as the brain becomes more active toward morning (Dement & Kleitman, 1957a, 1957b; Pivik & Foulkes, 1968; Wamsley et al., 2007; Zimmerman, 1970). Dreaming there-fore cannot be triggered by episodic physiological needs represented in the mind as wishes. The theory is also directly refuted by laboratory studies of lobotomized schizophrenics, who slept soundly even though they did not report dreams from REM awakenings throughout the night (Jus, Jus, Gautier, et al., 1973; Jus, Jus, Villeneuve, et al., 1973). It is also called into serious question by the fact that preschool children rarely if ever dream and dream only infrequently from ages 5 to 9 (Foulkes, 1982, 1999, 2017; Foulkes et al., 1990), as discussed in chapter 6.

In the face of these strong findings, the hypothesis that dreams are the guardians of sleep has been defended by claiming dreaming may involve the "backward projection" of the impulses arising in the dopaminergic system (located in the basal forebrain) to the inferior parietal lobes and visual asso-ciation cortex (Solms, 1997). Other dream researchers quickly noted there is little or no evidence that such a mechanism is responsible for dreaming (Antrobus, 2000). More recently, this possibility has been refuted by the evidence that a subsystem of the default network is the neural substrate that enables dreaming (Domhoff, 2011; Domhoff & Fox, 2015; Eichenlaub, Bertrand, & Ruby, 2014; Fox, Nijeboer, Solomonova, Domhoff, & Christoff, 2013; Uitermarkt, Bruss, Hwang, & Boes, 2020), as discussed in chapter 2.

In addition, Solms (1997) defends the guardian-of-sleep hypothesis on the basis of patients who reported the cessation of dreaming. Their replies to a questionnaire suggested they had disrupted sleep more often than did a control sample. However, those findings are not impressive in that 51% of the 101 patients with global loss of dreaming indicated that their sleep was not disrupted (Solms, 1997, pp. 164–165; Solms & Turnbull, 2002, p. 214). If dreaming is an important evolutionary adaptation necessary to preserve sleep, then virtually all patients reporting the complete loss of dreaming also should report an inability to sleep at all. Moreover, the two studies of over a dozen lobotomized patients in a sleep lab found they almost never recalled a dream after REM awakenings. They had normal REM/NREM cycles. They reported in the morning they had slept well, a claim supported by the EEG records of their sleep (Jus, Jus, Gautier, et al., 1973; Jus, Jus, Villeneuve, et al., 1973).

Although Freud's adaptive theory has been discarded by most dream researchers, it was the general starting point for the development of most of the adaptive theories, all of which were first proposed by clinical theorists influenced, directly or indirectly, by Freudian ideas. This generalization includes theories having to do with problem-solving and emotional regulation (see Dallett, 1973, for a historical account). These Freudian-influenced adaptive theories are discussed in chapter 10.

On the psychological level, the theory begins by explaining how memories of events from the previous day have a role in the process of dreaming. They create a connection to a disturbing wish: "a reference to the events of the *day just past* is to be discovered in every dream" (Freud, 1900, p. 127; emphasis in the original). By linking memories of seemingly trivial external events of the previous day to important internal physiological urges such as hunger, thirst, and sex, the theory thereby synthesized the emphasis on either external stimuli or internal physiological events in the nineteenth century, as reviewed in an early chapter in *The Interpretation of Dreams* (Freud, 1900, pp. 22–42). However, five studies refuted this conclusion decades ago when they found that only 44–58% of dreams contain any sign of day residue, as identified by the dreamer (Botman & Crovitz, 1989; Harlow & Roll, 1992; Hartmann, 1968; Marquardt, Bonato, & Hoffmann, 1996; Nielsen & Powell, 1992).

Nor could researchers, who collected statements from their participants about the events and concerns of the day, find significant involvement of these events in the dream reports they subsequently collected in laboratory and nonlaboratory settings (Roussy, 1998; Roussy, Brunette, et al., 2000; Roussy et al., 1996). More generally, a very wide range of studies focused on the incorporation of numerous different types of stimuli and daily events have found no more than a few percent of such efforts are incorporated, even when the stimuli are very intense (Domhoff, 2017, pp. 25–33). Assessing the full range of laboratory studies of stimulus incorporation, which were carried out over the 35 years between 1958 and the early 1990s, one of the original dream researchers concluded: "Probably the most general conclusion to be reached from a wide variety of disparate stimuli employed, and analyses undertaken, is that dreams are relatively autonomous, or 'isolated,' mental phenomena, in that they are not readily susceptible to either induction or modification by immediate pre-sleep manipulation, at least those within the realm of possibility in ethical human experimentation" (Foulkes, 1996a, p. 614).

As part of his emphasis on the large role of specific memories in shaping dream content, Freud (1900, p. 266) said that all significant speech acts in dreams can be traced to memories of conversations heard or sentences read because "the dream-work is also incapable of creating speech." But an analysis of hundreds of speech acts in laboratory dream reports showed they are new constructions appropriate to the unfolding dream context, including the use by bilingual participants of the language relevant to the person to whom they are speaking in the dream (Foulkes et al., 1993; B. Meier, 1993). This Freudian assertion in relation to speech-acts, along with the claims about memories of the previous day as the instigator of all dreams, are relatively minor, but they are an indication of the degree to which the entire empirical basis of the theory is doubtful.

Turning to the more general theoretical level, Freud (1900, p. 106; emphasis in the original) claimed that wish-fulfillment "is the meaning of *each and every* dream." As one key piece of evidence for this claim, he claimed the dreams of young children are "invaluable as proof" of this theory because the wishes are clearly expressed (Freud, 1900, p. 102). This is a highly unlikely assertion about children's dreams if the findings from the longitudinal and cross-section studies of young children summarized in chapter 6 are taken seriously.

In the case of adults, on the other hand, Freud (1900, chaps. 4, 6) concluded that most of the wishes in their dreams are disguised by four cognitive processes: *condensation* (two or more wishes expressed with one element), *displacement* (the expression of a wish via an unlikely element), *regard for representability* (does the element fit reasonably in the narrative being constructed), and *secondary revision* (the "final editing," so to speak, before the dream becomes "manifest" in the dreaming mind). Together, these four processes comprise the dream-work, which "imposes a censorship on the wish and by this censoring distorts its expression" (Freud, 1900, p. 113). The dream-work is said to be "peculiar to the life of dreams" and "far more remote from the model of waking thought than even the most determined belittlers of the psyche's feats of dream-formation have thought" (Freud, 1900, pp. 328–329). Freud (1900, chap. 4) devoted an entire chapter to the "dream distortion" purportedly created by the dream-work, which assumes that "two psychical forces" are "the originators of dream-formation in the individual; one of these forms the wish uttered by the dream, while the other imposes a censorship on the dream wish and by this censoring distorts its expression" (Freud, 1900, p. 113).

The deciphering of these dreams has to be carried out through the use of a process called "free association" during a therapeutic session, which requires patients to let their minds flow from waking thought to waking thought, without any self-censorship, in reaction to the dream in general or to each element of the dream that is mentioned by the therapist. In a book presenting a comprehensive assessment of all aspects of Freudian theory, based on empirical studies, not simply on dreams, the coauthors conclude there is no reliable empirical evidence that free associations lead inevitably to the meaning contained in dreams (S. Fisher & Greenberg, 1977, p. 66). This conclusion was confirmed by a failed large-scale effort to use the method as a research tool (Foulkes, 1978), which convinced the investigator, on the basis of "extensive experience in association gathering," that free association suffers from an "inherent arbitrariness" (Foulkes, 1996a, p. 617).

In addition, a psychoanalytically oriented study that used dream reports, free associations, and psychotherapy transcripts to identify the central personal conflicts of one male and two female patients, found free associations did not add to what independent judges could infer about the patients, based on the dream reports alone (Popp, Lubarsky, & Crits, 1992). Similarly, a study by Freudian-oriented researchers, in which dream reports were collected from participants in a laboratory study and then compared with material from their psychoanalytic sessions and structured interviews, led them to conclude nothing further was added by the clinical materials, including free associations (Greenberg, Katz, Schwartz, & Pearlman, 1992).

Suggestion, Demand Characteristics, Cognitive Dissonance, and Persuasion in Therapeutic Settings

Moreover, and contrary to Freudian claims about the method of free association being free of any suggestive influence by the psychotherapist, there is evidence that it has grave shortcomings for research purposes, precisely because it is part of the psychotherapy process. Several possible confounds therefore may arise because of the many possible influences the therapist may have on the patient. These influences can be generally summarized as suggestion, demand characteristics, cognitive dissonance, and persuasion. The plausibility of applying these concepts to dream interpretations in a therapeutic setting was demonstrated in experimental studies in the context of simulated therapeutic relationships. In these studies, many of the unwitting participants gradually became convinced that one or another "planted" memory, which allegedly had been the basis for their dreams, had actually happened in their pasts (Mazzoni & Loftus, 1998; Mazzoni, Loftus, Seitz, & Lynn, 1999). The influence of these social-psychological processes also was demonstrated on the basis of interviews, transcripts, and court records in a case study of a dream-oriented therapeutic group and its clients. The group took on some of the trappings of a cult-like setting, which in turn led to revealing court cases (Ayella, 1998).

This line of analysis is supported by the fact that Freud (1900, pp. 114–119) often reported he had arguments with his patients when they denied his interpretations of their dreams as having wishful contents, which he further claimed were based on infantile desires and memories. He told them they were using the defense mechanism of "resistance" and that their therapy consisted in good part of understanding and overcoming this resistance. However, what Freud and later psychoanalysts understand as a need for their patients to overcome their "resistance" can be understood from the vantage point of social psychology as a process of persuasion and conversion. The patients' beliefs, based on their own memories, are set at odds with their respect for the expertise and healing powers of the therapist they hope will help them, and they often resolve the resulting cognitive dissonance by agreeing with the therapist.

This analysis does not demonstrate that all psychoanalytic sessions are exercises in suggestion, demand characteristics, cognitive dissonance, and persuasion. But it does mean the burden of proof is on Freudian researchers, when working in a therapeutic setting, to demonstrate the therapeutic data they use are not confounded by any of these four processes.

The Lack of Evidence for the Dream-Work as the Basis for Dream Content

Freud (1900, pp. 397–401) believed that the dream-work carries out its efforts to disguise unacceptable wishes at the behest of the cognitive process of repression. He also claimed the process of repression led to the forgetting of most dreams (Freud, 1900, pp. 337–338). But there is no experimental evidence for the existence of this process (Loftus, Joslyn, & Polage, 1998; Loftus & Ketcham, 1994; Otgaar, Howe, Patihis, Lilienfeld, & Loftus, 2019). In addition, research on the relationship between frequency of dream recall and various personality and cognitive variables casts doubt on the notion of any process of denial or self-censorship being involved in dream forgetting (D. Cohen, 1979; D. Cohen & Wolfe, 1973; Goodenough, 1991).

In an overall assessment of Freudian claims about the dream-work, based on studies by a wide range of researchers, two psychologists doubted its existence. Numerous studies of dream content outside a clinical setting found more information about the dreamers' lives and personal concerns than would be expected if the dream-work disguises the underlying wishes (S. Fisher & Greenberg, 1977, chap. 2). Put another way, the emphasis on the disguise function of the dream-work leads to the expectation that most, if not all dream content, is *not* continuous with waking thoughts. Recalling the findings from quantitative studies of lab and nonlab dream reports in chapters 3 and 4, virtually every finding about dream content can be understood as a refutation of Freudian theory.

Within the context of the emphasis on the dream-work, claims about emotions in dreams become of interest as well. On the one hand, Freud (1900, pp. 304, 305) thought that sleep itself may dampen emotions, but he did not think this dampening could be the "whole story." He further conjectured the dream-work "often mutes the tonality" of the emotions as well. In terms of assessing Freudian dream theory, however, the most important point is that Freud (1900, p. 299) concluded "there are plenty of dreams" in which an "intense expression of affect will make its appearance accompanying a content that does not seem to offer any occasion for the release of affect." In these instances, the dream-work has transformed the contents but the emotions remain unaltered. Contrary to this conclusion, the general findings since the 1960s suggest the relatively few emotions in dream reports are consistent with dream content when they do occur. This is best shown in the lab study focused solely on emotions in dreams (Foulkes et al., 1988). On the basis of judgments by each dreamer and assessments by independent judges, the study, as generally summarized in a theoretically oriented book, concluded that the emotions in dream reports are "overwhelmingly appropriate to the dream content" (Foulkes, 1999, p. 68).

The Inadequacies of Empirical Studies Attempting to Support Freudian Dream Theory

In earlier decades, the most consistent defense of Freud's ideas about the role of symbolism in dreams was based on evidence from studies using subliminal stimulation, which were already discussed and rejected in chapter 5 (e.g., C. Fisher, 1954). In the case of later studies by psychoanalysts, they often relied on the production of waking mental images and free associations provided to them by participants after laboratory awakenings from both REM and NREM sleep, so they have nothing directly to do with dreams (Shevrin, 1996, 2003; Slipp, 2000). And as Shevrin and Eiser (2000, p. 1006) state, these studies do not "establish the disguising function" of dreams, which is central to Freudian dream theory.

Building on the past studies by psychoanalysts, a Freudian-oriented research psychologist carried out a series of independent subliminal studies using updated methodologies, which presented new evidence for unconscious processes influencing thought through the presentation of psychodynamically relevant stimuli (Erdelyi, 1985, 1996). However, non-Freudian experimental psychologists familiar with the subliminal literature remained highly critical of Freudian studies on methodological grounds (Fudin, 2006; Kihlstrom, 2004, p. 97). These conclusions have been reinforced by neuroimaging findings. These studies demonstrate, to use the same useful summary quotation cited in chapter 5, that subliminal stimuli can register in the visual cortex but they do not have any impact on thinking, however small, unless they are "further processed by cognitive control networks underlying working memory" (LeDoux, 2019, p. 272). Therefore, researchers would have to demonstrate the subliminal stimuli shown to participants have registered in memory networks.

The main proponent of Freudian dream theory in more recent decades, Mark Solms (1997, 2002), who recast the theory as one of "neuro-psychoanalysis," believes there is support for Freud's wish fulfillment hypothesis. It is based on the alleged involvement of dopamine in REM sleep as its primary neurochemical. This is because dopamine is supportive of the "appetitive interests," which he believes are akin to Freud's concept of the libido (Solms, 2000, 2002; Solms & Turnbull, 2002, pp. 116, 312). However, the highly complex neurochemistry of REM sleep primarily involves cholinergic, glutamatergic, and GABAergic neurons (Boucetta, Cisse, Mainville, Morales, & Jones, 2014), with the role of dopamine still uncertain at best, and perhaps nonexistent (J. Siegel, 2017a, pp. 9–10, 12).

Based on his own studies of the effects of lobotomies and antipsychotic drugs on the ventral medial region in the prefrontal cortex (through which the dopamine reward system links the basal forebrain and the prefrontal cortex), as well as his analysis of the literature on antipsychotic drugs, this proponent of Freudian theory further concluded that "the functional anatomy of dreaming is almost identical to that of schizophrenic psychosis" (Solms & Turnbull, 2002, p. 213). However, as discussed in chapter 1, there is strong neuroimaging evidence showing the neural substrate supporting dreaming is not similar to those underlying hallucinations in psychotic patients (P. Allen et al., 2012; Ford et al., 2009; Waters et al., 2016).

With the exception of the usual case studies psychoanalysts and clinical psychologists have been producing for over 100 years, which continue to be problematic for testing the theory for a variety of methodological reasons, there are essentially no studies supporting Freudian dream theory at the neurocognitive, developmental, or content levels (Domhoff, 2003, pp. 136–143; S. Fisher & Greenberg, 1977; Mazzoni & Loftus, 1998; Mazzoni et al., 1999). More recently, due to the meager findings on symbolism in systematic studies of dream content, the theory faces new challenges because of its emphasis on symbolism (Freud, 1916, chap. 10; 1933, pp. 22–26). The emphasis on symbolism is also doubtful because neuroimaging studies have discovered several of the brain areas needed to generate symbolism are relatively deactivated during dreaming, as discussed in chapter 5.

The Activation-Synthesis Theory of Dreaming

Activation-synthesis theory claims the brainstem has a major role in generating dreaming and shaping dream content; cognitive processes are a secondary issue at best. The theory asserts that dreaming is initiated by putatively random and chaotic firings, which arise from the brainstem during REM sleep. Dreaming is a cortical attempt to make sense of these brainstem events, which extend to the occipital lobe as well (Hobson, 2000; Hobson, Pace-Schott, & Stickgold, 2000a, 2000b). As a result, activation-synthesis theory focuses on seemingly unusual features said to be unique to dreaming. They include such events as frequent occurrences of flying under one's own power. More specifically, the theory states "the individual historical components of dream plot construction" are "diluted" by "chaotic cerebral activation processes," which lead to "visuomotor hallucinations, delusional beliefs, thought impairments, emotional storms, and memory defects" (Hobson & Kahn, 2007, p. 857). Dreams are said to be "making the best of a bad job in producing even partially coherent dream imagery from the relatively noisy signals sent up to it from the brainstem" (Hobson & McCarley, 1977, p. 1347).

More generally, dreaming is said to be a hallucinatory and delusional state, which has the same formal structure as a psychosis: "psychosis is, by definition, a mental state characterized by hallucinations and/or delusions." Dreaming is therefore "as psychotic a state as we ever experience while awake" (Hobson, 2002, pp. 98–99). It is a form of "delirium," which is characterized by disorientation, visual hallucinations, illogical thinking, loss of recent memory, and confabulation (i.e., fabricating explanations and stories without the intention to deceive) (Hobson, 2002, pp. 23, 98–106). This suggests that it is "at least possible that dream content is as much dross as gold, as much cognitive trash as treasure, and as much informational noise as a signal of something" (Hobson, 2002, p. 101). However, as discussed in chapter 2, neuroimaging studies show hallucinations and dreaming are supported by different neural substrates (Ford et al., 2009; Hoffman & Hampson, 2012; Waters et al., 2016; Zmigrod, Garrison, Carr, & Simons, 2016).

The idea of dreaming as similar to psychosis leads to the claim that dream content cannot be studied scientifically. An approach that "begins with dream content" is "fraught with conceptual, methodological, and analytic problems." This assumption leads to the conclusion it is better to start with the "details of brain physiology of REM sleep in animals" and then make "inferences back up to the psychological level (of dreaming) in humans" (Hobson, 1988, pp. 157–158). However, this conclusion is contradicted by cross-national and cross-cultural differences, by replicated findings on the handful of gender differences, and by the wide individual differences in what people dream about. The findings on consistency in dream content over decades and the continuity of dream content with waking thoughts about people and avocations also refute this assertion.

The Putative Brainstem Basis for Dreaming

Activation-synthesis theory originally began by locating the origins of REM periods in giant neurons in the pontine gigantocellular tegmental field (Hobson & McCarley, 1977; Hobson, McCarley, & Wyzinski, 1975). It further claimed that the neurochemical initiation of the periodic episodes of REM sleep is cholinergic in nature (in the form of acetylcholine). Finally, it concluded that neurons in the locus coeruleus are responsible for ending episodes of REM sleep (Hobson, Lydic, & Baghdoyan, 1986, pp. 378–379; Hobson & McCarley, 1977). Although the fact is not often remarked upon in the literature focused exclusively on dreaming and dream content, all of these claims were refuted within the space of a few years by those sleep researchers and physiologists who specialize in the study of REM sleep.

Experimental lesion studies soon showed the activation of the giant cells in the pons was not specific to REM sleep. Instead, these neurons are related to movement, as indicated by their very high levels of activity in waking. Activation-synthesis theorists missed this relationship because they used a head restraint while making their recordings (J. Siegel & McGinty, 1977; J. Siegel, McGinty, & Breedlove, 1977). Nor did the destruction of the entire gigantocellular tegmental region have any effect on REM sleep (Drucker-Colin & Pedrazo, 1981; Friedman & Jones, 1984a, 1984b; B. E. Jones, 1979; Sakai, Petitjean, & Jouvet, 1976; Sastre, Sakai, & Jouvet, 1981; Vertes, 1977, 1979). Moreover, "histochemical and pharmacological data" showed the neurochemistry of REM was not cholinergic in nature (B. E. Jones, 1986, p. 410). Still another surgical lesion study demonstrated the neurons in the locus coeruleus do not play any essential role in the cyclic appearance of REM and NREM sleep (B. E. Jones, Harper, & Halaris, 1977). The new evidence was succinctly synthesized as follows: *"The lateral pons is the brain region critical for REM sleep.* Medial pontine regions, including the gigantocellular tegmental field neurons, are not critical for REM sleep generation . . . the neurons whose interaction is critical for REM sleep constitute only a small percentage of the cells within the lateral pontine region" (J. Siegel & McGinty, 1986, p. 421; emphasis in the original).

Physiologist Barbara E. Jones (2000, p. 956) later concluded that she did "not know of any physiological evidence that the cortex has no control over the brainstem or over the central activity of dreams." She added, "corticofugal outputs reach the entire brainstem as well as the spinal cord, influencing the very neurons shown to be critical for the initiation and maintenance of REM sleep in the pontine reticular formation." Subsequent evidence confirmed the large degree of forebrain control of the REM generator, especially through the hypothalamus (Luppi, Clement, & Fort, 2013). The highly complex nature of this system was demonstrated in a study briefly mentioned in the discussion of Freudian dream theory. Using new methods of histochemical neuronal identification, it discovered cholinergic, glutamatergic, and GABAergic types of neurons are active in varying degrees within three functional subgroups in the part of the brainstem involved in the production of REM sleep. In addition, many cells previously thought to be cholinergic because of their discharge activity were found to be either GABAergic or glutamatergic through definitive histochemical identification (Boucetta et al., 2014). The sole emphasis on the cholinergic neurotransmitter acetylcholine in activation-synthesis theory thereby proved once again to be misguided.

One of the two co-creators of activation-synthesis theory later conceded "the apparent REM discharge selectivity of the large paramedian reticular cells was a function of the head restraint required to allow long-term recording." He also agreed that "the original attribution of cholinergic neuromodulation to the REM-on cells of the pontine reticular formation was incorrect" (Hobson, 2007, p. 75). However, even though some alterations were made in the neural foundations of their theory, activation-synthesis theorists continued to claim that dreaming is hallucinatory and delusional, and contains highly unusual content (Hobson, 2007; Hobson & Kahn, 2007, p. 857).

The Rejection of Developmental Studies of Dreaming

As activation-synthesis theorists fully recognized around the turn of the century, every aspect of their theory is challenged by the longitudinal and cross-sectional laboratory studies of children (Foulkes, 1982; Foulkes et al., 1990). In response, they asserted they could imagine dreaming even in neonates: "Similarly, we specifically suggest that the human neonate, spending as it does more than 50% of its time in REM sleep, is having indescribable but nevertheless real oneiric experiences. . . . For us, it is not at all difficult to imagine that an infant might be experiencing hallucino-sis, emotions, and fictive kinesthetic sensations during REM sleep" (Hob-son et al., 2000b, p. 803). This claim ignores the likelihood that dreaming depends upon the same cognitive capacities (concept formation, mental imagery, narrative skills, imagination, and an autobiographical self) mak-ing daydreams and many other waking cognitive processes possible. None of the necessary cognitive capabilities, with the exception of the ability to create concepts, has been found to be very well developed until ages 5–8, as discussed in chapter 6.

The strong claims by the activation-synthesis theorists are based in part on their analysis of a word-of-mouth sample, which they gathered from friends and friends-of-friends in the Boston/Cape Cod area. It consisted of eight children ages 4–5 (three girls, five boys) and six children ages 8–10 (one girl, five boys) (Resnick, Stickgold, Rittenhouse, & Hobson, 1994, p. 32). The dream reports were collected on 13 consecutive days by the parents of the young participants. The collection period began with reports after spontaneous awakenings for five mornings. Then, for the next five nights, the children had to repeat, "'I will remember my dreams' three times out loud just before going to sleep" (Resnick et al., 1994, p. 33). The next morning, they were awakened 15 minutes before their usual time of waking. For those five mornings, "parents were explicitly instructed to elicit as much detail as possible by guiding their children through the *who, what, when, where, why* questions and thereafter to base their questions on the children's reports" (Resnick et al., 1994, p. 33; emphasis in the original). However, the parents were also "reminded that it was important not to ask leading questions and to wait until the child finished a statement before asking another question."

In the final phase, the morning awakenings were supplemented through awakenings by the parents after three and five hours of sleep. These night awakenings led to very few dream reports. They were not included in the analysis because "in many of these attempts the parent was unable to rouse the child sufficiently to get any response and hence reflect a failure to awaken rather than a failure to recall" (Resnick et al., 1994, p. 40). By assuming that the children could not be sufficiently awakened, the activation-synthesis theorists ignored the possibility that the children had no dreams to report, a possibility demonstrated by the low frequency of recall in the longitudinal and cross-sectional laboratory studies of dreaming in young children (Foulkes, 1982; Foulkes et al., 1990).

Based on this questionable set of procedures, dreams were reported after 85 of 131 morning awakenings (64.9%), which is well above the lab figures until ages 9–11 (Foulkes, 1982; Foulkes et al., 1990). The investigators then concluded the morning dream reports they collected revealed "remarkable similarities to those of adults in terms of length, number of characters and settings, and the presence of dream bizarreness" (Resnick et al., 1994, p. 43). Once again turning to the findings from the longitudinal and cross-sectional studies of children, which showed girls have higher rates of recall than boys, the results in the Boston/Cape Cod study are all the more unlikely because 10 of the 14 participants were boys (71.4%).

Numerous waking experimental studies of children, which examine the influence of the nature of the questions on children's responses, demonstrate the type of questioning used in the activation-synthesis theorists' study invariably leads to compliant answers, which add up to demand characteristics (Ceci, Bruck, & Battin, 2000; Foley, 2013; see Lamb, Hershkowitz, Orbach, & Esplin, 2008, pp. 50–57, for a summary). This is especially the case when the same questions are repeated. Thus, the likely confounds and strong demand characteristics raise doubts about the results, despite the warning to parents about asking leading questions. These doubts begin with the fact that the sample was in effect primed to please the investigators and the parents of the participants. The reasons for doubts also include the social persuasion and social pressure implied by repeated parental questioning. The implicit expectation they would do a good job of remembering and reporting their dreams was reinforced on five of the days by the admonition to improve their recall through the pre-sleep repetition of a vow to remember their dreams.

These findings were not replicated in the best-controlled nonlab study of children's dreams. It, too, used morning awakenings by parents but reported very different results (Sandor, Szakadát, Kertész, & Bódizs, 2015). The study was based on a six-week cross-sectional study of 19 boys and 21 girls, ages 3 to 8.5. Based on 1,680 awakenings, which were carried out over the space of 42 mornings, the researchers first found a mean recall rate of only 21%. The even lower median recall rate was 15.5%. These figures are very similar to the results in the longitudinal and cross-sectional lab studies discussed in chapter 6. They are also similar to findings in the longitudinal studies in terms of their low frequencies of friendly and aggressive interactions (Sandor et al., 2015). The main differences occurred because the investigators had the parents ask several leading questions at the end of each of the morning interviews, such as "Did you see your dream as a motion picture or was it rather like a photo," and "Did you feel for example angry, sad, happy, surprised or scared or were you just calm?" (Sandor et al., 2015, pp. 11, 13). As noted in the previous paragraph, repeated questioning of this nature very often leads to the answers the children think the questioners want to hear from them.

Due to all the problems inherent in the Boston/Cape Cod study and the inability of independent researchers to replicate it, the conclusions drawn by the activation-synthesis theorists are not credible.

Bizarreness in Large Group Samples of Dream Reports

Activation-synthesis theorists focus on the alleged unusual features of dream content, such as sudden scene shifts and a purported sense of constant movement. Along with many other aspects of dream content, unusual features and sudden scene shifts are classified by activation-synthesis theorists as "bizarre" (Hobson & Kahn, 2007, p. 857). However, the other cofounder of the theory had found very few of the elements in dream content considered to be bizarre by activation-synthesis theory fully 25 years earlier (McCarley & Hoffman, 1981). As recounted in chapter 5, his coauthored study was based on 104 REM dream reports, obtained from a University of Cincinnati sleep-dream lab. Readers may recall that putatively bizarre instances, such as the sudden appearance or disappearance of a character, occurred in only 2.9% of the dream reports, and abrupt shifts in time in even fewer, 1%. There were no sudden appearances or disappearances of any objects. Flying under one's own power occurred in only 2% of the dream reports, and unusual environmental combinations were found in 17.3% (McCarley & Hoffman, 1981, pp. 904, 908, and figs. 4 and 6).

From the outset, the alleged bizarre features in dreams were said to be the result of the episodic firings originating in the brainstem during REM periods, which are also called "phasic events" (Hobson & McCarley,

1977). However, in the 1960s and early 1970s, laboratory dream researchers already had tried and failed to link a range of phasic activities to dream content through immediate awakenings at the moment these signs appeared (Antrobus, Kondo, & Reinsel, 1995; Foulkes, 1985; Pivik, 1986, 1991; Reinsel, Antrobus, & Wollman, 1985). Despite the general consensus on the futility of the search for causal relationships between phasic events and dream content, which was largely abandoned by the 1980s, activation-synthesis theorists nonetheless continued to characterize the relationship between phasic events in REM and the nature of dream content as "weak but consistently positive" (Hobson et al., 2000b, p. 799). They also continued to assume that the putative "chaotic cerebral activation processes" during REM sleep led to "visuomotor hallucinations, delusional beliefs, thought impairments, emotional storms, and memory defects" (Hobson & Kahn, 2007, p. 857). They did so without carrying out any laboratory research on this alleged relationship.

The emphasis on sudden scene shifts in dream reports serves as the main empirical basis for these continuing assertions about alleged bizarreness in dreaming. However, the claims about scene shifts were shown to be incorrect in the early 1990s. As documented in chapter 2, in the section concerning brief episodes of dreaming during drifting waking thought, one lab-based study found that there were more scene shifts in a sample of waking reports (Reinsel et al., 1992, pp. 169–170, 173). Simply stated, the studies by the activation-synthesis theorists lack the necessary control group (drifting waking thought) for any assertions about sudden scene shifts in dreams to be taken seriously.

Based on these and other findings, one of the coauthors of the necessary daytime control study, which compared REM reports with samples of drifting waking thought, criticized activation-synthesis theorists for continuing to "attribute bizarre cognition to chaotic pontine activation despite the fact that no experiments have supported this association." He further noted that he and his coworkers had reported instances in which there were bizarre dream elements when "phasic activity is minimal" and other instances in which there is no unusual dream imagery when there is high phasic activity (Antrobus, 2000, pp. 905–906; Antrobus et al., 1995). As Antrobus (2000, pp. 905–906) concluded, activation-synthesis theorists' "assumptions about how the pons determines the features of dreaming are completely without empirical support."

Emotions in Dream Reports

The original statement of activation-synthesis theory did not put a strong emphasis on emotions. It frankly stated that the theory "cannot yet account for the emotional aspects of the dream experience, but we assume they are produced by the activation of brain regions subserving affect in parallel with the activation of the better-known sensorimotor pathways." Further, these brain regions are "considered to be a part of the forebrain," and "strong emotion may or may not be associated" with the "distinctive formal properties of the dream" (Hobson & McCarley, 1977, p. 1336). Since neuroimaging studies show that most of the brain regions subserving emotions are relatively deactivated during sleep, including REM sleep, the basic underlying assumption underlying the activation-synthesis theory of emotions during dreaming has been refuted.

This relative lack of emphasis on emotions in the original version of the theory was in effect supported in the study by one of its cofounders, who joined with a coworker to code 104 REM dream reports for emotions, which were obtained from the sleep-dream lab at the University of Cincinnati (McCarley & Hoffman, 1981). As readers may recall, this study found little evidence for emotions in dream reports, as already discussed in chapter 8, so its results can be briefly summarized here. Using their own coding categories for anxiety, anger, sadness, and joy/happiness, the researchers found that at best 38% of the dream reports included at least one instance of any emotion and that they were primarily negative (McCarley & Hoffman, 1981, p. 908, and fig. 3).

However, based on a nonlab study a little over a decade later, other activation-synthesis theorists claimed there was "underreporting" in that study (Merritt, Stickgold, Pace-Schott, Williams, & Hobson, 1994, p. 46). This conclusion concerning underreporting is based on a study of 200 dream reports, which one of the activation-synthesis theorists had collected from students in his course on The Biopsychology of Waking, Sleeping and Dreaming, sponsored by the Harvard Extension School. Students were asked to keep a dream diary "to encourage active participation" in the material of the course (Williams, Merritt, Rittenhouse, & Hobson, 1992, p. 173). Whenever a dream was recalled during their daily lives, the 12 student participants were asked to write a report of it on a standardized report form. After documenting their reports, the participants were instructed to self-code each report on a line-by-line basis. They did so through the use of a list of eight emotion categories, which was included on the same sheet on which they wrote the dream report: anxiety/fear, anger, joy/elation, shame, sadness, surprise, and affection/eroticism. In addition, there was a category labeled "other." This open-ended category made it possible for participants to add any additional emotions they may have experienced.

These independently developed emotion categories are very similar to those in the Hall/Van de Castle (HVdC) coding system. However, the HVdC system includes shame in the Fear/Apprehension category, which is roughly the same as the "anxiety/fear" category in this study. In addition, the "joy/elation" category seems to parallel the HVdC category for Happiness/Joy, and the affection/eroticism category may partially overlap with Happiness/Joy. Surprise, as noted in chapters 3 and 8, was once part of the HVdC Confusion category but has subsequently been redefined as a cognitive category concerned with cognitive appraisals.

The 200 dream reports each contained 50 or more words. Emotions were found to be present in 95% of the reports (Merritt et al., 1994, pp. 46, 55). This figure is 2.5 times higher than what was found in the above study based on REM dream reports and coded by independent coders. However, this extremely high figure is very close to those reported in studies based on self-ratings by participants, as discussed in chapter 8 (Röver & Schredl, 2017; Schredl & Doll, 1998; Sikka et al., 2017; Sikka et al., 2014). Sixty-eight percent of the emotions were negative in the nonlab dreams from the 12 students, which is similar to what is found in most studies using independent coders or raters (Merritt et al., 1994, p. 56).

A third activation-synthesis study of emotions is based on a unique "lab/ nonlab" setting. Each of the nine participants, seven women and two men, was studied individually over three consecutive nights in her or his own bedroom at home in Oslo. The lead researcher in the study, a native of Norway, sat in a nearby room, using a portable polysomnograph to monitor their sleep stages and carry out awakenings during REM periods (Fosse, Stickgold, & Hobson, 2001). The study thereby combined aspects of lab and nonlab studies, and very likely overcame any putative biases concerning emotions that might be present in one setting or the other. After awakenings, which resulted in a total of 88 REM dream reports overall, the participants were asked to write their accounts of their dreams on a standardized report form. After documenting their reports, the participants coded each report line-by-line for any emotions they could identify. These self-codings were carried out on the basis of the same eight categories used in the above study (Fosse et al., 2001, p. 1).

Twenty-six percent of the reports in this study had no emotional elements, despite the immediacy of recall in a comfortable and familiar home setting (Fosse et al., 2001, pp. 1–2). The study also reported negative emotions were more prevalent than positive emotions in the majority of the dream reports (44% vs. 33%), with another 23% rated as neutral. Since the definition of positive emotions included affection/eroticism content, as well as "interest" and "motivation," and the definition of negative emotions was somewhat expanded as well, it is not possible to make exact comparisons with the HVdC findings (Fosse et al., 2001, pp. 2–3). However, even though this broadening of the number of categories leads to caution in assessing the *degree* to which negative emotions predominated in this study, the fact that this study broadened the categories of emotions makes the lack of any emotion in 26% of the dream reports all the more striking. It does not fit with the activation-synthesis theorists' nonlab study, based on the same list of emotions, nor with their post-1990 emphasis on the importance of emotions in dreams (Merritt et al., 1994).

Activation-synthesis theorists also carried out a fourth study of emotions. They rely on it exclusively in their later discussions of the theory (e.g., Hobson & Kahn, 2007). For this fourth study, the dream reports were collected in 1999 "as a graded class exercise" in the course on the Biopsychology of Waking, Sleeping and Dreaming at the Harvard Extension School. This requirement was included in order to "motivate students to comply with the complex instructions." However, the students were also reassured they "would be given a set of dream reports if they could not recall their own dreams." The mention of this alternative set of dream reports was meant "to discourage undesirable demand characteristics" (Kahn, Pace-Schott, & Hobson, 2002, p. 35). Thirty-five students provided 320 dream reports. Using the same coding categories that activation-synthesis theorists had been using since their 1994 study, the students self-coded their dream reports as including far more emotions than was the case in the activation-synthesis theorists' lab/nonlab study the year before (Fosse et al., 2001).

Somewhat more specifically, emotions were "almost always evoked" (Kahn et al., 2002, p. 34). In this study, however, the emotions self-codings were used in a more specific way. They were categorized in relation to the different types of characters appearing in the dream reports. In dream reports including characters known to the dreamer, "feelings were reported 81.4% of the time," compared to 69.3% in dream reports with generic or unknown characters (Kahn et al., 2002, pp. 40–41). In this study, unlike past studies by activation-synthesis theorists, the most frequent emotions were positive ones, caring (18.2%) and joy (14.2%), "followed by anger (11.6%), and anxiety (9.7%)" (Kahn et al., 2002, p. 45).

The overall findings on emotions by activation-synthesis theorists are not useful for five reasons, the first three of which overlap to some extent. First, their own findings are not consistent over the four studies. Second, they ignore this inconsistency by not making use of their two studies reporting lower frequencies of emotions in dream reports (Fosse et al., 2001; McCarley & Hoffman, 1981).

Third, the two studies they foreground are based on dream reports collected from students as one of the requirements for a passing grade in a course taught by one or another activation-synthesis theorist. This one fact alone suggests there are strong demand characteristics built into their nonlab studies. Fourth, the extremely high percentages of dream reports with at least one emotion in the classroom-based studies are based on self-codings by the dreamers themselves, which usually lead to much higher frequencies of emotions than do studies using independent coders, as discussed in chapter 8. It is therefore extremely likely that the findings by the activation-synthesis theorists are compromised by a wake-state bias as well as by demand characteristics. Fifth, and finally, the neuroimaging evidence discussed in chapter 8, which shows that only one of the association networks involved in supporting waking emotions is activated at certain times during sleep, raises major doubts as to whether this truncated neural network can support the high percentage of emotions that activation-synthesis-theorists insist are an important dimension of dreaming.

For all these reasons, any claims about emotions in dreams by activation-synthesis theorists cannot be taken at face value.

Summarizing the Criticisms of Activation-Synthesis Theory

The neurophysiological underpinnings of activation-synthesis theory have been convincingly refuted by neurophysiologists, and in particular by those researchers who focus on REM sleep. The alleged effects of phasic events on dream content have been proven wrong by numerous studies by other dream researchers. Then, too, the activation-synthesis theorists' one study of children proved to be inadequate on methodological grounds. The study of children was also called into question by contradictory findings in both lab and nonlab studies by other researchers. The activation-synthesis theorists' studies of adult dream content also have been met with skepticism on methodological grounds. These studies are directly challenged due to the ways in which they were conducted and also because better-controlled studies by other dream researchers led to different findings. To take the two most important examples, there is not as much bizarreness and emotion in dreams as activation-synthesis theorists claim.

Conclusions and Implications

Neither of the two well-known traditional comprehensive theories of dreams discussed in this chapter stands up well in the light of the wide range of empirical findings accumulated over several decades. Most, but not all, of their own studies have been criticized on methodological grounds as well. Moreover, Freudians have done studies casting doubt on some of the studies by other Freudians, and the activation-synthesis theorists have produced contradictory results on bizarreness and emotions, as well as other findings rejected by most dream researchers. There is no longer any reason to believe that either of these theories provides a sound basis for future theorizing. Nevertheless, some of the legacies of both theories are found in several of the adaptive theories of dreaming discussed in the next chapter.

Based in large part on four different types of descriptive empirical findings, which are inconsistent to varying degrees with every theory of dream function that has been proposed, this chapter argues that dreaming has no adaptive function. In addition, still other empirical findings refute specific aspects of one or another adaptive theory. The four general findings are already familiar to readers because they were discussed in previous chapters, so they can be briefly summarized for the purposes of this chapter's introductory section.

First, the extremely infrequent recall of the several dreams during sleep onset and at night seems to be much lower than would be expected if dream content contains useful information people need to consciously know in waking life. This result alone has contributed strongly to the promulgation of new adaptive theories since the late 1990s, which claim that dreaming has an impact on neurocognitive processes and/or waking behavior without any need for recall. Second, the likely absence of dreaming during the preschool years, and its infrequency and lack of complexity until ages 9–11, seem to limit any adaptive function to the adolescent and adult years. However, most adaptive theories, including those not requiring recall, currently assume there is dreaming at an early age, or by age 6 at the latest.

Third, the lesion studies, which show that people can lose dreaming without losing necessary cognitive functions, provide a strong challenge to most adaptational theories. Fourth, the consistency of dream content over months, years, and decades, and the fact that dream content changes only gradually when it does change, calls into question any adaptive theory claiming that dreams deal with new problems as they arise or help people to prepare for the future.

These numerous empirical findings provide the main basis for assessing the three most widely discussed general types of theories of dream function. These three general types of theories focus on (1) solving specific problems, (2) helping with the regulation or assimilation of emotions in waking life, and (3) mastering waking situations. The third of those categories, "mastery," is a very general one. It encompasses Freudian-derived theories, social-rehearsal theories not involving waking recall, and assertions about dreaming during REM sleep having a role in consolidating the brain's predictive-coding capabilities.

Although there are many reasons to reject all of the adaptive theories, there are good reasons to suggest dreaming is a type of "nonadaptation." According to evolutionary biologists, nonadaptations include genetic drift and preadaptations, which persist because they do not interfere with reproductive success. In the case of dreaming, it is more specifically a "by-product," which is a type of nonadaptation that is an unintended outcome of one or more adaptations. From a neurocognitive point of view, dreaming is a by-product of the evolutionary selection for the default network, which supports the imaginative capacities making it possible to rethink the past, imagine possible alternatives in new situations as they arise, and plan for the future.

Even though dreaming is most likely a by-product, there is an abundance of evidence showing human beings in virtually all cultures have used their waking imaginations to invent uses for dreams (Benjamin, 2014; D'Andrade, 1961; Gregor, 1981). Dreaming thereby has emergent cultural uses in a variety of contexts. Due to age-old beliefs about a link between dreams and the world of spirits, the most important of these invented functions involve religious and healing ceremonies (Bulkeley, 2008; Tylor, 1871/1958). In particular, dreams may become extremely important in times of cultural crisis, when cultural heroes claim that a dream or a dreamlike state was an important basis for the new religious beliefs they propound (LaBarre, 1972). In this way, the private world of an individual's dreams may lead to new cultural practices. These new practices may help to bring about group solidarity and also may aid individuals in coping with anxiety and other negative emotions. The uses of dreams are explored further after the several theories of dream function are discussed in detail.

The Problem-Solving Theory of Dream Function

The general idea that dreams sometimes provide solutions to unresolved problems was put forth by several early advocates of Freudian dream theory, who had come to believe the adaptive dimension of the theory is too narrow (Dallett, 1973, p. 409). In their view, dreams help people to cope with their personal problems. This idea was developed and elaborated upon by a psychoanalyst who was also a sleep-laboratory dream researcher (Ullman, 1959). Problem-solving theories were carried forward in slightly different forms by non-Freudian dream researchers through experimental studies (e.g., Cartwright, 1977, 1996; Cartwright, Agargun, Kirkby, & Friedman, 2006).

All versions of the problem-solving theory are called into question by several different empirical findings. First, the truncated neural substrate underlying dreaming may be limited in its ability to support many forms of waking thought, as discussed in chapter 2 and as demonstrated in the case of figurative thinking in chapter 5. Second, most people recall only a very small percentage of their dreams, which seemingly would be recalled more frequently if they were important in dealing with waking-life problems. Third, the small percentage of recalled dreams rarely contain even a hint of a solution to a problem, so the theory in effect ignores the extremely large number of instances it cannot explain.

In one of the largest and most carefully done studies, 76 college students were asked to choose a problem they hoped to resolve and then to write down any dreams they had about the problem within the next week or longer if necessary (D. Barrett, 1993). Only half of the participants recalled a dream they felt related to the problem, most of which simply enacted relationship dilemmas or educational/vocational decisions and did not seem to offer any solutions. There were only two instances in which both the dreamer and two independent judges agreed the dream reports contained the problem and offered a plausible solution. The language used in the written reports about those two dreams strongly suggests the possible solutions to their problems dawned on the participants as they were writing down and reflecting upon their dreams (Domhoff, 2003, pp. 160–161). Since conscious attention is usually if not always needed for problem solving, it is more likely that any realizations about a possible solution to a problem emerge while a person is thinking about the dream in the waking state. This point has been convincingly argued and demonstrated in an insightful critique of problem-solving theory by a cognitively trained research psychologist, who focuses on dreams (Blagrove, 1992, 1996, 2000; Blagrove, Edwards, van Rijn, Reid, & Malinowski, 2019).

The findings from studies of dream series, as overviewed in chapter 4, also present major problems for the problem-solving theories of dreams. As these studies demonstrate, most of the dream reports in the most recent years of a dream series dwell upon the same personal concerns appearing earlier in the series. Changes in dream content happen only gradually over time in a dream series, if they occur at all. This point is perhaps best made as far as problem-solving theories through the results at the end of chapter 7 regarding the Izzy and Jasmine dream series. Readers may recall those two dream series begin in the teenage years and stretch into early adulthood, a time period when individuals are facing new problems and undergoing many changes. Both series reveal the continuing appearance of the same people and avocations throughout. The findings with dream series also reinforce another conclusion, based on studies of individuals who are participants in sleep-dream lab studies or who make daily reports: dreams only rarely deal with actual daily waking events of any type (Domhoff, 2017, pp. 25–33; Foulkes, 1996a). This issue is also briefly overviewed in the critique of Freudian dream theory in chapter 9.

As a result of the doubts about the systematic studies relating to the problem-solving theory of dreams, the best evidence for it consists primarily of anecdotal examples concerning the writing of new poems, the creation of new story plots, and the emergence of new scientific insights. However, it is one thing for dreams to provide inspiration for poetry, stories, films, or musical creations, but it is another for dreams to contain solutions to problems confronting the dreamers. Even while acknowledging that recalled dreams may inspire waking creativity, the alleged dream may not have been a dream in some of the best-known examples. In one such instance, the new ideas also involved using opium before taking a nap, which led one wary investigator to do a detailed historical study that caused her to doubt every aspect of the story (E. Schneider, 1953). Another instance, based on a drug-induced state, may not have involved much sleep (I. Bell, 1992). In another, the creative thoughts were reveries while nodding off to sleep in front of the fireplace in the early evening. Moreover, the dreamer had been thinking intensely during the day about the theoretical problem in chemistry he had been trying to solve for at least a few weeks (Rudofsky & Wotiz, 1988). In examples such as that one, the drifting waking thought during the sleep-onset process may or may not have reached the point at which it could be considered "dreaming," as shown in detailed studies of the sleep-onset process that are discussed in chapter 2 (Foulkes et al., 1966; Foulkes & Vogel, 1965). Still other anecdotal examples are secondhand accounts from decades after the time of the alleged dream. They turned out to be inaccurate or else the reports are of dubious validity because no text of the alleged dream from many years earlier can be found (Baylor, 2001, pp. 89–90).

Beyond these famous instances, though, there are more recent examples of dreams providing intellectual and artistic inspiration (D. Barrett, 2001, 2017). In an interview study of several present-day computer programmers and physicists, they reported that they gained new insights from dreams while working intensely on their projects (D. Barrett, 2001, pp. 102–103, 107–109). However, it cannot be ruled out that the insights might have arisen in a drowsy state, mind-wandering during NREM 2 late in the sleep period, or transitioning into waking. It would therefore be necessary to carry out well-controlled studies and replicate them. Replication studies are especially necessary in these cases because any problem solving relating to dreams may occur while the person is thinking about the dream (Blagrove, 1992, 1996).

Emotions-Related Theories of the Adaptive Function of Dreaming

In still another variant to Freud's (1900) conclusion about the adaptive function of dreaming, a clinical psychologist suggested the dream state has a number of advantages over waking consciousness for dealing with emotional material (Breger, 1969, pp. 409–410). Based in part on his work with children, this theorist claimed that associational processes may be more fluid and that stored information may be more readily available during dreaming. Moreover, the pressures toward social acceptability may be at a minimum, so the overall context of dreaming may creatively open up memory systems. Finally, emotionally arousing situations during the day might be more readily compared during dreaming with strategies used previously in such situations (Breger, 1967, 1969). In a similar fashion, two psychoanalysts suggested that REM dreaming may provide the context for emotion regulation, by bringing together new and past experiences in order to facilitate new adaptive solutions (Greenberg, Katz, Schwartz, & Pearlman, 1992; Greenberg

& Pearlman, 1993; Greenberg, Pearlman, & Gampel, 1972). As part of their effort, they did experimental studies with rodents, which suggested that REM sleep facilitated information processing and memory consolidation (Pearlman & Becker, 1974; Pearlman & Greenberg, 1973).

The Memory-Consolidation Theory of Emotional Regulation

Although the Freudian-derived theories were largely abandoned, a new version of an emotions-regulation theory of dreaming was created at the turn of the twenty-first century by activation-synthesis theorists (Stickgold, 1998; Stickgold, James, & Hobson, 2000; Stickgold, Scott, Rittenhouse, & Hobson, 1999). This theory claims memory consolidation includes the regulation of emotions in the context of dreaming during REM periods. As with the Freudian-oriented emotions-regulation theorists, the new version built in part on studies of rodents. Some of the animal studies involved depriving rodents of REM sleep. The deprivation of REM sleep in these studies was accomplished by placing the rodents on a small pedestal, from which they would fall into water if they went into REM sleep (due to the muscle atonia accompanying REM sleep) (Carlyle Smith, 1985). These studies were heavily criticized because the stress caused by this procedure very likely confounded the results in a major way (Horne & McGrath, 1984; Vertes, 1995; Vertes & Eastman, 2000). The other and more lasting animal basis for this line of theorizing focused on the tracing of hippocampal brain patterns in rodents during NREM slow-wave sleep. Several studies found these brain patterns to be similar to the brain waves recorded during an earlier spatial behavioral task in the waking state (Wilson & McNaughton, 1994).

At the human level, this new version of the emotions-regulation theory claims that memories of daytime experiences are reprocessed and consolidated during dreams in REM. This assumption is supported by evidence of various changes in REM dream reports after participants view either an emotionally arousing picture, such as one of a wrecked automobile, or a more neutral picture. Dreaming is said to be "simply the conscious perception of the stream of images, thoughts, and feelings evoked in the brain by one or more of the many forms of off-line learning and memory processing that occur during sleep" (Stickgold & Wamsley, 2017, p. 514). This conclusion provides the basis for a further conclusion: "the function of dreaming may be reducible to a question of the function of the sleep-dependent memory processes that result in the conscious experience of dreaming," with an emphasis on REM dreams as aiding in emotional regulation (Stickgold & Wamsley, 2017, p. 513).

The emphasis on emotional memory processing during REM is first of all doubtful due to the continuing lack of convincing evidence for any type of memory consolidation during REM sleep in rodents. For example, one independent research team concluded, on the basis of its own research: "Altogether, these findings indicate an importance of NonREM rather than REM sleep for the encoding of information that is independent of the emotionality of the materials" (Kaida, Niki, & Born, 2015, p. 72). These conclusions are similar to those in other research studies by the same core research group (Born & Wilhelm, 2012; Rasch & Born, 2015; Rasch, Pommer, Diekelmann, & Born, 2009). Their analysis is supported by a study showing that "sleep did not stabilize memory" for a word list learned earlier in the day and then followed by another learning task. The study led to the conclusion that "the stabilizing effect of sleep against interference" by subsequent new learning "has been overestimated" (Pohlchen, Pawlizki, Gais, & Schönauer, 2020, p. 1).

In a review of the two decades of work on the replay during sleep of the waking brain-wave patterns accompanying memory consolidation in rodents in experimental settings, the researchers did not find evidence for memory consolidation during REM sleep (Z. Chen & Wilson, 2017). The sharp-wave ripples in the hippocampal region, which are the brain waves used to indicate memory consolidation, seem to behave differently in REM than at other times. Sharp-wave ripples are frequent in slow-wave sleep and quiet wakefulness, but in REM sleep the "firing-rate correlation was not related" to the relevant learning experience during the experiment (Z. Chen & Wilson, 2017, p. 3). Adding further complexity to the research literature on this topic, some of the neurons in the hippocampus active during REM sleep may play a role in forgetting (Izawa et al., 2019). At the least, then, the role of REM in memory consolidation remains uncertain in animal models.

The emphasis on emotional memory consolidation during REM in humans is first of all doubtful on the basis of studies of the many thousands of people who unexpectedly lost most of their REM sleep in the 1950s and 1960s, due to their use of the first generation of antidepressant medications. Despite the relative absence of REM sleep, they did not report any memory difficulties or other cognition issues. This finding is particularly striking in the case of those who took monoamine oxidase inhibitors, which all but abolish REM sleep (Schweitzer & Randgazzo, 2017; J. Siegel, 2021; Vertes & Eastman, 2000; Wyatt, Fram, Kupfer, & Snyder, 1971).

Similar results are reported in a study using antidepressants developed a few decades later, which reduce REM sleep by as much as 30%. In this well-controlled experimental study, the moderately depressed patients first took several cognitive tests. Then they were placed on one or the other of two medications. After a one-week trial, there was "no association of REM sleep diminution with decreases in memory performance or cognitive flexibility" (Göder et al., 2011, p. 544).

In addition, there is the case of a young man who lost all but a very few percent of his REM sleep due to shrapnel wounds during routine military training. This accident permanently damaged the lateral pontine region of the brainstem, as discovered by means of CT scans after he complained of headaches (Lavie, Pratt, Scharf, Peled, & Brown, 1984). Even without REM sleep, he earned a B.A. and a law degree, practiced law, prepared a new crossword puzzle each week for a major newspaper, and in general led a normal life. He was examined again in greater detail 34 years later at age 68. A full daytime neuropsychological assessment was made. He spent four nights in a laboratory setting while being studied with EEG recordings and a CT scan. He was found to be cognitively normal. He still was almost devoid of REM sleep. He also lacked the REM sleep concomitants that could be assessed, such as increased heart rate and muscle atonia (Efrat et al., 2018).

The neuroimaging studies of the neural substrates supporting emotions during the waking state, which are discussed in chapter 8, are also relevant to this discussion of emotions during REM dreaming. This is because the dorsolateral prefrontal cortex, which is a necessary part of the neural systems subserving the experiencing of emotions in waking (LeDoux, 2012, 2015; Lindquist et al., 2012), is deactivated during both REM and NREM sleep (Fox, Nijeboer, Solomonova, Domhoff, & Christoff, 2013; Uitermarkt, Bruss, Hwang, & Boes, 2020). Then, too, the emotions-regulation theory of REM is called into question by neuroimaging studies of emotion states. Four of the five association networks known to be the basis for emotion states during waking are relatively deactivated during all stages of sleep, as overviewed in chapter 8 (Polner-Clark, Wager, Satpute, & Barrett, 2016; Touroutoglou, Lindquist, Dickerson, & Barrett, 2015).

The low levels of emotions in dream reports collected from children, adolescents, and adults, inside and outside of laboratory settings, are consistent with these neuroimaging results. The findings from studies of dream content also cast doubt on the emotions-regulation theory in their own right. First, as discussed in chapter 8, there is a relative absence of emotions in the dreams of children and preadolescents, which means the theory could not likely be relevant until early adolescence (Foulkes, 1982; Strauch, 2005). Even during adulthood, the relevance of the theory is doubtful. The absence of emotions from at least 25–30% of the dream reports collected in several different lab studies, based on REM awakenings, are contrary to what the theory predicts. The REM-awakening study using a portable EEG in home settings is especially noteworthy. It not only discovered that 26% of the awakenings led to reports of no emotions, but it also found that the emotional intensity of the emotions was rated as low or medium in another 46% of the reports (Fosse et al., 2001, p. 2). It is also noteworthy that the strongest advocate of the emotions-regulation theory is a coauthor on this lab/nonlab study, which has not been factored into the emotions-regulation theory.

Equally problematic, the lab study that focused exclusively on the appropriateness of emotions to other aspects of dream content (Foulkes et al., 1988) found that 17% of the 88 REM dream reports were without emotions but would have had emotions in the same situation in waking life. This conclusion is based on judgments by both participants shortly after the REM awakenings and by independent judges. It seems difficult to explain why a sleep stage focused on regulating emotions would not be processing emotions in situations that would trigger emotions in waking life.

Finally, other evidence shows the alleged regulation of emotional memories during REM occurs just as well during slow-wave NREM sleep during naps, if not better: "Although prior evidence has connected negative emotional memory formation to REM sleep physiology, we found that non-REM delta activity and the amount of slow wave sleep (SWS) in the nap were robustly related to the selective consolidation of negative information" (Payne, Kensinger, Wamsley, Spreng, & Alger, 2015, p. 176). The researchers then said the "magnitude of the emotional memory benefit" conferred by REM sleep "can be economically achieved by taking a nap" (Payne et al., 2015, p. 176). Simply resting, according to another study, can have the same impacts (Humiston, Tucker, Summer, & Wamsley, 2019). These two sets of findings, whose coauthors include several of the advocates of the emotions-regulation theory of dreaming, provide evidence that REM dreaming is at most a minor factor in processing emotions.

After three decades of research on dreaming and emotions regulation during REM sleep, it is highly unlikely REM sleep involves the reprocessing and regulation of emotions. At best, the evidence is inconclusive. To summarize, there is no evidence for the presence of memory consolidation during REM sleep in rodents. The low levels of activation in the frontoparietal, dorsal attention, and salience/ventral networks during human dreaming raise serious doubts. The cognitive unimportance of the loss of REM sleep in humans remains difficult for the theory to explain. The infrequency of emotions in dream reports and their near-total absence from children's dream reports until ages 9–11 is not what the theory would predict. The equal benefits of taking a NREM nap or taking a waking rest, when considered in the context of the other negative evidence, suggests that REM sleep does not have an important role in regulating emotions.

The Emotions-Assimilation Theory of Dreaming

The emotions-assimilation theory of the function of dreaming is "somewhat divergent" from the theory analyzed in the previous subsection (Malinowski & Horton, 2015, p. 3). It places greater emphasis on the assimilation and integration of waking emotions into the memory system during dreaming: "emotions act as a marker for information to be selectively processed during sleep, including consolidation into long term memory structures and integration into pre-existing memory networks" (Malinowski & Horton, 2015, p. 3). It thereby puts more emphasis on dreaming itself. Nor does it emphasize REM sleep as the only sleep stage in which emotions assimilation during dreaming can occur.

Instead, this theory focuses on the psychological benefits of emotions assimilation, such as its usefulness in facilitating insights and creative thinking during waking. It includes a greater emphasis on the assimilation of positive emotions than related theories do. In addition, the emotions-assimilation theory assigns more importance to the presence of metaphors in dream content, because they aid in emotion assimilation. It also claims "hyperassociativity" during dreaming has a role similar to that of metaphors. (Hyperassociativity is indexed by various forms of bizarreness during dreaming, such as sudden topic changes and highly unusual constructions in dream content.)

This theory first suffers from several of the same problems facing the emotions-regulation theory. As discussed in chapter 8, the relative deactivation during sleep of all but one of the association networks involved in the generation of waking emotions is highly problematic for all theories concerning the importance of emotions during sleep, including this one. The findings on the low frequency of emotions in quantitative analyses of dream content by independent coders is consistent with the neuroimaging findings and adds to these doubts. The theory's strong emphasis on the assimilation of positive emotions is problematic because most studies of dream reports conclude that positive emotions are even more infrequent than negative emotions. Nor does the importance placed on emotional assimilation during dreaming fit well with the low levels of dreaming in the first nine or 10 years of life or with the lack of emotions in the infrequent dreaming that does occur.

In addition, the theory is doubtful because it relies heavily on studies of emotions that make use of self-ratings. However, as shown in a section of chapter 8, self-ratings likely overstate the frequency of emotions. Nor is there evidence for bizarreness (hyperassociativity) in dreaming, as the theory assumes. In addition, several types of hyperassociativity are equally frequent during drifting waking thought (Klinger, 2009; Reinsel, Antrobus, & Wollman, 1992). Further doubts arise because of the lack of evidence for any appreciable amount of metaphoric thinking during dreaming, as demonstrated in chapter 5. The theorists' own study of metaphoric thinking during dreaming is based on metaphoric interpretations emerging in the course of interviews with the four participants. In each case, two dream reports with apparent "high levels of wake-dream continuity" and two with seemingly "low levels of waking dream continuity," were discussed (Malinowski, Fylan, & Horton, 2014, p. 164). However, reflections on dreams during interviews are very likely based on waking insights, not on what is in the dreams themselves (e.g., Blagrove, 1992; Edwards, Ruby, Malinowski, Bennett, & Blagrove, 2013).

Mastery and Rehearsal Theories of Dream Function

Two Freudian-oriented research psychologists independently suggested dreams have a mastery function, which primarily concerned the most important interpersonal conflicts experienced by the dreamer. For Richard

M. Jones (1962), dreams function to keep the person's identity integrated, a form of ego synthesis. Put another way, they help to resolve inner crises. In a similar vein, a second Freudian-oriented research psychologist thought that dreamers' understanding of their central personal problems improve during the course of a dream, putting them in a position of relatively more mastery than at the beginning of the dream (Witkin, 1969; Witkin & Lewis, 1967). There were no direct follow-ups on these speculations and the introductory evidence for them.

However, based on the later adoption of a cognitive emphasis by several theorists, new mastery theories were put forward in the twenty-first century. Generally speaking, they are social-rehearsal theories that can emphasize either solidification of social-interaction skills or the mastery of threatening situations. The most fully developed and visible version of a social-rehearsal theory emphasizes learning new skills during dreaming for dealing with threats (Revonsuo, 2000a, 2000b; Revonsuo & Valli, 2000; Valli & Revonsuo, 2009). The other social-rehearsal theories focus primarily on improving social skills (Brereton, 2000; M. Franklin & Zyphur, 2005; Revonsuo, Tuominen, & Valli, 2015). There is also one other mastery type of theory, which involves the efficient consolidation of updates in the neural substrates involved in predictive coding (Friston, 2014).

The Threat-Simulation Theory of Dream Function

The threat-simulation theory of dreaming combines a cognitive emphasis on simulation with an emphasis on the origins of dreaming in the "ancestral environment," in which "human life was short and full of threats" (Revonsuo, 2000b, p. 877). It begins by assuming dreaming prepares people to cope more adequately with waking threats that endanger reproductive success, without necessarily being recalled. At the brain level, it claims that the "biological function of dreaming is to simulate threatening events, and to rehearse threat perception and threat avoidance," which prepare people to cope more adequately with waking threats that endanger reproductive success (Revonsuo, 2000b, p. 793). The theory thereby assumes the neural substrate enabling dreaming has the capacity to support implicit learning (incidental learning without awareness) during sleep, which threat-simulation theorists regard as the most difficult assumption in the theory to test (Revonsuo, 2000a, pp. 890, 1081; Valli & Revonsuo, 2009, p. 33).

In fact, most experimental psychologists doubt the importance of implicit learning even during waking, due to its limited scope, its occurrence in only brief durations, and its relevance for only a few types of information (see Ryals & Voss, 2015, pp. 44–45, for a summary of the various criticisms). Moreover, the small amount of implicit learning requires the participant to be paying attention to the main task at hand. One study reported a "significant negative correlation between mind-wandering and implicit learning" (M. Franklin, Smallwood, Zedelius, Broadway, & Schooler, 2016, p. 223). Implicit learning during dreaming therefore may be extremely difficult because dreaming shares qualities with mind-wandering. In the case of implicit *sequence* learning, which seems to be the most relevant type of implicit learning in terms of the sequential, quasi-narrative nature of most dreams, a researcher who has studied the sequential-learning issue in detail concludes that implicit learning cannot be separated from explicit learning (Shanks, 2003, p. 38). Nor is there any evidence for new learning during sleep beyond a conditioning study that paired odors with tones to condition sniffing responses in reaction to tones (Arzi et al., 2012).

In addition to its reliance on implicit memory in order to learn from the very complex process of dreaming, the theory also assumes that this implicit learning can be transferred to waking situations, which is known as "transfer of learning" or "transfer of training" in the psychological literature. Once again, findings in the twenty-first century suggest that the transfer of learning also seems to be limited in its scope, with little or no evidence for improvement on tasks outside of tightly related, overlapping domains (see Sala & Gobet, 2017, and Sala, Tatlidil, & Gobet, 2017, for meta-analyses and overviews of very large literatures). Nor is there any evidence for transfer of learning from sleep to waking, and it is unlikely that such a complex transfer could occur. One neuroimaging study reported that the transfer of learning involves the regions in the brain that support the central executive network (i.e., the frontoparietal control network), which relies heavily on the dorsolateral prefrontal cortex (Verghese, Garner, Mattingley, & Dux, 2016). This finding, which needs to be replicated before it can be fully accepted, raises doubts about the threat-simulation theory because numerous studies show the dorsolateral prefrontal cortex is deactivated in both REM and NREM sleep (Fox et al., 2013; Uitermarkt et al., 2020).

The threat-simulation theory of dreaming does not focus heavily on other animals, but it does suggest the early mammals had "long periods of sleep," with an emphasis on REM sleep. According to the theory, these REM periods may have allowed for the simulation of threat and survival skills during a time when the small early mammals had to compete for resources "with the much larger and more numerous reptiles" (Revonsuo, 2000b, p. 900n14). This speculation leads to the hypothesis that humans may have inherited threat scripts that are triggered by "ecologically valid cues" (Revonsuo, 2000b, p. 878). The evidence for this claim is based on the behaviors of decorticated cats, such as "hunting, stalking, running as if chasing imaginary prey," which happen during REM periods. These behaviors also happen during waking, but this finding is brushed aside because "the brain in REM is most like the brain in very alert wakefulness" (Revonsuo, 2000a, pp. 1070–1071). However, there are large differences between waking and REM, as shown by neuroimaging studies, and there is no reason to believe there is imagination in the brainstems of any decorticated mammals. Nor is there any evidence mammals have the cognitive capacities needed to dream, such as mental imagery, narrative ability, imagination, and an autobiographical self, which are discussed in chapter 6 (Foulkes, 1983, pp. 317–319, 325–327, 332–333; 2017).

Threat simulation during dreaming is said to work well in children and adolescence "as soon as their perceptual and motor skills are at a level that enables threat recognition and avoidance in the waking state" (Revonsuo, 2000b, p. 899n5). This assertion is primarily based on an inadequate earlier study of children's dream reports collected by parents and teachers, which is now known to be very risky due to suggestion and demand characteristics—among the many problems encountered when working with young children (Hall & Domhoff, 1963a). The study very likely greatly overstated the degree of aggression in children's dream reports. As shown in subsequent research in well-controlled lab and nonlab settings, the few bland dream reports from preschool children contain no aggression, misfortune, or failure, and the dream reports of elementary school children very rarely contain these negative elements (Foulkes, 1982; Foulkes et al., 1990; Sandor et al., 2015).

However, the more soundly based laboratory findings were not factored into threat simulation theory because "these data do not show that such experiences are not possible, at least occasionally or in specific subgroups of children who are living in less safe environments or who otherwise have been exposed to various threatening events" (Revonsuo, 2000b, p. 899n5). Moreover, the numerous waking studies of successful threat detection by children as young as ages 3–5 demonstrate they are already good at threat detection, including the detection of angry faces and perhaps even earlier on some specific issues, such as a threat response to snakes (LoBue, Matthews, Harvey, & Thrasher, 2014; LoBue & Rakison, 2013; LoBue, Rakison, & DeLoache, 2010). In other animals, one-trial escape learning after upsetting events and punishment has been well established for generations by means of experimental studies by research psychologists (see Postman, 1963, for a summary going back to the 1920s).

The theory also assumes that traumatic events lead to a more rapid development of the capacity to dream in children (Revonsuo, 2000b, p. 889n5), but this claim seems doubtful in terms of what is known about the gradual sequential unfolding of neural and cognitive development, as discussed in chapter 6 and in other sources (Fair et al., 2008, 2009; Nelson, 2005). The strong assertions on this issue by threat-simulation theorists were later amended to claim that threat-simulation dreams are possible at age 6 and thereafter (Valli & Revonsuo, 2009, p. 33). This small adjustment also remains unlikely in terms of the findings on the immaturity of the default network until ages 9–13 (Fair et al., 2007, 2008). The longitudinal and cross-sectional laboratory findings on the low frequencies of aggression, misfortune, and failure in the dream reports of children ages 6–9, also are at odds with this theory, as summarized in chapter 6 (Domhoff, 1996, pp. 91–95; Foulkes, 1982; Foulkes et al., 1990).

Turning now to empirical doubts based on studies of dream content, the indicators of threat used to test the theory on the basis of dream content encompass just about everything negative that can happen in dreams. In addition to physical attacks and threats of aggressive acts, they include threats to valuable material resources, social status, and events perceived as "subjective threats," such as mistakes and feelings of personal failure (Revonsuo & Valli, 2000, pp. 5, 23–25). Research by its proponents finds that 60–77% of dream reports have threatening events (Revonsuo & Valli, 2000, 2008). This finding replicates the following Hall and Van de Castle (1966) results in their normative samples of men and women almost exactly, but it overlooks some key points: only 44% of women's dream reports and 47% of men's dream reports have at least one aggression in them, only half of which or less are physical aggressions; 33% of women's reports and 36% of men's reports have at least one misfortune, most of which are minor events; 10% of women's reports and 15% of men's reports have at least one failure; and overall, 70.7% of women's reports and 73.8% of men's reports have at least one of these three types of elements (see Hall & Van de Castle, 1966, as updated using a spreadsheet that includes 491 of the 500 women's reports and all 500 of the men's reports).

Similarly, when threat-simulation theorists report dreamers are involved in a large majority of the life-threatening events (Revonsuo & Valli, 2000, p. 10), they are replicating the Hall/Van de Castle (HVdC) normative findings on both female and male dreamers. These normative findings show dreamers are involved in 80% of the aggressive interactions in their dreams. But threat-simulation theorists ignore related findings revealing that the dreamers are the victims in the majority of the aggressive interactions. Also, the dreamers only infrequently respond to their victimization, as shown in table 3.2 in chapter 3. It is difficult to believe that the interactions relating to aggressions in dreams contain useful lessons to transfer to the waking state. At the same time, the exclusive concern with threats in this theory cannot explain the considerable percentage of dream reports not containing threats (Zadra & Donderi, 2000b). This criticism is supported by the 25–30% of the dream reports in the normative sample not containing the negative elements of aggression, misfortune, or failure. This figure seems too large to dismiss as a small glitch in an evolutionarily evolved threat-oriented dreaming process.

The nonthreatening nature of many dreams is demonstrated in two studies of dream reports from students at the University of Cape Town, both of which were meant to examine threat-simulation theory. Although the university is located in the most violent and crime-ridden area in South Africa, the first study found that very few dream reports had realistic life-threatening elements and that effective responses by dreamers to threats of direct physical harm were rare (Malcolm-Smith & Solms, 2004; Malcolm-Smith, Solms, Turnbull, & Tredoux, 2008b). The second of the two studies included a comparison with 116 dream reports from students at Bangor University, which is in a low-crime area in North Wales. Nevertheless, the sample from Wales contained a higher percentage of dream reports including life-threatening events than did those from Cape Town (18.6% vs. 8.7%) (Malcolm-Smith et al., 2008b, pp. 1285–1286).

The weaknesses of the threat-simulation theory are also demonstrated in a study by another team of independent investigators, which used 212 previously collected recurrent dreams (Zadra, Desjardins, & Marcotte, 2006). Recurrent dreams are acknowledged by threat-simulation theorists to very often contain threat simulations (Revonsuo, 2000a, p. 1076). Unlike what threat-simulation theorists would expect, the study found that 81% of the instances of threats "belonged to the realm of fantasy or fiction" or would be "very unlikely to occur in the subject's waking life." Furthermore, a great majority of the recurrent dreams with a threat of any type in them ended with the threat being fulfilled (40%), the participant awakening (37%), or the imminence of the threat changing "abruptly" (17%). As a result, only 17% of the recurrent dreams with a threat in them had a positive outcome (Zadra, Desjardins, & Marcotte, 2006, p. 457).

The content findings in studies of dream series also demonstrate the shortcomings of this theory because dream content is so consistent over time and responds very slowly, if at all, to changing waking circumstances, as discussed in chapter 4. For example, in a dream series spanning just over five decades, which was not discussed in chapter 4, about 72% of the 904 dream reports included at least one of six recurring personal concerns throughout the series (Domhoff, 1996, pp. 142–145, 206). Dorothea, the pseudonym the dreamer gave herself, was eating, starting to eat, preparing a meal, buying or seeing food, watching someone eat, or mentioning that she is hungry in just over 20% of her dream reports. In many of these dreams she is at the family dinner table with her parents and one or more of her several siblings. In these dream reports she often portrays herself as being treated unfairly as to portions. The last dream documented by this well-educated professional woman, who was living in a middle-class retirement home in an idyllic American state where she enjoyed a daily swim in its pool, was recorded four days before her unexpected death at age 78. It was about having dinner in a familiar home setting from the time when she was growing up. Her mother served her siblings too much food, which seemed to leave her with nothing to eat.

In about 16% of the dream reports she lost an object, which was most often her purse, and in about 10% of her dream reports she was in a small or disorderly room, or people were barging into her room. Another 10% involved the dreamer and her mother doing something together. The dreamer was going to the toilet in about 8% of the dreams and she was late or worried about being late, or about missing a bus or train, in 6% (Dom-hoff, 1996, pp. 143–145). This repetition of several themes, along with the consistency with which she dreamt about her long-deceased mother, do not accord with the idea that dreams deal with new threat situations as they arise. Instead, they often deal with ongoing and past personal concerns. Dorothea's dream reports are available on Dreambank.net.

A more dramatic and poignant example of the repetitiveness in a dream series, which was documented three decades after the previous example, concerns 315 dream reports written down over a period of several years by an artist ("Merri"), when she was in her late 30s. The most frequent character in her dream reports, her older sister, had been killed by a hit-and-run driver while she was jogging, three years before the dream series began. She was in a hospital on life support for five days, with the dreamer and her brother by their sister's side when she died (Bulkeley, 2009, p. 98). Contrary to what a threat-simulation theory might expect, Merri's deceased sister appeared in 34.3% of the dream reports in this series and she is at the center of the small-world social network constructed on the basis of coappearances by dream characters in dream reports, a topic discussed more generally in chapter 4 (Han et al., 2015).

These dream reports express the dreamer's major personal concern: her preoccupation with the deceased sister she deeply admired for her many accomplishments. If these dreams represent threat simulations that eventually will allow her to move forward in waking life, they were very unsuccessful during the time span in which she recorded her dreams. The findings from the Merri series can be found on the "Information" and "Further Information" pages on Merri on DreamBank.net, and some of them appear in a published paper (Bulkeley, 2009).

In responding to the various studies of dream content that refute their theory, threat-simulation theorists discount the findings because the settings and circumstances in which the dream reports were collected allegedly did not include ecologically valid cues (Revonsuo, 2000a, pp. 1071–1073; Revonsuo & Valli, 2008). This rejection of solid, replicated findings thereby raises the possibility that the theory is unfalsifiable by studies of dream content, at least according to the theorists who created it, because of the repeated claim that ecologically valid cues were not present (Desjardins & Zadra, 2006; see Malcolm-Smith, Solms, Turnbull, & Tredoux, 2008a, for a similar suggestion). To remain viable at the level of dream content, the threat-simulation theory also would have to explain the 25–30% of dream reports not containing a threat, whether physical or subjective.

Although the threat-simulation theorists reject the theory-threatening findings on dream content by other investigators, they nonetheless have to deal sooner or later with the serious doubts about the four main underlying assumptions upon which the theory is based. First, threat-simulation theory claims there was at least a rudimentary form of dreaming during REM sleep in ancient mammals hundreds of million years ago. However, this assertion does not fit with the strong evidence that no other animals, including other great apes, have the cognitive capacities necessary for dreaming. The time frame for the alleged ecologically valid cues purportedly shaping dreaming therefore would have to be greatly narrowed. Second, there is no evidence for complex implicit learning during sleep. Third, there is no evidence for the transfer of implicit learning to waking situations. Fourth, studies of young children between ages 6 and 11 suggest that they do not have the cognitive capacities to generate threat-simulation dreams with any frequency or complexity until they are in their early teenage years.

Until threat-simulation theorists provide convincing evidence showing that their four questionable assumptions are at all plausible, the theory remains in the realm of pure speculation, despite all the equally doubtful empirical evidence they bring forth. Their discussions of negative elements in dreams, and their resort to the concept of ecologically valid cues as a way to dismiss the wide range of empirical studies discussed in this subsection, is not credible. However, their mistaken portrayal of dream content is a secondary issue if the four assumptions upon which their theorizing is based cannot be satisfactorily defended.

Other Social-Rehearsal Theories

The other social-rehearsal theories focus primarily on the learning of social skills through positive social interactions (Brereton, 2000; M. Franklin & Zyphur, 2005; Revonsuo et al., 2015). They first would have to deal with many of the same objections raised in the specific case of threat-simulation theory, such as the lack of evidence for implicit learning during sleep and for the transfer of training to the waking state. They also would have to explain why there are relatively few friendly interactions in dreams, as well as the fact that these friendly acts are rarely reciprocated, as shown in table

3.2 in chapter 3. They also would need to account for the high frequencies of aggression and misfortunes in dream reports.

Nor can these theories readily explain the replicated finding that dream content does not change very fast, if at all, from the teenage years into old age, as discussed in chapters 4 and 7. And, once again, the infrequency of dreaming in young children until ages 9–11 and the small number of social acts in these infrequent dream reports suggest that the few plausible social rehearsals in dreaming do not occur until ages 9–11 at the earliest.

Dreaming as an Adaptation for the Consolidation of Waking Predictive Codes

According to the predictive-coding theory, the brain is designed through evolutionary selection to minimize surprises by constantly matching new sensory inputs with established neural patterns. If the brain detects a mismatch, it sends an "error message" to higher brain centers to update the neural codes and thereby enhances the chances for survival (Friston, 2010). Since the brain receives no new information from the environment during sleep, the role of REM sleep is to "optimize" the model by removing the "redundancies" that are "accrued during wakefulness" (Friston, 2014, p. 139). Two of the theory's main proponents claim that "the mechanics of predictive coding provide a compelling three-way link" between brain waves that have their origins in the brain stem, the occurrence of rapid eye movements, and "the fictive percepts of dreaming" (Hobson & Friston, 2012, p. 97).

To test their hypothesis concerning rapid eye movements, the predictive-coding theorists reanalyzed a "short six-minute study" of video footage that contained 43 instances of rapid eye movement bursts. These 43 instances were obtained from 11 participants who were examined using fMRI. The researchers first carried out "24 independent studies" focused on the individual participants. They infer from their results with several of these individuals that the participants' eye movements were scanning a "virtual sensorium," which is supported by a "hot zone" in the "posterior left hemisphere" (Hong, Fallon, & Friston, 2021, p. 2). In addition, three of the brain areas predictive-coding theorists assign to the default network—the precuneus, posterior cingulate cortex, and retrosplenial cortex—were said to be deactivated when the rapid eye movements were occurring, so they assert that the default network is not active during dreaming (Hong, Fallon, & Friston, 2021, p. 11).

Since the theory has an evolutionary basis that long precedes humans, its emphasis on dreaming cannot be sustained. Studies claiming to show dreaming in other animals were found to be flawed, and no other animals have the ability to produce personal memories or narrative thought flow, or to develop an autobiographical self (Foulkes, 1983; Tulving, 2005). The human brain is much larger and has far more cortical connections than do other primates, and more anterior-posterior connections in the default network; the necessary capacity for a full imagination developed gradually within a group context over the past few hundred thousand years (Fuentes, 2020; LeDoux, 2019). Further, the lack of evidence for memory consolidation during REM sleep, as discussed earlier in the chapter, puts the onus on predictive-coding theorists to show that *any* type of consolidation or updating can occur, with or without dreaming, during REM sleep.

As for the hypothesis that the rapid eye movements during REM sleep in humans are scanning dream scenarios, it was first refuted in a study in which the four participants were immediately awakened from 75 REM periods that suddenly showed electrooculogram (EOG) evidence of one of four distinctive eye-movement patterns. A blind judge could not correctly match participant-based subsets of the 56 dream reports that described clear visual events with the accompanying eye-movement patterns at better than a chance level. Nor could plausible correspondences be discerned in a nonblind comparison of several visually distinctive dream reports with the participant's eye-movement patterns just before the awakening (Moskowitz & Berger, 1969, pp. 613–614).

Later systematic studies also rejected the scanning hypothesis (Christensen et al., 2019; Jacobs, Feldman, & Bender, 1972; Pivik, 1991, p. 227). These cumulative results are further supported by a comparison of eye movements during REM sleep and waking, which found that "REMs were significantly slower than waking saccades of comparable amplitude," among several differences between eye movements in REM sleep and waking, so "one wonders why REMs behave so differently if they are simply the unmodified motor concomitant of visual imagery in dreaming" (Aserinsky, Lynch, Mack, Tzankoff, & Hurn, 1985, pp. 1, 9). Moreover, the primary visual cortex, which is of central concern for predictive-coding theorists, is relatively deactivated during REM sleep, and damage to it has no impact on visual dream imagery (e.g., Braun, et al., 1998; Solms & Turnbull, 2002, pp. 209–210). Taken together, these various studies raise two further questions about the scanning hypothesis. Since visual imagery appears to be continuous throughout REM sleep dreams, it seems unlikely that eye movements that happen only periodically during the course of a REM period could be tracking the ongoing visual imagery. Nor can the scanning hypothesis account for the visual imagery that occurs in dreams outside of REM sleep.

The emphasis on the scanning hypothesis by predictive-coding theorists, despite the studies cited in the previous two paragraphs, is based in good part on studies of REM Behavior Disorder (RBD), in which the eye movements, twitching of limbs, and major body movements, including walking, are assumed to be the acting out of an ongoing REM dream (e.g., Arnulf, 2012). However, aside from the risks of confounds in using patients in dream studies, as discussed in chapter 1 and demonstrated in chapter 2, most of the motor events in RBD (66%–83%) are "minor, elementary, or jerky limb movements," and only 1.8% have been described as possibly acting out the dream content, which makes them "exceedingly rare" (Blumberg & Plumeau, 2016, p. 35).

In one of the most detailed studies, "there were only 19 apparently goal-oriented behaviours in nine patients" who were studied over a period of 77 nights and reported their dreams after spontaneous awakenings (Leclair-Visonneau, Oudiette, Gaymard, Leu-Semenescu, & Arnulf, 2010, p. 1742). Based on an analysis of 162 episodes of rapid eye movements that occurred during these dreams, the researchers reported they had correctly predicted there would be eye movements in over half of these instances (58.3%), but they were wrong on the remaining 41.7% of their predictions; further, they reported that the eye movements were determined to be "concordant with the direction of the RBD-associated motor behaviour" in 82% of the cases with eye movements (Leclair-Visonneau, Oudiette, Gaymard, Leu-Semenescu, & Arnulf, 2010, p. 1742). While these results appear to be better than chance, such rare events in studies of a small number of patients are not a solid basis for making strong claims in the face of the systematic evidence to the contrary, based on representative samples of participants who were studied in a variety of well-controlled research studies.

The claim by predictive-coding theorists that the default network is not involved in dreaming is first of all weakened by the theorists' statement that there were merely "modest" declines in the activation levels in the precuneus and the posterior cingulate cortex in 12 of the 24 independent studies of the 11 individual participants; there were also small decreases in activation levels in the retrosplenial cortex in 6 of the 24 independent studies (Hong, Fallon, & Friston, 2021, p. 7). Whatever the degree of decline, the important point is that the precuneus is not generally considered to be part of the default network by neurocognitive researchers, and the posterior cingulate cortex is in the default network's zone of integration (Andrews-Hanna, Smallwood, & Spreng, 2014, p. 35). Furthermore, previous neuroimaging studies already had concluded that both of these areas are relatively deactivated during sleep. Therefore, they are not among the areas in the default network that are discussed in chapter 2 as part of the neural substrate that supports dreaming. As for the six instances of relative deactivation of the retrosplenial cortex, four of them involved only 1 of the 11 participants.

Most problematic of all, "in the group analysis, REM-locked fMRI signal decreases were not seen," which means the modest declines in some of the researchers' individual studies do not reach the level of statistical significance in the overall sample (Hong, Fallon, and Friston, 2021, p. 7). As they therefore frankly state in their discussion of the limitations in their study, the fact that they were "using uncorrected statistical thresholds" and did not find "statistically significant decreases at the group level" means that their "group and individual levels both call for replication" (Hong, Fallon, and Friston, 2021, p. 15). Until such a replication occurs, these results are not credible, and two of them, relating to the precuneus and the posterior cingulate cortex, are irrelevant in any case.

Finally, several established findings on dreaming and dream content cannot be explained by predictive-coding theory. First, dreaming is infrequent and not very complex until ages 9–11, as shown in chapter 6, so the theory cannot explain why preschool and elementary school children function so well during waking. Second, the loss of REM sleep dreaming by people who took the first generation of antidepressant drugs did not lead to cognitive impairments, as already discussed earlier in this chapter. Third, few events from the previous day appear in dream reports, as discussed in chapter 9, so there is very little new information to be updated. Fourth, predictive-coding theory is challenged by the consistency of dream content in the numerous dream series studied to date, as documented in chapter 4, chapter 7, and earlier in this chapter, which does not leave much time during REM sleep dreaming for updating.

Based on the findings in these various systematic studies using several different methodologies, the attempt to extend the waking-state predictive-coding theory to REM sleep and dreaming is not plausible.

Conclusions Regarding Adaptive Theories

As shown by the discussions of each of the three main types of theories proposing a possible adaptive function for dreaming, no empirical evidence supports any of them. At the same time, there are many replicated findings leading to grave doubt about all of these theories. Since these several different types of evidence cast strong doubt on all of these adaptive theories, the next section of this chapter turns to the evidence concerning the several ways in which people have creatively made use of dreaming and dream content in waking life. They do so in an attempt to cope with uncertainty, anxiety, grief, guilt, and the inevitability of death.

Dreaming as a Culturally and Individually Useful By-Product

As noted at the outset of this chapter, an abundance of evidence suggests that dreaming and the content of dreams have been put to use historically and cross-culturally in human societies. The most detailed cross-cultural study of a wide range of indigenous societies is based on a sample of 63 societies in the Yale Human Relations Area database. The researcher found four factors correlated with a high concern with dreams: supernaturals give people powers through dreams; shamans and priests use their own dreams in curing and in making prophecies; certain types of dreams are required before a person can assume certain roles (e.g., warrior, shaman); and special techniques are needed to induce dreams (D'Andrade, 1961, p. 321). In addition, it is hunting and gathering societies, which cannot rely on agriculture and/or animal husbandry for their food supply, that are most likely to make extensive use of dreams. This finding led to the insight that the uncertainty of a food supply may be a factor in explaining why people turn to dreams for guidance (D'Andrade, 1961, pp. 324–325).

This idea can be generalized to claim that dreams become important when there is any situation of crisis or uncertainty. This generalization encompasses societies in danger because of large-scale threats or violent intrusions by nearby societies. It also includes the many situations in which individuals perceive themselves to be in conditions of personal uncertainty and crisis, as discussed shortly. Then, too, shamans, who are important far beyond hunting and gathering societies, are in some ways similar in their functions to psychodynamic therapists. Both shamans and psychodynamic therapists make a transition to another realm—the spirit world for the shaman, the world of the unconscious for psychodynamically oriented therapists.

Either way, whether in indigenous societies or Westernized nation-states, dreams are thought of as coming from "somewhere else," as something that "happens" to people outside the realm of normal cognitive functioning. Shamans and psychotherapists also share the goal of learning the causes of an illness, which are generally attributed to angry malevolent spirits in indigenous societies and to unconscious forbidden wishes in Western civilizations, as is often discovered through dreams. Finally, shamans and psychodynamically oriented psychotherapists share the goal of curing illnesses, and shamans are sometimes called the "first professionals" by cultural anthropologists (Singh, 2018).

Within this context, and as already mentioned in the introduction to this chapter, dreams are often seen as the main source of new religions. This insight was first stated in the nineteenth century by one of the founding figures of anthropology, Edward B. Tylor (1871/1958). It was subsequently elaborated in greater detail by anthropologists (e.g., LaBarre, 1972; Lincoln, 1935). Relevant cross-cultural evidence demonstrates that dreams are sometimes an important part of a person's conversion to a new religion (Bulkeley, 1995, 2008).

Dreams are very often intermixed with music in religious and healing ceremonies. This finding is of interest because music is a far more important "accidental by-product of traits that evolved for other purposes" than are dreams (McDermott, 2009, p. 164). This conclusion about music being a by-product is also attested to by other researchers (Patel, 2008, pp. 367–400). They include one of the deans of musical studies, who wrote that "many of the capacities for music emerge independently at different times in hominin evolution, but only with modern humans are we likely to find an integrated capacity for music together with language" (Cross, 2016, pp. 11–12). Music is very frequently at the center of peak experiences for individuals and groups, and music as an expression of "religious faith is a worldwide phenomenon," including the work of great European composers such as Bach, Beethoven, Handel, and Hayden (Gabrielsson, Whaley, & Sloboda, 2016, p. 226). Music also can be helpful for stroke victims who have lost speech (Patel, 2010, pp. 15–20; 2014) and in calming people in postoperative hospital settings (Hole, Hirsch, Ball, & Meads, 2015). Music is far more important in most people's cultures and personal lives than is dreaming, but these two evolutionary by-products are often used together in some ceremonial events.

The close relationship between music and dreams in religious and healing ceremonies has been documented in the greatest detail in studies of the Temiar people in the highlands of Malaysia, who were one of the few small indigenous societies still almost fully intact in the early 1980s (see Benjamin, 2014, for the definitive account on music and dreams in this culture, and M. Roseman, 1991, for a detailed study focused primarily on music in the same small society). It therefore can be concluded that both music and dreaming have emergent cultural uses, which have been invented by human beings in many different societies in the course of their histories.

The way in which the connection between dreams and religion may have developed is seen in dreams about deceased loved ones, who sometimes are seen and perceived as alive in a dream experienced by a grieving survivor. The dreamers are amazed, shocked, or delighted because they realize they are talking to a deceased person. Such dreams have an even greater impact when the deceased loved one provides reassurances that all is well or gives solace or advice to the dreamer (see D. Barrett, 1992, for the seminal study of the dream reports of deceased loved ones, and Domhoff, 1996, pp. 199–202, for a summary and further evidence from student participants). When such dreams are told and retold by many people, they become evidence for life after death and the existence of a spirit world.

Several dreams of this type were documented on the morning after they occurred in the dream series kept by the widower. Ed did not consider himself to be a very religious person, although he attended religious ceremonies on important days in the calendar of his religion. However, he wrote in his reflections that in a few of the dreams it felt like he actually had been "visited" by Mary:

> A few dreams have me wondering if I had actually "supernatural-like" experiences. I swear that Mary really did "visit" me. The first was my very first dream of her. I had this just a week or two after she died. She came to me in a dream and told me she wants me to be happy. (Domhoff, 2015, p. 4)

Dreams also may be seen as important for both an individual and a society when the transition from youth to adulthood occurs. In many societies, as part of this particular rite of passage, it is necessary for the young initiates into the adult world to have a revelatory dream to determine their future role in the society (see Van Gennep, 1909/1960, for the classic source on rites of passage). This dream can be induced by social isolation, sleep deprivation, fasting, and/or the use of a hallucinogen. At a strictly individual level, dreams may become important during times of personal illness or family conflict, or when a loved one dies. For example, in questionnaire surveys and case studies concerning paranormal experiences, from one-third to two-thirds of the participants claimed that they had had a paranormal experience in a dream that foretold serious health issues, a major accident, or even the death of a relative (Van de Castle, 1977, 476–481). A summary of past research nearly four decades later concluded that "spontaneous case collections have shown that ESP is most likely to occur during dreaming, followed closely by daydream-like states" (Richard Broughton, 2015, p. 145).

Dreams can also have a lifelong impact on the lives of a few individuals (Bulkeley, 1994; Knudson, 2001, 2003; Knudson & Minier, 1999; Kuiken & Sikora, 1993). One of the most striking and often-cited examples seems ideal in this context because it involves both music and creativity. In this instance, a professor of music told one of his colleagues in psychology that a dramatic dream led to his finally going to college at about age 22, with a focus on music. He added that he later wrote a prize-winning musical composition based on the same dream. As he recounted, he had been a self-taught rock guitarist for three or four years after his graduation from high school. Then he attended a very different type of performance out of curiosity, a Beethoven piano concerto. The performance inspired him enough to buy a recording of it. After playing the recording "at least a thousand times," he unexpectedly had a dream in which the leaves in swaying trees were like musical notes and he was hearing the music as well. At that point his life began to change. He went to college, took music lessons, and earned a doctorate in music. It was about 15 years after he completed graduate school that he won the aforementioned academic prize for a musical composition that had its origins in his profoundly inspiring dream (Knudson, 2001, p. 170).

Conclusions and Implications

Drawing upon a wide range of studies, this chapter concludes that there is no solid empirical evidence for any of the theories of dream function put forth since 1900. Moreover, there is much evidence that inadvertently casts strong doubt on these theories. Each of these theories is called into question by one or more of the four types of descriptive findings summarized in the introduction to this chapter: the very small percentage of dreams that are recalled or remembered for very long if they are recalled; the absence of any immediate or major effects from lesions eliminating dreaming; the immaturity of the default network in children, which makes any adaptive function for dreaming highly unlikely until early adolescence; and the consistency found in studies of dream series.

In addition, there are various types of evidence casting further doubt on specific adaptational theories, such as the absence of evidence for repression; the low levels of reciprocity to friendly or aggressive acts; the questionable nature of free association as a method for studying dreams; the absence of evidence for memory consolidation during dreaming in REM sleep; the lack of evidence for implicit learning during sleep; the absence of evidence for any putative implicit learning during sleep being transferred to waking thought; and the absence of evidence that the functioning of synapses during REM sleep can support the consolidation of daytime updates in predictive codings.

At the same time, the neurocognitive theory of dreaming can explain the widespread belief in the importance of dreams in both cultural ceremonies and individual lives by pointing to the evidence for the inventiveness of the human mind. As in the case of an even more important by-product, namely music, dreaming has its most important emergent function in human history as a central aspect of religion and as part of the healing practices once related closely to religion in the history of Western civilization. Dreams can impact the personal lives of individuals as well, usually for short periods of time, but sometimes for decades or for a lifetime.

11 The Neurocognitive Theory Compared to Other Dream Theories

As this book demonstrates, the neurocognitive theory of dreaming does not face any of the problems confronting the Freudian, activation-synthesis, and adaptive theories. It has a neurocognitive basis rooted in the findings on the default network. Future detailed neuroimaging findings will continue to refine this picture. Future lesion studies may make it possible to be more specific concerning the exact locations leading to the loss of dreaming or to alterations in dreaming. Nevertheless, the work to date has outlined the general picture.

It is therefore the activated portions of the default network that provide the neural basis subserving the cognitive processes that generate dreaming. However, for dreaming to occur, the default network must be mature and intact, and it must be cut off from the external world and incoming stimuli by gates in the thalamus. There also has to be continuing relative deactivation of the frontoparietal control, dorsal attention, and salience/ventral networks. These conclusions could be incorporated by adaptation theories, and perhaps by Freudian theorists, but they are inconsistent with the major emphasis activation-synthesis theory puts on the importance of random stimuli from the brainstem in generating dreaming and dream content.

The neurocognitive theory of dreaming also has a sound developmental dimension because it incorporates the evidence concerning the gradual maturation of the default network, as well as the longitudinal and cross-sectional studies of the changes in the frequency, complexity, and content of dream reports collected in sleep-dream laboratories from more than 100 children ages 3–15 (Foulkes, 1982; Foulkes et al., 1990; Strauch, 2004, 2005; Strauch & Lederbogen, 1999). The theory also makes use of the parallel findings on the development of waking cognition in young children, preadolescents, and adolescents, which provide convincing supporting evidence for the dream-lab results. However, none of the other theories of dreaming has incorporated these replicated findings as yet.

Much still needs to be learned about the development of dreaming, and refinements very likely will be made. However, there is little doubt about the default network maturing only gradually and not beginning to be adultlike until ages 9–13. The work on the development of cognition in children is also firmly established. There are plausible parallels between the developmental pattern of dream content and the emergence of various cognitive capacities, such as the production of mental imagery and the use of narrative. To the degree that the longitudinal and cross-sectional lab findings on the frequency, complexity, content, and emotionality of dreaming and dream content are judged to be credible, then the absence of a developmental dimension in the Freudian, activation-synthesis, and adaptation theories can be seen as a major shortcoming in them.

Few, if any, of the other theories can accommodate the systematic findings on dream content. The findings on continuity, the rarity of symbolism, and the relative lack of emotions are very difficult for Freudian theory. The everyday nature of most dream content, the minor role of unusual elements, and the relative infrequency of emotions are contrary to what activation-synthesis theorists expect. The fact that dream content can be studied in a systematic fashion also goes against their claims. As for adaptation theories, the minor role of incorporations from daily waking life, the slowness with which dream content changes over time, and the frequent enactments of several basic concerns over years and decades, are not what most adaptation theories would expect.

Dream series provide dream researchers with an unexpected source of ideal unobtrusive and nonreactive archival data. The numerous studies of dream series have led to replicated results that are similar to those from representative samples of dream reports analyzed with the same quantitative content methods. The similarity of the findings, using both representative samples of dream reports and dream series, is a form of convergent validity that increases the likelihood that the results from both types of studies are valid. Studies of dream series also have extended the scientific findings on dream content to include impressive consistency over months and years, as well as further evidence for the continuity between the frequency with which characters and avocations appear in dream reports and the intensity of conscious personal concerns about those characters and avocations in waking life. No other theory of dreams has incorporated the consistency dimension, and only a few of them could accommodate the findings on continuity.

Although none of the other dream theories makes use of the systematic findings on dream content in adults or children, there are several non-laboratory studies of adult dream content by activation-synthesis theorists and one of children. However, their nonlab studies are all inadequate and questionable due to one or more of several methodological problems. Their rating scales and coding categories are not of proven reliability or validity. Their results are very different from those reported by other researchers on many topics, and in particular those concerning bizarreness and emotions. These differences are due to their inclusion of scene changes as bizarre and to the demand characteristics they build into their studies of emotions. As a result, their atypical results have been frequently doubted or directly questioned by other dream researchers (Antrobus, 2000; Foulkes, 1996a, 1996b; Reinsel et al., 1992; Revonsuo & Salmivalli, 1995).

The neurocognitive theory of dreaming, unlike several other theories of dreaming, does not assume most published studies of dreaming and dream content can be taken at face value. Nor is there much, if any, consideration of the issues of a wake-state bias and demand characteristics in studies relying on self-ratings. This general contrast may be especially important on the issues of bizarreness and emotions in dream content. In the case of studies of emotions, the self-ratings are likely prone to the confounds created by a wake-state bias and perhaps especially when they are instructed to include any implicit emotions. Unless new methods are found to study emotions in dreams, it is very likely that the study of emotions will remain the most divisive issue facing dream researchers.

In addition to continuing to insist that bizarreness and emotions are hallmarks of dreaming, several theories assume that findings from well-controlled studies, based on large samples of nonclinical participants, can be used interchangeably with studies focused on a wide range of atypical mental states during both waking and sleep. Those atypical states include psychiatric symptoms, drug states, PTSD nightmares, and self-awareness during sleep. This back-and-forth, grab-bag approach ignores the distinction between foundational and translational models.

Meaning, Lawfulness, and Adaptive Function

Although the meaning of "meaning" is a fuzzy topic, the neurocognitive theory of dreaming leads to the conclusion that there is a considerable degree of meaning in dream content. There is, for example, coherency in dream reports that simulates what are agreed to be meaningful events in waking life. There are also several correspondences between dream content and waking life, which are meaningful because it is generally agreed that individual differences, gender differences, and cross-cultural differences are meaningful in waking life. Similarly, the dream reports in a dream series from an individual are meaningful because the consistencies in them parallel the consistencies in the individual's waking personal concerns and avocational interests. Finally, the developmental regularities in the frequency and complexity of dream reports from children from ages 5 to 12 are meaningful because they parallel the meaningful regularities in brain maturation and cognitive development.

On the other hand, Freudian dream theory claims that much, if not all, of the meaning in dreams is hidden. Meaning therefore can only be found through free association, which has been shown to be a dubious method for many reasons. Alternatively, Freudian theorists often claim that the hidden meaning in dream reports can be deciphered through the interpretation of alleged dream symbolism, which has been found to be rare in blind analyses of dream reports, a finding supported by neuroimaging evidence.

As for activation-synthesis theorists, they claim there is little if any meaning in dreams because they are triggered by random stimuli that are generated in the brainstem. They sometimes grant that there is some degree of coherency in dream content but add that a large percentage of dream reports contain bizarre elements, which are meaningless except as evidence for dreaming as a form of delirium. They also sometimes agree that the content in dream reports is similar to what the dreamer thinks or does in waking life, at least to some degree, but they accord little or no meaning to these similarities because dreams are considered to be cognitive trash in this theory (Hobson, 2002, p. 101). The adaptive theories, on the other hand, are able to accommodate the concept of meaning, and often depend upon it, as in the case of the problem-solving and social-rehearsal theories.

In addition to claiming dreams are meaningful, the neurocognitive theory concludes that there is evidence for dream reports containing more lawfulness than is usually thought to be the case. This point is best seen in the small-world nature of character networks in six different dream series that have the same properties as waking social networks (Han, 2014; Han & Schweickert, 2016; Han et al., 2015). The claim about dreams being lawful is not problematic for adaptation theories. However, Freudian dream theory cannot accommodate the findings on which this conclusion is based. Any lawfulness that Freudian dream theorists assume to be present in dreams cannot be detected without uncovering the ways in which the dream-work transforms the wishes underlying dream content into a more benign but often opaque, seemingly meaningless narrative. As already noted, the evidence for the theory's main concept related to lawfulness, the dream-work, is meager. To the degree that Freudians can put forth any evidence for this concept, it is based on the flawed method of free association or on unlikely claims about symbolism.

As for activation-synthesis theory, the emphasis on randomness is a claim about lawfulness, which is built into the theory by definition. There can be no lawfulness in dream reports generated by random events in the brainstem. This assumption is refuted by the results discussed in this book, which have been replicated numerous times. Thus, specific gender differences and specific cross-cultural differences can be "predicted" in any future study if sample sizes are large enough to detect them. Then, too, the fact of individual differences in any new dream series can be "predicted" with confidence. Since activation-synthesis theories have explicitly stated in at least one study in the past 15 years that there are no gender or individual differences in dream reports, their randomness theory is in effect refuted at the content level by those two empirically incorrect claims alone (Hobson & Kahn, 2007, p. 854).

Finally, as stated in chapter 10, the neurocognitive theory of dreaming makes no assumptions about the alleged adaptive function(s) of dreaming. It suggests that human imagination has created both cultural and individual uses for dreams in the course of history, often relating to religion, the treatment of illness, artistic creativity, and self-help measures. It thereby differs from Freudian theory, with its unlikely claim about dreams being the guardians of sleep. As also shown in chapter 10, the neurocognitive theory of dreaming also differs from the adaptive theories, which are based on unsupported claims about memory consolidation during dreaming in REM sleep, high levels of metaphoric thinking and hyperassociativity in dreams, implicit learning during dreaming, the transfer of new strategies acquired during dreaming to waking thought, or the consolidation of updated predictive codes during REM sleep dreaming.

Looking toward the Future

Based on the comparisons of the neurocognitive theory of dreaming with other theories of dreaming in this chapter, the new theory has the virtue that it can be tested, amended, extended, or refuted, due to the numerous hypotheses and replicated findings it has synthesized from a variety of independent research fields. At the same time, the other theories do not appear to have feasible agendas for further development. Few of their claims have been replicated by independent investigators, and most of their basic assertions have been called into question or refuted by a range of replicated studies.

The neurocognitive theory of dreaming could be tested at the level of dream content using various combinations of smartphone apps and home-based sleep-awakening devices that detect sleep stages. Individually or combined, those two devices make it possible to report dreams when they are recalled at any time of the day or night. They also make it possible to carry out awakenings throughout the night from a research office, while individual participants from ages 9 to 90 are asleep in the familiarity and comfort of their own bedrooms. Many basic questions about the frequency of dreaming, the factors involved in dream recall, and the representativeness of nonlab waking samples of dream reports could be studied anew in ways never before possible.

In addition, smartphone apps can be used by children as young as ages 7–8 and hence could be used to document dream recall and the content of dream reports. Smartphone apps, or periodic collections of written Most Recent Dreams (MRDs) in classrooms, make it feasible to study changes in the frequency, complexity, and substance of dream reports in a faster, easier, and more confidential way than was possible in the past. For example, a visit to every classroom in a middle school every two or three months during the school year to collect MRD reports would provide good samples for three one-year longitudinal studies and a three-year cross-sectional study. Such a study would include most of the time period (ages 9–13) during which dream content gradually becomes more adultlike. Then, too, MRD studies of fifth and sixth graders on the same day across cities, states, or countries might provide a valuable database on the frequency, length, and content of dream reports at the time when children begin to dream in a more adultlike fashion. In addition, greater efforts could be made to locate dream journals kept for a month or more during adolescence. The dream reports in them could be studied with Hall/Van de Castle (HVdC) content categories, individually tailored word strings, and those generic word strings making use of regular expressions to reduce false positives.

The content analyses of large samples of dream reports could be facilitated in several ways. First, the HVdC coding categories could be automated through an online interface that would display each dream report and provide menus or buttons for the entry of the relevant codings. The codings selected could be automatically entered into a spreadsheet at a central location. Codings could be carried out by multiple coders and then automatically checked for reliability. Final codings could be fed into statistical analysis tools similar to those already available on dreamresearch.net. These statistical analyses could be extended to include the extant approximate randomization program for h and p values, as well as programs correcting for multiple testing and for detecting autocorrelation in a dream series. Eventually, codings could be carried out by crowd-sourcing strategies. Although it is unlikely that generic word strings could ever carry out reliable and complete searches for social interactions and other complex aspects of dream content, the few currently available generic word strings for relatively simple types of content (e.g., sensory mentions, mentions of religion-related elements, mentions of nature terms) could be greatly improved by carefully checking for false positives after the results are available.

Based on the gender differences in the frequency and completeness of dream reports, which are documented in several different chapters, it would be methodologically sensible to concentrate on girls and women in order to more readily obtain large samples with complete dream reports, whether using MRDs, two-week diaries, or dream series. This focus on girls and women seems especially plausible because the findings presented in this book suggest studies of gender differences are not likely to make any further contributions to the construction of a neurocognitive theory of dreaming. Dream reports could still be collected in mixed-gender settings when necessary, but the focus of the data analysis would be on the girls' and women's reports.

Finally, there are now neuroimaging technologies that might make it easier to test hypotheses concerning the neural substrates and association networks subserving dreaming. In particular, parts of the theory's agenda perhaps could be tested in dream research laboratories through the use of functional near-infrared spectroscopy (fNIRS), a relatively recent neuroimaging technology in terms of its technical sophistication and feasibility. It does not require the presence of a technician, tolerates motion well, and costs in the $60,000 to $300,000 range. This range is approximately one-fourth to one-tenth as much as equivalent fMRI technologies, depending on size and the quality of resolution (Cui, Bray, Bryant, Glover, & Reiss, 2011; Kamran, Mannan, & Jeong, 2016). Moreover, portable versions of the fNIRS technology are not much bigger than a large kitchen microwave. The usefulness of this technology has been validated in at least one comparison with fMRI scans in adults (Cui et al., 2011) and for use with infants by comparing its results with those from the simultaneous use of fMRI (Bulgarelli et al., 2018). It has been used in two small pilot studies of the default network in adults (Durantin, Dehais, & Delorme, 2015; Harrivel, Weissman, Noll, & Peltier, 2013) and to locate the nascent default network in infants (Bulgarelli et al., 2020). It also has been used to study the early stages of sleep with a handful of sleep-deprived participants who were sitting in chairs (Nguyen et al., 2018).

Because fNIRS tolerates head motion better than fMRI does, its widest use in neurocognitive studies has been in studies of the prefrontal cortex in infants (Grossmann, 2013; Piazza, Hasenfratz, Hasson, & Lew-Williams, 2020). Its disadvantages concern its less detailed spatial resolution and its decreased signal-to-noise ratio, as well as its inability to reach below the cortex (Cui et al., 2011; Kamran et al., 2016; McKendrick, Parasuraman, & Ayaz, 2015). However, these possible limitations did not prove to be a problem in two studies of the default network in adults, which were mentioned in the previous paragraph (Durantin et al., 2015; Harrivel et al., 2013). Still, it might be necessary to make the bulky optode sensors, usually embedded in a cap, more comfortable for sleeping in a reclining position.

Using fNIRS technology, it might be possible to replicate the serendipitous finding of brief episodes of dreaming during drifting waking thought in individual participants reclining in a solitary laboratory setting (Foulkes, 1985; Foulkes & Fleisher, 1975; Foulkes & Scott, 1973). If brain patterns could distinguish mind-wandering from episodes of dreaming, perhaps the neural substrate supporting dreaming could be studied in more detail than is possible during experimental awakenings at night, when participants may take several seconds to awaken and respond (see Domhoff & Fox, 2015, for suggestions on what a study of this kind might include). The fNIRS technology also might make it possible to study a wide range of atypical dreamers in their home settings (Domhoff, 2020). For example, the two extremes on dream recall (those who have never recalled a dream and those who recall three or four dreams each day) could be compared in greater detail. (See Eichenlaub, Bertrand, & Ruby, 2014, for a study of frequent recallers; Pagel, 2003, for a study of 16 people who have never recalled a dream; and Vallat, Eichenlaub, Nicolas, & Ruby, 2018, for a replication of the findings on frequent recallers.)

The availability of portable fNIRS devices also provides a feasible and unobtrusive way to study the development of dreaming in young children in apartment-like lab settings. It also might make it possible to do neuro-imaging studies of children who are precocious or delayed dreamers, or to do home-based longitudinal studies of the development of the default network, the frequency of dream recall, and the nature of dream content in children ages 7–13. All it would take is a few good participants.

If smartphone apps, home sleep-awakening devices, the two relatively underutilized methods for collecting large samples of dream reports (MRDs and dream series), and the new neuroimaging technologies were used in a sustained way, backed by adequate resources, then the fate of the neurocognitive theory of dreaming, as well as the fate of the other theories discussed in this book, could be decided within a decade.

Acknowledgments

I am extremely grateful to my editor on this project, Philip Laughlin, for his careful consideration of an interdisciplinary manuscript on an enigmatic topic, for his patience in making sure it was thoroughly reviewed both before and after extensive revisions, and for his moral support and reassurances when I was struggling to improve both the focus and flow of the manuscript. I once again thank my longtime research assistant and frequent coauthor, Adam Schneider, for his graphics, tables, and formatting, as well as for his frank feedback and astute judgments on both substantive and editorial issues.

I thank dream researcher Pilleriin Sikka for helping me in understanding her fine series of studies on emotions and for helping me to avoid mistakes, but she is not responsible for any misjudgments and wrong interpretations I may have made in drawing on her work. I thank neuroscientist Joseph LeDoux for checking the several paragraphs I wrote in chapter 8 on the basis of his work on emotions. I thank my frequent coauthor on topics unrelated to dreams, social psychologist Richard L. Zweigenhaft, for his useful editorial suggestions on the introduction and chapter 1. I am very grateful to one of the reviewers of this manuscript for saving me from a mistake and for several useful suggestions and also to several of the other reviewers for providing useful suggestions and for insisting that the manuscript should be made much more comprehensible to readers. I am grateful to Jerry M. Siegel, an expert on the neurobiology of REM sleep, for helping me once again to understand the new advances in the understanding of REM sleep and its many concomitants. I also thank neuroimaging researcher B. T. Thomas Yeo for sending me the basic material that made figure 2.1 possible, and dream researchers Mark Blagrove and Anthony Zadra for their helpful comments on the subsection on predictive-coding theory.

Most of all, I thank my wife, Elizabeth K. Gray, a sociologist and word-smith par excellence, for taking the time from her own reading, research-ing, and bountiful backyard orchard in order to read and help me rewrite the two chapters that proved most difficult for me to compose: chapter 5, on figurative thinking and symbolism, and chapter 8, on emotions. I would not have made it through the manuscript that became this book without her help.

References

Abdallah, C., Averill, C., Ramage, A., Averill, L., Young-McCaughan, S., Fox, P., & Peterson, A. (2019). Salience network disruption in U.S. Army soldiers with posttraumatic stress disorder. *Chronic Stress, 3*, 1–10. https://doi.org/10.1177/2470547019850467.

Abraham, A. (2013). The world according to me: Personal relevance and the medial prefrontal cortex. *Frontiers in Human Neuroscience, 7*, 341–344.

Abraham, A. (2016). The imaginative mind. *Human Brain Mapping, 37*, 4197–4211.

Abraham, A. (2018). *The neuroscience of creativity.* New York: Cambridge University Press.

Achenbach, T. M., & Rescorla, L. A. (2007). *Multicultural understanding of child and adolescent psychopathology.* New York: Guilford Press.

Addis, D., Pan, L., Vu, M.-A., Laiser, N., & Schacter, D. (2009). Constructive episodic simulation of the future and the past: Distinct subsystems of a core brain network mediate imagining and remembering. *Neuropsychologia, 47*, 2222–2238.

Addis, D., Wong, A. T., & Schacter, D. (2007). Remembering the past and imagining the future: Common and distinct neural substrates during event construction and elaboration. *Neuropsychologia, 45*, 1363–1377.

Alexandra, G., & Chua, E. (2018). Transcranial direct current stimulation over the prefrontal cortex alters encoding and judgments of learning based on fluency. *Journal of Cognitive Neuroscience, 31*, 1710–1725. https://doi.org/10.1162/jocn_a_01449.

Allen, E., Laroi, F., McGuire, P., & Aleman, A. (2008). The hallucinating brain: A review of structural and functional neuroimaging studies of hallucinations. *Neuroscience & Biobehavioral Reviews, 32*, 175–191. https://doi.org/10.1016/j.neubiorev.2007.07.012.

Allen, P., Modinos, G., Hubl, D., Shields, G., Shotbolt, P., Plaze, M., & Hoffman, R. (2012). Neuroimaging auditory hallucinations in schizophrenia: From neuroanatomy to neurochemistry and beyond. *Schizophrenia Bulletin, 38*, 695–703. https://doi.org/10.1093/schbul/sbs066.

Allport, G. (1942). *The use of personal documents in psychological science*. New York: Social Science Research Council.

Andrews-Hanna, J., & Grilli, M. (2021). Mapping the imaginative mind: Charting new paths forward. *Current Directions in Psychological Science, 30*, 82–89. https://doi.org/10.1177/0963721420980753.

Andrews-Hanna, J., Irving, Z., Fox, K., Spreng, R. N., & Christoff, K. (2018). The neuroscience of spontaneous thought: An evolving interdisciplinary field. In K. Fox & K. Christoff (Eds.), *The Oxford handbook of spontaneous thought: Mind-wandering, creativity, and dreaming* (pp. 143–163). New York: Oxford University Press.

Andrews-Hanna, J., Reidler, J., Huang, C., & Buckner, R. (2010). Evidence for the default network's role in spontaneous cognition. *Journal of Neurophysiology, 104*, 322–335.

Andrews-Hanna, J., Reidler, J., Sepulcre, J., Poulin, R., & Buckner, R. (2010). Functional-anatomic fractionation of the brain's default network. *Neuron, 65*, 550–562.

Andrews-Hanna, J., Smallwood, J., & Spreng, R. N. (2014). The default network and self-generated thought: Component processes, dynamic control, and clinical relevance. *Annals of the New York Academy of Sciences, 1316*, 29–52.

Antrobus, J. (2000). How does the dreaming brain explain the dreaming mind? *Behavioral and Brain Sciences, 23*(6), 904–907.

Antrobus, J., Kondo, T., & Reinsel, R. (1995). Dreaming in the late morning: Summation of REM and diurnal cortical activation. *Consciousness & Cognition, 4*, 275–299.

Arkin, A. (1981). *Sleep talking: Psychology and psychophysiology*. Mahwah, NJ: Erlbaum.

Armitage, R. (1995). The distribution of EEG frequencies in REM and NREM sleep stages in healthy young adults. *Sleep, 18*, 334–341. https://doi.org/10.1093/sleep/18.5.334.

Arnulf, I. (2012). REM sleep behavior disorder: Motor manifestations and pathophysiology. *Movement Disorders, 27*, 677–689. https://doi.org/10.1002/mds.24957.

Arzi, A., Shedlesky, L., Ben-Shaul, M., Nasser, K., Oksenberg, A., Hairston, I., & Sobel, N. (2012). Humans can learn new information during sleep. *Natural Neuroscience, 15*, 1460–1465. https://doi.org/10.1038/nn.3193.

Aserinsky, E., Lynch, J., Mack, M., Tzankoff, S., & Hurn, E. (1985). Comparison of eye motion in wakefulness and REM sleep. *Psychophysiology, 22*(1), 1–10. https://doi.org/10.1111/j.1469-8986.1985.tb01551.x.

Atance, C., & Metcalf, J. (2013). Future thinking in young children. In M. Taylor (Ed.), *The Oxford handbook of the development of imagination* (pp. 305–324). New York: Oxford University Press.

Avidan, A. (2017). Non-rapid eye movement parasomnias: Clinical spectrum, diagnostic features, and management. In M. Kryger, T. Roth, & W. Dement (Eds.), *Principles and practices of sleep medicine* (6th ed., pp. 981–1001). Philadelphia: Elsevier.

Avila-White, D., Schneider, A., & Domhoff, G. W. (1999). The most recent dreams of 12–13-year-old boys and girls: A methodological contribution to the study of dream content in teenagers. *Dreaming, 9*, 163–171.

Avneon, M., & Lamy, D. (2018). Reexamining unconscious response priming: A liminal-prime paradigm. *Consciousness and Cognition, 59*, 87–103. https://doi.org/10.1016/j.concog.2017.12.006.

Ayella, M. (1998). *Insane therapy: Portrait of a psychotherapy cult*. Philadelphia: Temple University Press.

Baird, B., Castelnovo, A., Gosseries, O., & Tononi, G. (2018). Frequent lucid dreaming associated with increased functional connectivity between frontopolar cortex and temporoparietal association areas. *Science Reports, 8*(17798). https://doi.org/10.1038/s41598-018-36190-w.

Barrett, D. (1979). The hypnotic dream: Its relation to nocturnal dreams and waking fantasies. *Journal of Abnormal Psychology, 88*, 584–591.

Barrett, D. (1991). Flying dreams and lucidity: An empirical test of their relationship. *Dreaming, 1*, 129–134.

Barrett, D. (1992). Through a glass darkly: Images of the dead in dreams. *Omega, 24*, 97–108.

Barrett, D. (1993). The "committee of sleep": A study of dream incubation for problem solving. *Dreaming, 3*, 115–122.

Barrett, D. (2001). *The Committee of Sleep: How artists, scientists, and athletes use dreams for creative problem-solving—and how you can too*. New York: Crown/Random House.

Barrett, D. (2017). Dreams and creative problem-solving. *Annals of the New York Academy of Sciences, 1406*(1), 64–67. https://doi.org/10.1111/nyas.13412.

Barrett, D. (2020). Dreams about COVID-19 versus normative dreams: Trends by gender. *Dreaming, 30*, 216–221. https://doi.org/10.1037%2Fdrm0000149.

Barrett, L. F. (2013). The conceptual act theory: A precis. *Emotion Review, 6*, 292–297.

Barrett, L. F. (2017). The theory of constructed emotion: An active inference account of interoception and categorization. *Social Cognitive and Affective Neuroscience, 12*, 1–23.

Barsalou, L. (1982). Context-independent and context-dependent information in concepts. *Memory & Cognition, 10*, 82–93. https://doi.org/10.3758/BF03197629.

Barsalou, L. (1991). Deriving categories to achieve goals. In G. H. Bower (Ed.), *The psychology of learning and motivation: Advances in research and theory* (Vol. 27, pp. 1–64). San Diego: Academic Press.

Barsalou, L. (2003). Situated simulation in the human conceptual system. *Language and Cognitive Processes, 18*, 513–562.

Barsalou, L. (2009). Simulation, situated conceptualization, and prediction. *Philosophical Transactions of the Royal Society of London, Part B, 364*, 1281–1289. https://doi.org/10.1098/rstb.2008.0319.

Bauer, P. (2013). Memory. In P. Zelazo (Ed.), *The Oxford handbook of developmental psychology, Vol. 1: Body and mind* (pp. 503–541). New York: Oxford University Press.

Bauer, P., Burch, M., Scholin, S., & Güler, O. E. (2007). Using cue words to investigate the distribution of autobiographical memories in childhood. *Psychological Science, 18*, 910–916.

Baumeister, R., Bratslavsky, E., Finkenauer, C., & Vohs, K. (2001). Bad is stronger than good. *Review of General Psychology, 5*, 323–370.

Baylor, G. (2001). What do we really know about Mendeleev's dream of the periodic table? A note on dreams of scientific problem solving. *Dreaming, 11*, 89–92.

Baylor, G., & Cavallero, C. (2001). Memory sources associated with REM and NREM dream reports throughout the night: A new look at the data. *Sleep, 24*, 165–170.

Beaty, R., Chen, Q., Qiu, J., Silvia, P., & Schacter, D. (2018). Brain networks of the imaginative mind: Dynamic functional connectivity of default and cognitive control networks relates to openness to experience. *Human Brain Mapping, 39*, 811–821.

Beaty, R., & Silvia, P. J. (2013). Metaphorically speaking: Cognitive abilities and the production of figurative language. *Memory & Cognition, 41*, 255–267.

Beaty, R., Silvia, P. J., & Benedek, M. (2017). Brain networks underlying novel metaphor production. *Brain and Cognition, 111*, 163–170.

Beaulieu-Prevost, D., & Zadra, A. (2005a). Dream recall frequency and attitude towards dreams: A reinterpretation of the relation. *Personality and Individual Differences, 38*, 919–927.

Beaulieu-Prevost, D., & Zadra, A. (2005b). How dream recall frequency shapes people's beliefs about the content of their dreams. *North American Journal of Psychology, 7*, 253–264.

Beaulieu-Prevost, D., & Zadra, A. (2007). Absorption, psychological boundaries, and attitude towards dreams as correlates of dream recall: Two decades of research seen through a meta-analysis. *Journal of Sleep Research, 16*, 51–59.

Beaulieu-Prévost, D., & Zadra, A. (2015). When people remember dreams they never experienced: A study of the malleability of dream recall over time. *Dreaming, 25*, 18–31.

Belicki, K., Gulko, N., Ruzycki, K., & Aristotle, J. (2003). Sixteen years of dreams following spousal bereavement. *Omega, 47*, 93–106.

Bell, A., & Hall, C. (1971). *The personality of a child molester: An analysis of dreams.* Chicago: Aldine.

Bell, I. (1992). *Robert Louis Stevenson: Dreams of exile.* Edinburg: Mainstream Publishing.

Benedek, M., Beaty, R., Jauk, E., Koschutnig, K., Fink, A., Silvia, P. J., & Neubauera, A. (2014). Creating metaphors: The neural basis of figurative language production. *Neuroimage, 90,* 99–106.

Benjamin, G. (2014). *Temiar religion, 1964–2012: Enchantment, disenchantment, and re-enchantment in Malaysia's uplands.* Singapore: NUS Press.

Benjamini, Y., & Hochberg, Y. (1995). Controlling the false discovery rate: A practical and powerful approach to multiple testing. *Journal of the Royal Statistical Society: Series B (Methodological), 57,* 289–300.

Benjamini, Y., & Yekutieli, D. (2001). The control of the false discovery rate in multiple testing under dependency. *The Annals of Statistics, 29,* 1165–1188.

Bernstein, D. M., & Belicki, K. (1995–1996). On the psychometric properties of retrospective dream content questionnaires. *Imagination, Cognition and Personality, 15,* 351–364.

Bernstein, D. M., & Roberts, B. (1995). Assessing dreams through self-report questionnaires: Relations with past research and personality. *Dreaming, 5,* 13–27.

Bertolero, M., Yeo, B., Bassett, D., & D'Esposito, M. (2018). A mechanistic model of connector hubs, modularity and cognition. *Nature of Human Behavior, 10,* 765–777. https://doi.org/10.1038/s41562-018-0420-6.

Bickart, K., Dickerson, B., & Barrett, L. F. (2014). The amygdala as a hub in brain networks that support social life. *Neuropsychologia, 63,* 235–248.

Biderman, N., & Mudrik, L. (2108). Evidence for implicit—but not unconscious—processing of object-scene relations. *Psychological Science, 29,* 266–277. https://doi.org/10.1177/0956797617735745.

Bischof, M., & Basset, C. (2004). Total dream loss: A distinct neuropsychological dysfunction after bilateral PCA stroke. *Annals of Neurology, 56,* 583–586.

Blagrove, M. (1992). Dreams as a reflection of our waking concerns and abilities: A critique of the problem-solving paradigm in dream research. *Dreaming, 2,* 205–220.

Blagrove, M. (1996). Problems with the cognitive psychological modeling of dreaming. *Journal of Mind and Behavior, 17,* 99–134.

Blagrove, M. (2000). Dreams have meaning but no function. *Behavioral and Brain Sciences, 23,* 910.

Blagrove, M. (2007). Dreaming and personality. In D. Barrett & P. McNamara (Eds.), *The new science of dreaming: Content, recall, and personality correlates* (Vol. 2, pp. 115–158). Westport, CT: Praeger.

Blagrove, M., & Akehurst, L. (2000). Personality and dream recall frequency: Further negative findings. *Dreaming, 10,* 139–148.

Blagrove, M., Edwards, C., van Rijn, E., Reid, A., & Malinowski, J. (2019). Insight from the consideration of REM dreams, non-REM dreams, and daydreams. *Psychology of Consciousness: Theory, Research, and Practice, 6*, 138–162. https://doi.org/10.1037/cns0000167.

Blake, Y., Terburg, D., Balchin, R., van Honk, J., & Solms, M. (2019). The role of the basolateral amygdala in dreaming. *Cortex, 113*, 169–183. https://doi.org/10.1016/j .cortex.2018.12.016.

Bliwise, D., & Scullin, M. (2017). Normal aging. In M. Kryger, T. Roth, & W. Dement (Eds.), *Principles and practices of sleep medicine* (6th ed., pp. 25–38). Philadelphia: Elsevier.

Blumberg, M. S., & Plumeau, A. M. (2016). A new view of "dream enactment" in REM sleep behavior disorder. *Sleep Medicine Reviews, 30*, 34–42.

Born, J., & Wilhelm, I. (2012). System consolidation of memory during sleep. *Psychological Research, 76*, 192–203.

Botman, H., & Crovitz, H. (1989). Dream reports and autobiographical memory. *Imagination, Cognition and Personality, 9*, 213–214.

Boucetta, S., Cisse, Y., Mainville, L., Morales, M., & Jones, B. E. (2014). Discharge profiles across the sleep–waking cycle of identified cholinergic, GABAergic, and glutamatergic neurons in the pontomesencephalic tegmentum of the rat. *Journal of Neuroscience, 34*, 4708–4727.

Braun, A., Balkin, T., Wesensten, N., Carson, R., Varga, M., Baldwin, P., & Herscovitch, P. (1997). Regional cerebral blood flow throughout the sleep-wake cycle: An H_2 ^{15}O PET study. *Brain, 120*, 1173–1197.

Braun, A., Balkin, T., Wesensten, N., Gwadry, F., Carson, R., Varga, M., & Herscovitch, P. (1998). Dissociated pattern of activity in visual cortices and their projections during human rapid eye movement sleep. *Science, 279*, 91–95.

Breger, L. (1967). Function of dreams. *Journal of Abnormal Psychology, 72*(5, Pt. 2), 1–28. https://doi.org/10.1037/h0025040.

Breger, L. (1969). Children's dreams and personality development. In J. Fisher & L. Breger (Eds.), *The meaning of dreams* (pp. 64–100). California Mental Health Research Symposium, No. 3. Sacramento: Department of Mental Hygiene.

Brenman, M. (1949/1967). Dreams and hypnosis. In C. S. Moss (Ed.), *The hypnotic investigation of dreams* (pp. 455–465). New York: Wiley & Sons.

Brereton, D. (2000). Dreaming, adaptation, and consciousness: The social mapping hypothesis. *Ethos, 28*, 379–409.

Broughton, Richard. (2015). Psi and biology: An evolutionary perspective. In E. Cardeña, J. Palmer, & D. Marcusson-Clavertz (Eds.), *Parapsychology: A handbook for the 21st century* (pp. 139–148). Jefferson, NC: MacFarland & Co.

Broughton, Roger. (1968). Sleep disorders: Disorders of arousal? *Science, 159,* 1070–1078.

Buckley, J. (1970). *The dreams of young adults.* Doctoral dissertation, Wayne State University.

Buckner, R., Andrews-Hanna, J., & Schacter, D. (2008). The brain's default network: Anatomy, function, and relevance to disease. *Annals of the New York Academy of Sciences, 1124,* 1–38. https://doi.org/10.1196/annals.1440.011.

Buckner, R., Krienen, F., & Yeo, B. (2013). Opportunities and limitations of intrinsic functional connectivity MRI. *Nature Neuroscience, 16,* 832–837. https://doi.org/10.1038/nn.3423.

Bulgarelli, C., Blasi, A., Arridge, S., Powell, S., de Klerk, C., Southgate, V., Brigadoi, S., Penny, W., Tak, S., & Hamilton, A. (2018). Dynamic causal modelling on infant fNIRS data: A validation study on a simultaneously recorded fNIRS-fMRI dataset. *Neuroimage, 175,* 413–424.

Bulgarelli, C., Klerk, C. C. J. M., Richards, J., Southgate, V., Hamilton, A., & Blasi, A. (2020). The developmental trajectory of fronto-temporoparietal connectivity as a proxy of the default mode network: A longitudinal fNIRS investigation. *Human Brain Mapping, 41,* 2717–2740. https://doi.org/10.1002/hbm.24974.

Bulkeley, K. (1994). *The wilderness of dreams: Exploring the religious meanings of dreams in modern Western culture.* Albany: State University of New York Press.

Bulkeley, K. (1995). Conversion dreams. *Pastoral Psychology, 44*(1), 3–11.

Bulkeley, K. (2008). *Dreaming in the world's religions: A comparative history.* New York: New York University Press.

Bulkeley, K. (2009). The religious content of dreams: A new scientific foundation. *Pastoral Psychology, 58,* 93–106.

Bulkeley, K. (2012). Dreaming in adolescence: A "blind" word search of a teenage girl's dream series. *Dreaming, 22,* 240–252.

Bulkeley, K. (2014). Digital dream analysis: A revised method. *Consciousness and Cognition, 29,* 159–170.

Bulkeley, K. (2018). The meaningful continuities between dreaming and waking: Results of a blind analysis of a woman's 30-year dream journal. *Dreaming, 28,* 337–350. https://doi.org/10.1037/drm0000083.

Bulkeley, K., & Graves, M. (2018). Using the LIWC program to study dreams. *Dreaming, 28,* 43–58. https://doi.org/10.1037/drm0000071.

Bullmore, E., & Sporns, O. (2009). Complex brain networks: Graph theoretical analysis of structural and functional systems. *Nature Reviews Neuroscience, 10,* 186–198.

Burle, B., Spieser, L., Roger, C., Casini, L., Hasbroucq, T., & Vidala, F. (2015). Spatial and temporal resolutions of EEG: Is it really black and white? A scalp current density view. *International Journal of Psychophysiology, 197*, 210–220. https://doi.org/10.1016/j.ijpsycho.2015.05.004.

Bussey, K. (2013). Gender development. In M. Ryan & N. Branscombe (Eds.), *Sage handbook of gender and psychology* (pp. 81–99). Los Angeles: Sage.

Cai, B., Zhang, G., Hu, W., Zhang, A., Zille, P., Zhang, Y., & Wang, Y.-P. (2019). Refined measure of functional connectomes for improved identifiability and prediction. *Human Brain Mapping, 40*, 4843–4858. https://doi.org/10.1002/hbm.24741.

Calvo-Merino, B., Jones, A., Haggard, P., & Bettina, F. (2017). Embodiment and expertise effects on aesthetics judgments. *Meetings of the Society for Cognitive Neuroscience, San Francisco* (March 25–28).

Campbell, I., Grimm, K., de Bie, E., & Feinberg, I. (2012). Sex, puberty, and the timing of sleep EEG measured adolescent brain maturation. *Proceedings of the National Academy of Sciences, 109*, 5740–5743. https://doi.org/10.1073/pnas.1120860109.

Carey, B. (2020, October 6). Pandemic escape? Maybe not in your dreams. *New York Times*, p. 7.

Cartwright, R. (1977). *Night life: Explorations in dreaming.* Englewood Cliffs, NJ: Prentice-Hall.

Cartwright, R. (1996). Dreams and adaptation to divorce. In D. Barrett (Ed.), *Trauma and dreams* (pp. 179–185). Cambridge, MA: Harvard University Press.

Cartwright, R., Agargun, M., Kirkby, J., & Friedman, J. (2006). Relation of dreams to waking concerns. *Psychiatry Research, 141*, 261–270.

Ceci, S. J., Bruck, M., & Battin, D. B. (2000). The suggestibility of children's testimony. In D. Bjorklund (Ed.), *False-memory creation in children and adults: Theory, research, and implications* (pp. 169–201). Mahwah, NJ: Erlbaum.

Cesario, J., Johnson, D., & Eisthen, H. (2020). Your brain is not an onion with a tiny reptile inside. *Current Directions in Psychological Science, 29*, 255–260.

Chang, C., Liu, Z., Chen, M., Liu, X., & Duyn, J. (2013). EEG correlates of time-varying BOLD functional connectivity. *Neuroimage, 72*, 227–236.

Chen, A., Oathes, D., Chang, C., Bradley, T., & Zhou, Z.-W. (2013). Causal interactions between fronto-parietal central executive and default-mode networks in humans. *Proceedings of the National Academy of Sciences, 110*, 19944–19949.

Chen, Z., & Wilson, M. (2017). Deciphering neural codes of memory during sleep. *Trends in Neuroscience, 40*, 260–275. https://doi.org/10.1016/j.tins.2017.03.005.

Chow, H. M., Horovitz, S., Picchioni, D., Balkin, T., & Braun, A. (2013). Rhythmic alternating patterns of brain activity distinguish rapid eye movement sleep from other states of consciousness. *Proceedings of the National Academy of Sciences, 110,* 10300–10305.

Christensen, J., Aubin, S., Nielsen, T., Ptito, M., Kupers, R., & Jennum, P. (2019). Rapid eye movements are reduced in blind individuals. *Journal of Sleep Research, 28*(e12866), 1–10. doi:10.1111/jsr.12866.

Christoff, K. (2014). Thinking. In K. Ochsner & S. Kosslyn (Eds.), *The Oxford handbook of cognitive neuroscience* (pp. 318–333). New York: Oxford University Press.

Christoff, K., Gordon, A., Smallwood, J., Smith, R., & Schooler, J. (2009). Experience sampling during fMRI reveals default network and executive system contributions to mind wandering. *Proceedings of the National Academy of Sciences, 106,* 8719–8724.

Cicogna, P., Natale, V., Occhionero, M., & Bosinelli, M. (1998). A comparison of mental activity during sleep onset and morning awakening. *Sleep, 21*(5), 462–470.

Cipolli, C., Calasso, E., Maccolini, S., Pani, R., & Salzarulo, P. (1984). Memory processes in morning recall after multiple night awakenings. *Perceptual and Motor Skills, 59,* 435–446.

Cipolli, C., Mazzetti, M., Palagini, L., & Feinberg, I. (2015). Time-of-night variations in the story-like organization of dream experience developed during rapid eye movement sleep. *Journal of Sleep Research, 24,* 234–240. https://doi.org/10.1111/jsr .12251.

Cohen, D. (1974). Toward a theory of dream recall. *Psychological Bulletin, 38,* 122–138.

Cohen, D. (1979). *Sleep and dreaming: Origins, nature, and function.* New York: Pergamon Press.

Cohen, D., & Wolfe, G. (1973). Dream recall and repression: Evidence for an alternative hypothesis. *Journal of Consulting and Clinical Psychology, 41,* 349–355.

Cohen, J. (1960). A coefficient of agreement for nominal scales. *Educational and Psychological Measurement, 20,* 37–46.

Cohen, J. (1977). *Statistical power analysis for the behavioral sciences.* New York: Academic Press.

Cohen, J. (1990). Things I have learned (so far). *American Psychologist, 45,* 1304–1312.

Cohen, J. (1994). The earth is round ($p < .05$). *American Psychologist, 49,* 997–1003.

Côté, L., Lortie-Lussier, M., Roy, M.-J., & De Koninck, J. (1996). Continuity and change: The dreams of women throughout adulthood. *Dreaming, 6,* 187–199.

Coughlin, C. (2016). *Development of episodic prospection: Factors underlying improvements in middle and late childhood.* Doctoral dissertation, University of California, Davis.

Coughlin, C., Robins, R., & Ghetti, S. (2019). Development of episodic prospection: Factors underlying improvements in middle and late childhood. *Child Development, 90*, 1109–1122. https://doi.org/10.1111/cdev.13001.

Cross, I. (2016). The nature of music and its evolution. In S. Hallam, I. Cross, & M. Thaut (Eds.), *Oxford handbook of music psychology* (2nd ed., pp. 3–17). New York: Oxford University Press.

Crugnola, C., Maggiolini, A., Caprin, C., Martini, C., & Giudici, F. (2008). Dream content of 10- to 11-year-old preadolescent boys and girls. *Dreaming, 18*, 201–218.

Cui, X., Bray, S., Bryant, D., Glover, G., & Reiss, A. (2011). A quantitative comparison of NIRS and fMRI across multiple cognitive tasks. *Neuroimage, 54*, 2808–2821.

Curot, J., Valton, L., Denuelle, M., Vignal, J.-P., Maillard, L., Pariente, J., Trébuchon, A., Bartolomei, F., & Barbeau, E. J. (2018). Déjà-rêvé: Prior dreams induced by direct electrical brain stimulation. *Brain Stimulation, 11*, 875–885.

Dale, A., Lortie-Lussier, M., & De Koninck, J. (2015). Ontogenetic patterns in the dreams of women across the lifespan. *Consciousness and Cognition, 37*, 214–224.

Dale, A., Lortie-Lussier, M., & De Koninck, J. (2016). *Ontogenetic patterns in the dreams of men across the lifespan*. Ottawa: Department of Psychology, University of Ottawa.

Dale, A., Lortie-Lussier, M., Wong, C., & De Koninck, J. (2016). Dreams of Canadian students: Norms, gender differences, and comparison with Americans. *Journal of Cross-Cultural Psychology, 47*(6). https://doi.org/10.1177/0022022116655788.

Dallett, J. (1973). Theories of dream function. *Psychological Bulletin, 79*(6), 408–416.

D'Andrade, R. (1961). Anthropological studies in dreams. In F. Hsu (Ed.), *Psychological anthropology* (pp. 296–332). Homewood, IL: Dorsey Press.

D'Argembeau, A. (2020). Imagination and self-referential thinking. In A. Abraham (Ed.), *The Cambridge handbook of the imagination* (pp. 354–372). New York: Cambridge University Press.

Darwin, C. (1872/1998). *The expression of the emotions in man and animals* (3rd ed.). London: HarperCollins.

De Gennaro, L., Ferrara, M., Curcio, G., & Cristiani, R. (2001). Antero-posterior EEG changes during the wakefulness-sleep transition. *Clinical Neurophysiology, 112*, 1901–1911.

De Gennaro, L., Vecchio, F., Ferrara, M., Curcio, G., Rossini, P., & Babiloni, C. (2004). Changes in fronto-posterior functional coupling at sleep onset in humans. *Journal of Sleep Research, 13*, 209–217.

Dement, W., & Kleitman, N. (1957a). Cyclic variations in EEG during sleep and their relation to eye movements, body motility, and dreaming. *Electroencephalography and Clinical Neurophysiology, 9*, 673–690.

Dement, W., & Kleitman, N. (1957b). The relation of eye movements during sleep to dream activity: An objective method for the study of dreaming. *Journal of Experimental Psychology, 53*, 339–346.

Denkova, E., Nomi, J., Uddin, L., & Jha, A. (2019). Dynamic brain network configurations during rest and an attention task with frequent occurrence of mind wandering. *Human Brain Mapping, 40*, 4564–4576. https://doi.org/10.1002/hbm.24721.

Desjardins, S., & Zadra, A. (2006). Is the threat simulation theory threatened by recurrent dreams? *Consciousness and Cognition, 15*, 470–474.

Dixon, M., Andrews-Hanna, J., Spreng, R. N., Christoff, K., & Irving, Z. (2017). Interactions between the default network and dorsal attention network vary across default subsystems, time, and cognitive states. *Neuroimage, 147*, 632–649.

Dixon, M., Thiruchselvam, R., Todd, R., & Christoff, K. (2017). Emotion and the prefrontal cortex: An integrative review. *Psychological Bulletin, 143*, 1033–1081. https://doi.org/10.1037/bul0000096.

Dixon, M., Vega, A., Andrews-Hanna, J., Spreng, R. N., & Christoff, K. (2018). Heterogeneity within the frontoparietal control network and its relationship to the default and dorsal attention networks. *Proceedings of the National Academy of Sciences, 115*, E1598–E1607.

Domhoff, G. W. (1969). Home dreams and laboratory dreams: Home dreams are better. In M. Kramer (Ed.), *Dream psychology and the new biology of dreaming* (pp. 119–217). Springfield, IL: C. C. Thomas.

Domhoff, G. W. (1996). *Finding meaning in dreams: A quantitative approach*. New York: Plenum.

Domhoff, G. W. (1999). New directions in the study of dream content using the Hall and Van de Castle coding system. *Dreaming, 9*, 115–137. Retrieved from http://www.wkap.nl.

Domhoff, G. W. (2000, June 10). *A new neurocognitive theory of dreams*. Paper presented at the Annual Meeting of the Association for Psychological Science, Miami Beach, FL.

Domhoff, G. W. (2001). A new neurocognitive theory of dreams. *Dreaming, 11*, 13–33.

Domhoff, G. W. (2003). *The scientific study of dreams: Neural networks, cognitive development, and content analysis*. Washington, DC: American Psychological Association.

Domhoff, G. W. (2005). The content of dreams: Methodologic and theoretical implications. In M. Kryger, T. Roth, & W. Dement (Eds.), *Principles and practices of sleep medicine* (4th ed., pp. 522–534). Philadelphia: Elsevier Saunders.

Domhoff, G. W. (2011). The neural substrate for dreaming: Is it a subsystem of the default network? *Consciousness and Cognition, 20*, 1163–1174.

Domhoff, G. W. (2015). Dreaming as embodied simulation: A widower dreams of his deceased wife. *Dreaming, 25*, 232–256.

Domhoff, G. W. (2017). The invasion of the concept snatchers: The origins, distortions, and future of the continuity hypothesis. *Dreaming, 27*, 14–39.

Domhoff, G. W. (2018a). *The emergence of dreaming: Mind-wandering, embodied simulation, and the default network.* New York: Oxford University Press.

Domhoff, G. W. (2018b). Measurement and statistical issues relating to the study of dream content. Dreamresearch.net. https://dreams.ucsc.edu/Library/methodological _appendix.html.

Domhoff, G. W. (2020). New neuroimaging technologies and online methods for dream content analysis make it possible to study dreaming in non-disruptive and low-budget ways in sleep medicine clinics. In J. Pagel (Ed.), *Parasomnias and dreams* (pp. 293–319). Hauppauge, NY: Nova Science Publishers.

Domhoff, G. W., & Fox, K. (2015). Dreaming and the default network: A review, synthesis, and counterintuitive research proposal. *Consciousness and Cognition, 33,* 342–353.

Domhoff, G. W., & Gerson, A. (1967). Replication and critique of three studies on personality correlates of dream recall. *Journal of Consulting Psychology, 31*(4), 431.

Domhoff, G. W., & Kamiya, J. (1964). Problems in dream content study with objective indicators: I. A comparison of home and laboratory dream reports. *Archives of General Psychiatry, 11*, 519–524.

Domhoff, G. W., Meyer-Gomes, K., & Schredl, M. (2005–2006). Dreams as the expression of conceptions and concerns: A comparison of German and American college students. *Imagination, Cognition & Personality, 25*, 269–282.

Domhoff, G. W., & Schneider, A. (1999). Much ado about very little: The small effect sizes when home and laboratory collected dreams are compared. *Dreaming, 9*, 139–151.

Domhoff, G. W., & Schneider, A. (2008a). Similarities and differences in dream content at the cross-cultural, gender, and individual levels. *Consciousness and Cognition, 17*, 1257–1265.

Domhoff, G. W., & Schneider, A. (2008b). Studying dream content using the archive and search engine on DreamBank.net. *Consciousness and Cognition, 17*, 1238–1247.

Domhoff, G. W., & Schneider, A. (2015a). Assessing autocorrelation in studies using the Hall and Van de Castle coding system to study individual dream series. *Dreaming, 25*, 70–79.

Domhoff, G. W., & Schneider, A. (2015b). Correcting for multiple comparisons in studies of dream content: A statistical addition to the Hall/Van de Castle coding system. *Dreaming, 25*, 59–69.

Domhoff, G. W., & Schneider, A. (2020). From adolescence to young adulthood in two dream series: The consistency and continuity of characters and major personal interests. *Dreaming, 30*(2), 140–161. https://doi.org/10.1037/drm0000133.

Dorus, E., Dorus, W., & Rechtschaffen, A. (1971). The incidence of novelty in dreams. *Archives of General Psychiatry, 25*, 364–368.

dos Santos, R., Osório, F., Crippa, J., & Hallak, J. (2016). Classical hallucinogens and neuroimaging: A systematic review of human studies: Hallucinogens and neuroimaging. *Neuroscience and Biobehavioral Reviews, 71*, 715–728.

Dresler, M., Wehrle, R., Spoormaker, V., Koch, S., Holsboer, F., Steiger, A., & Czisch, M. (2012). Neural correlates of dream lucidity obtained from contrasting lucid versus non-lucid REM sleep: A combined EEG/fMRI case study. *Sleep, 35*, 1017–1020.

Drucker-Colin, R., & Pedrazo, J. (1981). Kainic acid lesions of gigantocellular tegmental field (FTG) does not abolish REM sleep. *Brain Research, 229*, 147–161.

Dudley, L., & Fungaroli, J. (1987). The dreams of students in a women's college: Are they different? *ASD Newsletter, 4*(6), 6–7.

Dudley, L., & Swank, M. (1990). A comparison of the dreams of college women in 1950 and 1990. *ASD Newsletter, 7*, 3.

Dunlap, W., Cortina, J., Vaslow, J., & Burke, M. (1996). Meta-analysis of experiments with matched groups or repeated measures designs. *Psychological Methods, 1*(2), 170–177.

Durantin, G., Dehais, F., & Delorme, A. (2015). Characterization of mind wandering using fNIRS. *Frontiers in Systems Neuroscience, 26*. https://doi.org/10.3389/fnsys.2015.00045.

Ebdlahad, S., Nofzinger, E., James, J. A., Buysse, D., Price, J., C., & Germain, A. (2013). Comparing neural correlates of REM sleep in posttraumatic stress disorder and depression: A neuroimaging study. *Psychiatry Research, 214*, 422–428.

Edlow, B., Takahashi, E., Wu, O., Benner, T., Dai, G., Bu, L., & Folkerth, R. (2012). Neuroanatomic connectivity of the human ascending arousal system critical to consciousness and its disorders. *Journal of Neuropathology and Experimental Neurology, 71*, 531–546.

Edwards, C., Ruby, P., Malinowski, J., Bennett, P., & Blagrove, M. (2013). Dreaming and insight. *Frontiers in Psychology, 4*. https://doi.org/10.3389/fpsyg.2013.00979.

Efrat, M., Hayat, H., Sharon, O., Andelman, F., Katzav, S., Lavie, P., & Nir, Y. (2018). Near-total absence of REM sleep co-occurring with normal cognition: An update of the 1984 paper. *Sleep Medicine, 52*, 134–137.

Eichenlaub, J. B., Bertrand, O., & Ruby, P. (2014). Brain reactivity differentiates subjects with high and low dream recall frequencies during both sleep and wakefulness. *Cerebral Cortex, 24*, 1206–1215. https://doi.org/10.1093/cercor/bhs388.

Eisbach, A. O. (2013a). *"Catching" a wandering mind: Developmental changes in the reporting of off-task thoughts*. Paper presented at the Society for Research on Child Development, Seattle, WA.

Eisbach, A. O. (2013b). What children understand about the flow of mental life. In M. Taylor (Ed.), *The Oxford handbook of the development of imagination* (pp. 359–375). New York: Oxford University Press.

Ellis, P. D. (2010). *The essential guide to effect sizes: Statistical power, meta-analysis, and the interpretation of research results*. Cambridge: Cambridge University Press.

Erdelyi, M. H. (1985). *Psychoanalysis: Freud's cognitive psychology*. New York: Macmillan.

Erdelyi, M. H. (1996). *The recovery of unconscious memories: Hypermnesia and reminiscence*. Chicago: University of Chicago Press.

Fair, D., Cohen, A. L., Dosenbach, N., Church, J. A., Miezin, F. M., Barch, D. M., & Schlaggar, B. L. (2008). The maturing architecture of the brain's default network. *Proceedings of the National Academy of Sciences, 105,* 4028–4032.

Fair, D., Cohen, A. L., Power, J., Dosenbach, N., Church, J. A., Miezin, F. M., & Peterson, S. (2009). Functional brain networks develop from a "local to distributed" organization. *PLoS Biology, 5,* 1–14. https://doi.org/10.1371/journal.pcbi.1000381.

Fair, D., Dosenbach, N., Church, J. A., Cohen, A. L., Brahmbhatt, S., Miezin, F. M., & Schlaggar, B. L. (2007). Development of distinct control networks through segregation and integration. *Proceedings of the National Academy of Sciences, 104,* 13507–13512.

Fan, F., Liao, X., Lei, T., Tao, S., & He, Y. (2021). Development of the default-mode network during childhood and adolescence: A longitudinal resting-state fMRI study. *Neuroimage, 226*(117581), 1–12. https://doi.org/10.1016/j.neuroimage.2020.117581.

Feinberg, I., & Campbell, I. (2010). Sleep EEG changes during adolescence: An index of a fundamental brain reorganization. *Brain and Cognition, 72* 56–65. https://doi.org/10.1016/j.bandc.2009.09.008.

Feinstein, J. (2015). Personal e-mail communication to G. William Domhoff, October 19.

Feinstein, J., Buzza, C., Hurlemann, R., Follmer, R., Dahdaleh, N., Coryell, W., & Wemmie, J. (2013). Fear and panic in humans with bilateral amygdala damage. *Nature Neuroscience, 16,* 270–272.

Ferguson, G. A. (1981). *Statistical analysis in psychology and education*. New York: McGraw-Hill.

Fife, D. (2020). The eight steps of data analysis: A graphical framework to promote sound statistical analysis. *Perspectives on Psychological Science, 15,* 1054–1075.

Fisher, C. (1954). Dreams and perception: The role of preconscious and primary modes of perception in dream formation. *Journal of the American Psychoanalytic Association, 2*, 389–445.

Fisher, C. (1966). Dreaming and sexuality. In R. Loewenstein, L. Newman, M. Schur, & A. Solnit (Eds.), *Psychoanalysis: A general psychology* (pp. 537–569). New York: International University Press.

Fisher, C., Kahn, E., Edwards, A., & Davis, D. (1973). A psychophysiological study of nightmares and night terrors: Physiological aspects of the Stage 4 night terror. *Journal of Nervous and Mental Disease, 157*, 75–98.

Fisher, S., & Greenberg, R. (1977). *The scientific credibility of Freud's theories and therapy.* New York: Basic Books.

Foley, M. A. (2013). Children's source monitoring of memories for imagination. In M. Taylor (Ed.), *The Oxford handbook of the development of imagination* (pp. 94–112). New York: Oxford University Press.

Ford, J., Roach, B., Jorgensen, K., Turner, J., Brown, G. G., Notestine, R., Bischoff-Grethe, A., Greve, D., Wible, C., Lauriello, J., Belger, A., Mueller, B. A., Calhoun, V., Preda, A., Keator, D., O'Leary, D. S., Lim, K. O., Glover, G., Potkin, S. G., & Mathalon, D. H. (2009). Tuning in to the voices: A multisite fMRI study of auditory hallucinations. *Schizophrenia Bulletin, 35*, 58–66. https://doi.org/10.1093/schbul/sbn140.

Fosse, R., Hobson, J. A., & Stickgold, R. (2003). Dreaming and episodic memory: A functional dissociation? *Journal of Cognitive Neuroscience, 15*, 1–9.

Fosse, R., Stickgold, R., & Hobson, J. A. (2001). The mind in REM sleep: Reports of emotional experience. *Sleep, 24*, 947–955.

Fosse, R., Stickgold, R., & Hobson, J. A. (2004). Thinking and hallucinating: Reciprocal changes in sleep. *Psychophysiology, 41*, 298–305.

Foulkes, D. (1978). *A grammar of dreams.* New York: Basic Books.

Foulkes, D. (1979). Home and laboratory dreams: Four empirical studies and a conceptual reevaluation. *Sleep, 2*, 233–251.

Foulkes, D. (1982). *Children's dreams: Longitudinal studies.* New York: Wiley.

Foulkes, D. (1983). Cognitive processes during sleep: Evolutionary aspects. In A. Mayes (Ed.), *Sleep mechanisms and functions in humans and animals: An evolutionary perspective* (pp. 313–337). Wokington, UK: Van Nostrand Reinhold.

Foulkes, D. (1985). *Dreaming: A cognitive-psychological analysis.* Hillsdale, NJ: Erlbaum.

Foulkes, D. (1996a). Dream research: 1953–1993. *Sleep, 19*, 609–624.

Foulkes, D. (1996b). Misrepresentation of sleep-laboratory dream research with children. *Perceptual and Motor Skills, 83*, 205–206.

Foulkes, D. (1999). *Children's dreaming and the development of consciousness*. Cambridge, MA: Harvard University Press.

Foulkes, D. (2017). Dreaming, reflective consciousness, and feelings in the preschool child. *Dreaming, 27,* 1–13.

Foulkes, D., & Fleisher, S. (1975). Mental activity in relaxed wakefulness. *Journal of Abnormal Psychology, 84,* 66–75.

Foulkes, D., Hollifield, M., Sullivan, B., Bradley, L., & Terry, R. (1990). REM dreaming and cognitive skills at ages 5–8: A cross-sectional study. *International Journal of Behavioral Development, 13,* 447–465.

Foulkes, D., Meier, B., Strauch, I., Kerr, N., Bradley, L., & Hollifield, M. (1993). Linguistic phenomena and language selection in the REM dreams of German-English bilinguals. *International Journal of Psychology, 28*(6), 871–891.

Foulkes, D., & Schmidt, M. (1983). Temporal sequence and unit comparison composition in dream reports from different stages of sleep. *Sleep, 6,* 265–280.

Foulkes, D., & Scott, E. (1973). An above-zero baseline for the incidence of momentarily hallucinatory mentation. *Sleep Research, 2,* 108.

Foulkes, D., Spear, P., & Symonds, J. (1966). Individual differences in mental activity at sleep onset. *Journal of Abnormal and Social Psychology, 71,* 280–286.

Foulkes, D., Sullivan, B., Hollifield, M., & Bradley, L. (1989). Mental rotation, age, and conservation. *Journal of Genetic Psychology, 150,* 449–451.

Foulkes, D., Sullivan, B., Kerr, N., & Brown, L. (1988). Appropriateness of dream feelings to dreamed situations. *Cognition and Emotion, 2,* 29–39.

Foulkes, D., & Vogel, G. (1965). Mental activity at sleep onset. *Journal of Abnormal Psychology, 70,* 231–243.

Fox, K. (2018). Neural origins of self-generated thought: Insights from intracranial electrical stimulations and recordings in humans. In K. Fox & K. Christoff (Eds.), *The Oxford handbook of spontaneous thought: Mind-wandering, creativity, and dreaming* (pp. 165–179). New York: Oxford University Press.

Fox, K., Andrews-Hanna, J., Mills, C., Dixon, M., Markovic, J., Thompson, E., & Christoff, K. (2018). Affective neuroscience of self-generated thought. *Annals of the New York Academy of Sciences, 1426,* 25–51.

Fox, K., Dixon, M., Nijeboer, S., Girn, M., Floman, J., Lifshitz, M., & Christoff, K. (2016). Functional neuroanatomy of meditation: A review and meta-analysis of 78 functional neuroimaging investigations. *Neuroscience Behavioral Review, 65,* 208–228.

Fox, K., Girn, M., Parro, C. C., & Christoff, K. (2018). Functional neuroimaging of psychedelic experience: An overview of psychological and neural effects and their relevance to research on creativity, daydreaming, and dreaming. In R. E. Jung & O. Vartanian (Eds.), *The Cambridge handbook of the neuroscience of creativity* (pp. 92–113). New York: Cambridge University Press.

Fox, K., Nijeboer, S., Solomonova, E., Domhoff, G. W., & Christoff, K. (2013). Dreaming as mind wandering: Evidence from functional neuroimaging and first-person content reports *Frontiers in Human Neuroscience, 7*(Article 412), 1–18.

Fox, K., Spreng, R. N., Ellamil, M., Andrews-Hanna, J., & Christoff, K. (2015). The wandering brain: Meta-analysis of functional neuroimaging studies of mind-wandering and related spontaneous thought processes. *Neuroimage, 111*, 611–621.

Frank, J. (1950). Some aspects of lobotomy (prefrontal leucotomy) under psychoanalytic scrutiny. *Psychiatry, 13*, 35–42.

Franklin, M., Smallwood, J., Zedelius, C. M., Broadway, J. M., & Schooler, J. W. (2016). Unaware yet reliant on attention: Experience sampling reveals that mind-wandering impedes implicit learning. *Psychonomic Bulletin & Review, 23*, 223–229. https://doi.org/10.3758/s13423-015-0885-5.

Franklin, M., & Zyphur, M. (2005). The role of dreams in the evolution of the human mind. *Evolutionary Psychology, 3*, 59–78.

Franklin, R., Allison, D., & Gorman, B. (1997). *Design and analysis of single-case research*. Mahwah, NJ: Erlbaum.

Fransson, P., & Marrelec, G. (2008). The precuneus/posterior cingulate cortex plays a pivotal role in the default mode network: Evidence from a partial correlation network analysis. *Neuroimage, 42*, 1178–1184.

Freud, S. (1900). *The interpretation of dreams*. London: Oxford University Press.

Freud, S. (1900/1909). *The interpretation of dreams* (Vol. 5). London: Hogarth Press.

Freud, S. (1916). *Introductory lectures on psychoanalysis* (Vol. 15). London: Hogarth Press.

Freud, S. (1933). *New introductory lectures on psychoanalysis* (Vol. 22). London: Hogarth Press.

Frick, A., Hansen, M., & Newcombe, N. (2013). Development of mental rotation in 3- to 5-year-old children. *Cognitive Development, 28*, 386–399.

Frick, A., Möhring, W., & Newcombe, N. S. (2014). Development of mental transformation abilities. *Trends in Cognitive Sciences, 18*, 536–542.

Friedman, L., & Jones, B. E. (1984a). Computer graphics analysis of sleep-wakefulness state changes after pontine lesions. *Brain Research Bulletin, 13*, 53–68.

Friedman, L., & Jones, B. E. (1984b). Study of sleep-wakefulness states by computer graphics and cluster analysis before and after lesions of the pontine tegmentum in the cat. *Electroencephalography and Clinical Neurophysiology, 57*, 43–56.

Friston, K. (2010). The free-energy principle: A unified brain theory? *Nature Reviews Neuroscience, 11,* 127–138.

Friston, K. (2014). Free energy and sleep. In N. Tranquillo (Ed.), *Dream consciousness: Allan Hobson's new approach to the brain and its mind* (pp. 137–142). New York: Springer.

Fudin, R. (2006). Critique of Sohlberg and Birgegard's (2003) report of persistent complex effects of subliminal psychodynamic activation messages. *Perceptual and Motor Skills, 103*(2), 551–564.

Fuentes, A. (2020). The evolution of a human imagination. In A. Abraham (Ed.), *The Cambridge handbook of the imagination* (pp. 13–29). New York: Cambridge University Press.

Gabrielsson, A., Whaley, J., & Sloboda, J. (2016). Peak experiences in music. In S. Hallam, I. Cross, & M. Thaut (Eds.), *The Oxford handbook of music psychology* (2nd ed., pp. 745–758). New York: Oxford University Press.

Gackenbach, J. (1988). The psychological content of lucid versus nonlucid dreams. In J. Gackenbach & S. LaBerge (Eds.), *Conscious mind, sleeping brain: Perspectives on lucid dreaming* (pp. 181–220). New York: Plenum.

Garcia, L. (2004). Escaping the Bonferroni iron claw in ecological studies. *Oikos, 105,* 657–660.

Garrison, K. A., Zeffiro, T. A., Scheinost, D., Constable, R. T., & Brewer, J. A. (2015). Meditation leads to reduced default mode network activity beyond an active task. *Cognitive, Affective & Behavioral Neuroscience, 15,* 712–720.

Gerbner, G. (Ed.) (1969). *The analysis of communication content.* New York: Wiley.

Germain, A., Jeffrey, J., Salvatore, I., Herringa, R. J., & Mammen, O. (2013). A window into the invisible wound of war: Functional neuroimaging of REM sleep in returning combat veterans with PTSD. *Psychiatry Research: Neuroimaging, 211,* 176–179.

Gibbs, R. (1994). *The poetics of mind: Figurative thought, language, and understanding.* New York: Cambridge University Press.

Gibbs, R. (1999). Speaking and thinking with metonymy. In K. Panther & G. Radden (Eds.), *Metonymy in language and thought* (pp. 61–75). Philadelphia: Benjamins.

Gibbs, R. (2006). *Embodiment and cognitive science.* New York: Cambridge University Press.

Gifuni, A., Kendal, A., & Jollant, F. (2017). Neural mapping of guilt: A quantitative meta-analysis of functional imaging studies. *Brain Imaging and Behavior, 11,* 1164–1178.

Göder, R., Seeck-Hirschner, M., Stingele, K., Huchzermeier, C., Kropp, C., Palaschewski, M., & Koch, J. (2011). Sleep and cognition at baseline and the effects of

REM sleep diminution after 1 week of antidepressive treatment in patients with depression. *Journal of Sleep Research, 20,* 544–551. https://doi.org/10.1111/j.1365 -2869.2011.00914.x.

Goodenough, D. (1991). Dream recall: History and current status of the field. In S. Ellman & J. Antrobus (Eds.), *The mind in sleep: Psychology and psychophysiology* (2nd ed., pp. 143–171). New York: Wiley & Sons.

Gopnik, A. (2009). *The philosophical baby: What children's minds tell us about truth, love, and the meaning of life.* New York: Farrar, Straus, and Giroux.

Gordon, E. M., Lee, P. S., Maisog, J. M., Foss-Feig, J., Billington, M. E., Van-Meter, J., & Vaidya, C. J. (2011). Strength of default mode resting-state connectivity relates to white matter integrity in children. *Developmental Science, 14,* 738–751.

Grady, J. (1999). A typology of motivation for conceptual metaphor: Correlation vs. resemblance. In R. Gibbs & G. Steen (Eds.), *Metaphor in cognitive linguistics* (pp. 79–100). Philadelphia: Benjamins.

Greenberg, R., Katz, H., Schwartz, W., & Pearlman, C. (1992). A research-based reconsideration of the psychoanalytic theory of dreaming. *Journal of the American Psychoanalytic Association, 40,* 531–550.

Greenberg, R., & Pearlman, C. (1993). An integrated approach to dream theory: Contributions from sleep research and clinical practice. In A. Moffitt, M. Kramer, & R. Hoffmann (Eds.), *The functions of dreaming* (pp. 363–380). Albany: State University of New York Press.

Greenberg, R., Pearlman, C., & Gampel, D. (1972). War neuroses and the adaptive function of REM sleep. *British Journal of Medical Psychology, 45,* 27–33. https://doi.org /10.1111/j.2044-8341.1972.tb01416.x.

Greenwald, A. G. (1992). New Look 3: Unconscious cognition reclaimed. *American Psychologist, 47*(6), 766–779.

Greenwald, A. G., Draine, S. C., & Abrams, R. L. (1996). Three cognitive markers of unconscious semantic activation. *Science, 273*(5282), 1699–1702.

Gregor, T. (1981). "Far, far away my shadow wandered . . .": The dream symbolism and dream theories of the Mehinaku Indians of Brazil. *American Ethnologist, 8,* 709–720.

Griffith, R., Miyago, O., & Tago, A. (1958). The universality of typical dreams: Japanese vs. Americans. *American Anthropologist, 60,* 1173–1179.

Gross, M. (1988). *Ein inhaltsanalytischer Vergleich von Heim- und Laborträumen* (A content analysis comparison of home and laboratory dreams). Master's thesis, University of Zurich, Switzerland.

Grossmann, T. (2013). The role of medial prefrontal cortex in early social cognition. *Frontiers in Human Neuroscience, 5*(7), 340.

Gusnard, D. A., Akbudak, E., Shulman, G. L., & Raichle, M. E. (2001). Medial prefrontal cortex and self-referential mental activity: Relation to a default mode of brain function. *Proceedings of the National Academy of Sciences, 98*, 4259–4264.

Gusnard, D. A., & Raichle, M. E. (2001). Searching for a baseline: Functional imaging and the resting human brain. *Nature Reviews Neuroscience, 2*, 685–694.

Hall, C. (1953). A cognitive theory of dream symbols. *Journal of General Psychology, 48*, 169–186.

Hall, C. (1964). Slang and dream symbols. *Psychoanalytic Review, 51*, 38–48.

Hall, C. (1966). *Studies of dreams collected in the laboratory and at home*. Santa Cruz, CA: Institute of Dream Research. PDF available for download in the Dream Library at dreamresearch.net.

Hall, C. (1969a). Content analysis of dreams: Categories, units, and norms. In G. Gerbner (Ed.), *The analysis of communication content* (pp. 147–158). New York: Wiley.

Hall, C. (1969b). Normative dream content studies. In M. Kramer (Ed.), *Dream psychology and the new biology of dreaming* (pp. 175–184). Springfield, IL: Charles C. Thomas.

Hall, C. (1984). A ubiquitous sex difference in dreams, revisited. *Journal of Personality and Social Psychology, 46*, 1109–1117.

Hall, C., & Domhoff, G. W. (1963a). Aggression in dreams. *International Journal of Social Psychiatry, 9*, 259–267.

Hall, C., & Domhoff, G. W. (1963b). A ubiquitous sex difference in dreams. *Journal of Abnormal and Social Psychology, 66*, 278–280.

Hall, C., & Domhoff, G. W. (1964). Friendliness in dreams. *Journal of Social Psychology, 62*, 309–314.

Hall, C., Domhoff, G. W., Blick, K., & Weesner, K. (1982). The dreams of college men and women in 1950 and 1980: A comparison of dream contents and sex differences. *Sleep, 5*, 188–194.

Hall, C., & Lind, R. (1970). *Dreams, life and literature: A study of Franz Kafka*. Chapel Hill: University of North Carolina Press.

Hall, C., & Van de Castle, R. (1966). *The content analysis of dreams*. New York: Appleton-Century-Crofts.

Han, H. J. (2014). *Structural and longitudinal analysis of cognitive social networks in dreams*. Doctoral dissertation, Purdue University.

Han, H. J., & Schweickert, R. (2016). Continuity: Knowing each other, emotional closeness, and appearing together in dreams. *Dreaming, 26*, 299–307. https://doi.org/10.1037/drm0000038.

Han, H. J., Schweickert, R., Xi, Z., & Viau-Quesnela, C. (2015). The cognitive social network in dreams: Transitivity, assortativity, and giant component proportion are monotonic. *Cognitive Science, 40,* 671–696.

Harlow, J., & Roll, S. (1992). Frequency of day residue in dreams of young adults. *Perceptual & Motor Skills, 74,* 832–834.

Harrivel, A., Weissman, D., Noll, D., & Peltier, S. (2013). Monitoring attentional state with fNIRS. *Frontiers in Human Neuroscience, 13*(7), 861. https://doi.org/10.3389/fnhum.2013.00861.

Hartmann, E. (1968). The day residue: Time distribution of waking events. *Psychophysiology, 5,* 222.

Hartmann, E. (1984). *The nightmare.* New York: Basic Books.

Hartmann, E., Elkin, R., & Garg, M. (1991). Personality and dreaming: The dreams of people with very thick or very thin boundaries. *Dreaming, 1,* 311–324.

Hartmann, E., Russ, D., Oldfield, M., Falke, R., & Skoff, B. (1980). Dream content: Effects of L-DOPA. *Sleep Research, 9,* 153.

Hasan, J., & Broughton, R. (1994). Quantitative topographic EEG mapping during drowsiness and sleep. In R. Ogilvie & J. Harsh (Eds.), *Sleep onset: Normal and abnormal processes* (pp. 219–235). Washington, DC: American Psychological Association.

Haux, L., Engelmann, J. M., Herrmann, E., & Tomasello, M. (2017). Do young children preferentially trust gossip or firsthand observation in choosing a collaborative partner? *Social Development, 26,* 466–474. https://doi.org/10.1111/sode.12225.

Hayashi, M., Katoh, K., & Hori, T. (1999). Hypnagogic imagery and EEG activity. *Perceptual and Motor Skills, 88,* 676–678.

He, W., Sowman, P., Brock, J., Etchell, A., Stam, C., & Hillebrand, A. (2019). Increased segregation of functional networks in developing brains. *Neuroimage, 200,* 607–620. https://doi.org/10.1016/j.neuroimage.2019.06.055.

Hedges, L., & Schauer, J. (2019). More than one replication study is needed for unambiguous tests of replication. *Journal of Educational and Behavioral Statistics, 44,* 543–570. https://doi.org/10.3102/1076998619852953.

Hobson, J. A. (1988). *The dreaming brain.* New York: Basic Books.

Hobson, J. A. (2000). The ghost of Sigmund Freud haunts Mark Solms's dream theory. *Behavioral and Brain Sciences, 23*(6), 951–952.

Hobson, J. A. (2002). *Dreaming: An introduction to the science of sleep.* New York: Oxford University Press.

Hobson, J. A. (2007). Current understanding of cellular models of REM expression. In D. Barrett & P. McNamara (Eds.), *The new science of dreaming: Biological aspects* (Vol. 1, pp. 71–84). Westport, CT: Praeger.

Hobson, J. A., & Friston, K. (2012). Waking and dreaming consciousness: Neurobiological and functional considerations. *Progress in Neurobiology, 98,* 82–98.

Hobson, J. A., & Kahn, D. (2007). Dream content: Individual and generic aspects. *Consciousness and Cognition, 16,* 850–858.

Hobson, J. A., Lydic, R., & Baghdoyan, H. (1986). Evolving concepts of sleep cycle generation: From brain centers to neuronal populations. *Behavioral and Brain Sciences, 9,* 371–448.

Hobson, J. A., & McCarley, R. (1977). The brain as a dream state generator: An activation-synthesis hypothesis of the dream process. *American Journal of Psychiatry, 134,* 1335–1348.

Hobson, J. A., McCarley, R., & Wyzinski, P. (1975). Sleep cycle oscillation: Reciprocal discharge by two brainstem neuronal groups. *Science, 189,* 55–58.

Hobson, J. A., Pace-Schott, E., & Stickgold, R. (2000a). Dream science 2000: A response to commentaries on dreaming and the brain. *Behavioral and Brain Sciences, 23,* 1019–1034.

Hobson, J. A., Pace-Schott, E., & Stickgold, R. (2000b). Dreaming and the brain: Toward a cognitive neuroscience of conscious states. *Behavioral and Brain Sciences, 23*(6), 793–842.

Hoffman, R., & Hampson, M. (2012). Functional connectivity studies of patients with auditory verbal hallucinations. *Frontiers in Human Neuroscience, 6,* 6. https://doi.org/10.3389/fnhum.2012.0000.

Hole, J., Hirsch, M., Ball, E., & Meads, C. (2015). Music as an aid for postoperative recovery in adults: A systematic review and meta-analysis. *Lancet, 60,* 169–176.

Holland, D., & Kipnis, A. (1994). Metaphors for embarrassment and stories of exposure: The not-so-egocentric self in American culture. *Ethos, 22,* 316–342.

Holm, S. (1979). A simple sequentially rejective multiple test procedure. *Scandinavian Journal of Statistics, 6,* 65–70.

Holyoak, K., & Stamenkovic, D. (2018). Metaphor comprehension: A critical review of theories and evidence. *Psychological Bulletin, 144,* 641–671. https://doi.org/10.1037/bul0000145.

Hong, C., Fallon, J., & Friston, K. (2021). fMRI evidence for default mode network deactivation associated with rapid eye movements in sleep. *Brain Sciences, 11*(1528), 1–21. https://doi.org/10.3390/brainsci11111528.

Hori, T., Hayashi, M., & Hibino, K. (1992). An EEG study of the hypnagogic hallucinatory experience. *International Journal of Psychology, 27*, 420.

Hori, T., Hayashi, M., & Morikawa, T. (1990). Typography and coherence analysis of hypnagogic EEG. In J. Horne (Ed.), *Sleep '90* (pp. 10–12). Bochum, Germany: Pontenagel Press.

Hori, T., Hayashi, M., & Morikawa, T. (1994). Topographic EEG changes and the hypnagogic experience. In R. Ogilvie & J. Harsh (Eds.), *Sleep onset: Normal and abnormal processes* (pp. 237–253). Washington, DC: American Psychological Association.

Horikawa, T., & Kamitani, Y. (2017). Hierarchical neural representation of dreamed objects revealed by brain decoding with deep neural network features. *Frontiers in Computational Neuroscience, 11*. https://doi.org/10.3389/fncom.2017.00004.

Horikawa, T., Tamaki, M., Miyawaki, Y., & Kamitani, Y. (2013). Neural decoding of visual imagery during sleep. *Science, 340*, 639–642.

Horn, A., Ostwalda, D., Reisert, M., & Blankenburg, F. (2013). The structural–functional connectome and the default mode network of the human brain. *Neuroimage, 102*, 142–151.

Horne, J., & McGrath, M. (1984). The consolidation hypothesis for REM sleep function: Stress and other confounding factors: A review. *Biological Psychology, 18*, 165–184. https://doi.org/10.1016/0301-0511(84)90001-2.

Horovitz, S. (2008). Low frequency BOLD fluctuations during resting wakefulness and light sleep: A simultaneous EEG-fMRI study. *Human Brain Mapping, 29*, 671–682.

Horovitz, S., Braun, A., Carr, W., Picchioni, D., Balkin, T., Fukunaga, M., & Duyn, J. (2009). Decoupling of the brain's default mode network during deep sleep. *Proceedings of the National Academy of Sciences, 106*, 11376–11381.

Horton, C. (2009). Applying memory theory to dream recall: Are dreams and waking memories the same? In M. Kelley (Ed.), *Applied memory* (pp. 275–303). Hauppauge, NY: Nova Science Publishers.

Horton, C. (2017). Consciousness across sleep and wake: Discontinuity and continuity of memory experiences as a reflection of consolidation processes. *Frontiers in Psychiatry, 8*. https://doi.org/10.3389/fpsyt.2017.00159.

Howard, M. (1978). *Manifest dream content of adolescents.* Doctoral dissertation, Iowa State University.

Humiston, G., Tucker, M., Summer, T., & Wamsley, E. (2019). Resting states and memory consolidation: A preregistered replication and meta-analysis. *Science Reports, 9*(19345).

Hurovitz, C., Dunn, S., Domhoff, G. W., & Fiss, H. (1999). The dreams of blind men and women: A replication and extension of previous findings. *Dreaming, 9*, 183–193.

Hursch, C., Karacan, I., & Williams, R. (1972). Some characteristics of nocturnal penile tumescence in early middle-aged males. *Comprehensive Psychiatry, 13,* 539–548.

Hyde, J. S. (2014). Gender similarities and differences. *Annual Review of Psychology, 65,* 373–398.

Inman, C. S., Manns, J. R., Bijanki, K. R., Bass, D. I., Hamann, S., Drane, D. L., Fasano, R. E., Kovach, C. K., Gross, R. E., & Willie, J. T. (2018). Direct electrical stimulation of the amygdala enhances declarative memory in humans. *Proceedings of the National Academy of Sciences, 115,* 98–103. https://doi.org/10.1073/pnas.1714058114.

Iorio, I., Sommantico, M., & Parrello, S. (2020). Dreaming in the time of COVID-19: A quali-quantitative Italian study. *Dreaming, 30,* 199–215. https://doi.org/10.1037/drm0000142te.

Iranzo, A. (2017). Other parasomnias. In M. Kryger, B. J. Roth, & W. Dement (Eds.), *Principles and practices of sleep medicine* (6th ed., Vol. 6, pp. 1011–1019). Philadelphia: Elsevier.

Izawa, S., Chowdhury, S., Miyazaki, T., Mukai, Y., Ono, D., Inoue, R., & Yamanaka, A. (2019). REM sleep–active MCH neurons are involved in forgetting hippocampus-dependent memories. *Science, 365,* 1308–1313. https://doi.org/10.1126/science.aax9238.

Jacobs, L., Feldman, M., & Bender, M. (1972). Are the eye movements of dreaming sleep related to the visual images of dreams? *Psychophysiology, 9,* 393–401.

Jalbrzikowski, M., Liu, F., Foran, W., Roeder, K., & Luna, B. (2020). Functional connectome fingerprinting accuracy in youths and adults is similar when examined on the same day and 1.5-years apart. *Human Brain Mapping,* online ahead of print, 1–13. https://doi.org/10.1002/hbm.25118.

Jenkins, A., & Mitchell, J. (2011). Medial prefrontal cortex subserves diverse forms of self-reflection. *Social Neuroscience, 6,* 211–218.

Ji, J. L., Spronk, M., Kulkarni, K., Anticevic, A., & Cole, M. W. (2019). Mapping the human brain's cortical-subcortical functional network organization. *Neuroimage, 185,* 35–57. https://doi.org/10.1016/j.neuroimage.2018.10.006.

Jones, B. E. (1979). Elimination of paradoxical sleep by lesions of the pontine gigantocellular tegmental field in the cat. *Neuroscience Letters, 13,* 285–293.

Jones, B. E. (1986). The need for a new model of sleep cycle generation. *Behavioral and Brain Sciences, 9,* 409–411.

Jones, B. E. (2000). The interpretation of physiology. *Behavioral and Brain Sciences, 23*(6), 955–956.

Jones, B. E. (2020). Arousal and sleep circuits. *Neuropsychopharmacology, 45,* 6–20. https://doi.org/10.1038/s41386-019-0444-2.

Jones, B. E., Harper, S., & Halaris, A. (1977). Effects of locus coeruleus lesions upon cerebral monamine content, sleep-wakefulness states and the response to amphetamine in the cat. *Brain Research, 124,* 473–496.

Jones, R. M. (1962). *Ego synthesis in dreams.* Cambridge, MA: Schenkman.

Jus, A., Jus, K., Gautier, J., Villeneuve, A., Pires, P., Lachance, R., & Villeneuve, R. (1973). Dream reports after reserpine in chronic lobotomized schizophrenic patients. *Vie medicale au Canada francais, 2,* 843–848.

Jus, A., Jus, K., Villeneuve, A., Pires, A., Lachance, R., Fortier, J., & Villeneuve, R. (1973). Studies on dream recall in chronic schizophrenic patients after prefrontal lobotomy. *Biological Psychiatry, 6,* 275–293.

Kaboodvand, N., Bäckman, L., Nyberg, L., & Salami, A. (2018). The retrosplenial cortex: A memory gateway between the cortical default mode network and the medial temporal lobe. *Human Brain Mapping, 39,* 2020–2034.

Kahn, D., Pace-Schott, E., & Hobson, J. A. (2002). Emotion and cognition: Feeling and character identification in dreaming. *Consciousness & Cognition, 11,* 34–50.

Kaida, K., Niki, K., & Born, J. (2015). Role of sleep for encoding of emotional memory. *Neurobiology of Learning and Memory, 121,* 72–79.

Kamiya, J. (1961). Behavioral, subjective, and physiological aspects of drowsiness and sleep. In D. W. Fiske & S. R. Maddi (Eds.), *Functions of varied experience* (pp. 145–174). Homewood, IL: Dorsey.

Kamran, M., Mannan, M., & Jeong, M. (2016). Cortical signal analysis and advances in functional near-infrared spectroscopy signal: A review. *Frontiers in Human Neuroscience, 10*(261), 1–12. https://doi.org/10.3389/fnhum.2016.00261.

Karagianni, M., Papadopoulou, A., Kallini, A., Dadatsi, A., & Abatzoglou, G. (2013). Dream content of Greek children and adolescents. *Dreaming, 23,* 91–96.

Kass, W., Preiser, G., & Jenkins, A. (1970). Inter-relationship of hallucinations and dreams in spontaneously hallucinating patients. *Psychiatric Quarterly, 44*(3), 488–499.

Kaufmann, C., Wehrle, R., Wetter, T. C., Holsboer, F., Auer, D., Pollmacher, T., & Czisch, M. (2006). Brain activation and hypothalamic functional connectivity during human non-rapid eye movement sleep: An EEG/fMRI study. *Brain, 129,* 655–667.

Keller, B. (2012). Detecting treatment effects with small samples: The power of some tests under the randomization model. *Psychometrika, 77,* 324–338.

Kellerman, A., & Mercy, J. (1992). Men, women, and murder: Gender-specific differences in rates of fatal violence and victimization. *Journal of Trauma, 33,* 1–5.

Kerr, N. (1993). Mental imagery, dreams, and perception. In D. Foulkes & C. Cavallero (Eds.), *Dreaming as cognition* (pp. 18–37). New York: Harvester Wheatsheaf.

Kerr, N., Foulkes, D., & Jurkovic, G. (1978). Reported absence of visual dream imagery in a normally sighted subject with Turner's syndrome. *Journal of Mental Imagery, 2*, 247–264.

Kihlstrom, J. (2002). Demand characteristics in the laboratory and the clinic: Conversations and collaborations with subjects and patients. *Prevention & Treatment, 5*. http://psycnet.apa.org/index.cfmfa=buy.optionToBuy&id=2003-04137-04003.

Kihlstrom, J. (2004). Availability, accessibility, and subliminal perception. *Consciousness and Cognition, 13*(1), 92–100.

Kimmins, C. W. (1920/1937). *Children's dreams*. London: Longmans, Green, and Co.

Klingenberg, B. (2008). Regression models for binary time series with gaps. *Computational Statistics and Data Analysis, 52*, 4076–4090.

Klinger, E. (2009). Daydreaming and fantasizing: Thought flow and motivation. In K. Markman, W. Klein, & J. Suhr (Eds.), *Handbook of imagination and mental simulation* (pp. 225–239). New York: Psychology Press.

Klinger, E., & Cox, W. (1987–1988). Dimensions of thought flow in everyday life. *Imagination, Cognition, and Personality, 7*, 105–128.

Knudson, R. M. (2001). Significant dreams: Bizarre or beautiful. *Dreaming, 11*, 167–177.

Knudson, R. M. (2003). The significant dream as emblem of uniqueness: The fertilizer does not explain the flower. *Dreaming, 13*, 121–134.

Knudson, R. M., & Minier, S. (1999). The on-going significance of significant dreams: The case of the bodiless head. *Dreaming, 9*, 235–245.

Kosslyn, S. (1994). *Image and brain: The resolution of the imagery debate*. Cambridge, MA: MIT Press.

Kosslyn, S., Margolis, J., Barrett, A., Goldknopfan, E., & Daly, P. (1990). Age differences in imagery abilities. *Child Development, 61*, 995–1010.

Kosslyn, S., Thompson, W. L., & Ganis, G. (2006). *The case for mental imagery*. New York: Oxford University Press.

Kövecses, Z. (2017). Levels of metaphor. *Cognitive Linguistics, 28*, 321–347. https://doi.org/10.1515/cog-2016-0052.

Kret, M., & de Gelder, B. (2012). A review on sex differences in processing emotional signals. *Neuropsychologia, 50*, 1211–1221. https://doi.org/10.1016/j.neuropsychologia.2011.12.022.

Kret, M., Pichon, S., Grèzes, J., & de Gelder, B. (2011). Men fear other men most: Gender specific brain activations in perceiving threat from dynamic faces and bodies—An fMRI study. *Frontiers in Psychology, 2,* Article 3, 1–11. https://doi.org/10.3389/fpsyg.2011.00003.

Krippendorff, K. (2004). *Content analysis: An introduction to its methodology.* Thousand Oaks, CA: Sage.

Krippner, S., Jaeger, C., & Faith, L. (2001). Identifying and utilizing spiritual content in dream reports. *Dreaming, 11,* 127–147.

Kucyi, A., Hodaie, M., & Davis, K. D. (2012). Lateralization in intrinsic functional connectivity of the temporoparietal junction with salience- and attention-related brain networks. *Journal of Neurophysiology, 108,* 3382–3392. https://doi.org/10.1152/jn.00674.2012.

Kuiken, D., & Sikora, S. (1993). The impact of dreams on waking thoughts and feelings. In A. Moffitt, M. Kramer, & R. Hoffmann (Eds.), *The functions of dreaming* (pp. 419–476). Albany: State University of New York Press.

LaBarre, W. (1972). *The ghost dance: Origins of religion.* New York: Dell Publishing.

LaBerge, S. (1980). *Lucid dreaming: An exploratory study of consciousness during sleep.* Doctoral dissertation, Stanford University.

LaBerge, S., Nagel, L., Dement, W., & Zarcone, V. (1981). Lucid dreaming verified by volitional communication during REM sleep. *Perceptual and Motor Skills, 52,* 727–732.

Lacey, S., & Lawson, R. (2013). *Multisensory imagery.* New York: Springer Science.

Lakoff, G. (1987). *Women, fire, and dangerous things.* Chicago: University of Chicago Press.

Lakoff, G. (1993). How metaphor structures dreams. *Dreaming, 3,* 77–98.

Lakoff, G. (1997). How unconscious metaphorical thought shapes dreams. In D. Stein (Ed.), *Cognitive science and the unconscious* (pp. 89–120). Washington, DC: American Psychiatric Press.

Lakoff, G., & Johnson, M. (1980). *Metaphors we live by.* Chicago: University of Chicago Press.

Lakoff, G., & Turner, M. (1989). *More than cool reason.* Chicago: University of Chicago Press.

Lamb, M. E., Hershkowitz, I., Orbach, Y., & Esplin, P. (2008). *Tell me what happened: Structured investigative interviews of child victims and witnesses.* New York: Wiley-Blackwell.

Landau, M., Meier, B., & Keefer, L. (2010). A metaphor-enriched social cognition. *Psychological Bulletin, 136,* 1045–1067. https://doi.org/10.1037/a0020970.

Landry, M., Lifshitz, M., & Raz, A. (2017). Brain correlates of hypnosis: A systematic review and meta-analytic exploration. *Neuroscience & Biobehavioral Reviews, 81,* 75–98.

Larson-Prior, L., Zempel, J., Nolan, T., Prior, F., Snyder, A., & Raichle, M. E. (2009). Cortical network functional connectivity in the descent to sleep. *Proceedings of the National Academy of Sciences, 106,* 4489–4494.

Latta, C. (1998). *The manifest dream content of premenarcheal and postmenarcheal girls.* Doctoral dissertation, California School of Professional Psychology. (ProQuest, UMI Dissertations Publishing, 9836011.)

Lavie, P., Pratt, H., Scharf, B., Peled, R., & Brown, J. (1984). Localized pontine lesion: Nearly total absence of REM sleep. *Neurology, 34*(1), 118–120.

Leaper, C. (2013). Gender development during childhood. In P. Zelazo (Ed.), *The Oxford handbook of developmental psychology, Vol. 2: Self and other* (pp. 326–377). New York: Oxford University Press.

Leaper, C., & Farkas, T. (2015). The socialization of gender during childhood and adolescence. In J. Grusec & P. Hastings (Eds.), *Handbook of socialization: Theory and research* (2nd ed., pp. 541–565). New York: Guilford Press.

Leclair-Visonneau, L., Oudiette, D., Gaymard, B., Leu-Semenescu, S., & Arnulf, I. (2010). Do the eyes scan dream images during rapid eye movement sleep? Evidence from the rapid eye movement sleep behaviour disorder model. *Brain, 133,* 1737–1746.

LeDoux, J. (1991). Emotion and the limbic system concept. *Concepts in Neuroscience, 2,* 169–199.

LeDoux, J. (2003). The emotional brain, fear, and the amygdala. *Cell and Molecular Neurobiology, 23,* 727–738.

LeDoux, J. (2012). Evolution of human emotion: A view through fear. *Progress Brain Research, 195,* 431–442.

LeDoux, J. (2015). *Anxious.* New York: Viking.

LeDoux, J. (2019). *The deep history of ourselves: The four-billion year story of how we got conscious brains.* New York: Viking.

LeDoux, J. (2021). What emotions might be like in other animals. *Current Biology, 31,* R1–R6.

Lee, M., Hacker, C., Snyder, A., Corbetta, M., Zhang, D., & Leuthardt, E. (2012). Clustering of resting state networks. *PLoS One, 7,* e40370. https://doi.org/40310 .41371/journal.pone.0040370.

Leech, R., & Sharp, D. J. (2014). The role of the posterior cingulate cortex in cognition and disease. *Brain, 137,* 12–32.

Levin, R., Fireman, G., & Rackley, C. (2003). Personality and dream recall frequency: Still further negative findings. *Dreaming, 13,* 155–162.

Levin, R., & Nielsen, T. (2007). Disturbed dreaming, posttraumatic stress disorder, and affect distress: A review and neurocognitive model. *Psychological Bulletin, 133,* 482–528.

Lewis, J. (2008). Dream reports of animal rights activists. *Dreaming, 18,* 181–200.

Lieberman, M. D. (2013). *Social: Why our brains are wired to connect.* New York: Broadway Books.

Lieberman, M. D., Straccia, M. A., Meyer, M. L., Du, M., & Tan, K. M. (2019). Social, self, (situational), and affective processes in medial prefrontal cortex (MPFC): Causal, multivariate, and reverse inference evidence. *Neuroscience and Biobehavioral Reviews, 99,* 311–328. https://doi.org/10.1016/j.neubiorev.2018.12.021.

Lincoln, J. S. (1935). *The dream in primitive cultures.* Baltimore: Williams and Wilkins.

Lindquist, K., Satpute, A., Wager, T., Weber, J., & Barrett, L. F. (2016). The brain basis of positive and negative affect: Evidence from a meta-analysis of the human neuroimaging literature. *Cerebral Cortex, 26,* 1910–1922. https://doi.org/10.1093/cercor/bhv001.

Lindquist, K., Wager, T., Kober, H., Bliss-Moreau, E., & Barrett, L. F. (2012). The brain basis of emotion: A meta-analytic review. *Behavioral and Brain Sciences, 35,* 121–143.

Linke, A., & Cusack, R. (2015). Flexible information coding in human auditory cortex during perception, imagery, and STM of complex sounds. *Journal of Cognitive Neuroscience, 27,* 1322–1333.

LoBue, V., Matthews, K., Harvey, T., & Thrasher, C. (2014). Pick on someone your own size: The detection of threatening facial expressions posed by both child and adult models. *Journal of Experimental Child Psychology, 118,* 134–142. https://doi.org/10.1016/j.jecp.2013.07.016.

LoBue, V., & Rakison, D. (2013). What we fear most: A developmental advantage for threat-relevant stimuli. *Developmental Review, 33,* 285–303. https://doi.org/10.1016/j.dr.2013.07.005.

LoBue, V., Rakison, D., & DeLoache, J. (2010). Threat perception across the life span: Evidence for multiple converging pathways. *Current Directions in Psychological Science, 19,* 375–379. https://doi.org/10.1177/0963721410388801.

Loftus, E., Joslyn, S., & Polage, D. (1998). Repression: A mistaken impression? *Development & Psychopathology, 10*(4), 781–792.

Loftus, E., & Ketcham, K. (1994). *The myth of repressed memory.* New York: St. Martin's Press.

Lortie-Lussier, M., Cote, L., & Vachon, J. (2000). The consistency and continuity hypotheses revisited through the dreams of women at two periods of their lives. *Dreaming, 10,* 67–76.

Luppi, P., Clement, O., & Fort, P. (2013). Paradoxical (REM) sleep genesis by the brainstem is under hypothalamic control. *Current Opinion in Neurobiology, 23*, 1–7.

MacDonald, S., & Culham, J. (2015). Do human brain areas involved in visuomotor actions show a preference for real tools over visually similar non-tools? *Neuropsychologia, 77*, 35–41.

Madsen, M. K., Stenbæk, D., Arvidsson, A., Armand, S., Marstrand-Joergensen, M., Jonansen, S., Linnet, K., Ozenne, B., Knudson, G., & Fisher, P. (2021). Psilocybin-induced changes in brain network integrity and segregation correlate with plasma psilocin level and psychedelic experience. *European Neuropsychopharmacology, 50*, 121–132. https://doi.org/10.1016/j.euroneuro.2021.06.001.

Madsen, P. L., Holm, S., Vorstrup, S., Friberg, L., Lassen, N. A., & Wildschidtz, G. (1991). Human regional cerebral blood flow during rapid-eye-movement sleep. *Journal of Cerebral Blood Flow Metabolism, 11*, 502–507.

Madsen, P. L., Schmidt, F., Holm, S., Vorstrup, S., Lassen, N. A., & Wildschidtz, G. (1991). Cerebral oxygen metabolism and cerebral blood flow in man during light sleep (stage 2). *Brain Research, 557*(1–2), 217–220.

Malcolm-Smith, S., & Solms, M. (2004). Incidence of threat in dreams: A response to Revonsuo's threat simulation theory. *Dreaming, 14*, 220–229.

Malcolm-Smith, S., Solms, M., Turnbull, O., & Tredoux, C. (2008a). Shooting the messenger won't change the news. *Consciouness and Cognition, 17*, 1297–1301.

Malcolm-Smith, S., Solms, M., Turnbull, O., & Tredoux, C. (2008b). Threat in dreams: An adaptation? *Consciousness and Cognition, 17*, 1281–1291.

Malinowski, J., Fylan, F., & Horton, C. (2014). Experiencing "continuity": A qualitative investigation of waking life in dreams. *Dreaming, 24*, 161–175.

Malinowski, J., & Horton, C. (2014). Memory sources of dreams: The incorporation of autobiographical rather than episodic experiences. *Journal of Sleep Research, 23*, 441–447. https://doi.org/10.1111/jsr.12134.

Malinowski, J., & Horton, C. (2015). Metaphor and hyperassociativity: The imagination mechanisms behind emotion assimilation in sleep and dreaming. *Frontiers in Psychology, 6*(1132). https://doi.org/10.3389/fpsyg.2015.01132.

Mandler, J. (2004). *The foundations of mind: Origins of conceptual thought.* New York: Oxford University Press.

Mandler, J. (2012). On the spatial foundations of the conceptual system and its enrichment. *Cognitive Science, 36*, 421–451.

Mansouri, F., Koechlin, E., Rosa, M., & Buckley, M. (2017). Managing competing goals—a key role for the frontopolar cortex. *Nature Reviews Neuroscience, 18*, 645–657.

Maquet, P., Peters, J., Aerts, J., Delfiore, G., Dequerldre, C., Luxen, A., & Franck, G. (1996). Functional neuroanatomy of human rapid-eye-movement sleep and dreaming. *Nature, 383*, 163–166.

Margulies, D. S., Ghosh, S. S., Goulas, A., Falkiewicz, M., Huntenburg, J. M., Langs, G., & Smallwood, J. (2016). Situating the default-mode network along a principal gradient of macroscale cortical organization. *Proceedings of the National Academy of Sciences, 113*, 12574–12579.

Marquardt, C. J. G., Bonato, R. A., & Hoffmann, R. F. (1996). An empirical investigation into the day-residue and dream-lag effects. *Dreaming, 6*, 57–65.

Marquis, L.-P., Paquette, T., Blanchette-Carrière, C., Dumel, G., & Nielsen, T. (2017). REM sleep theta changes in frequent nightmare recallers. *Sleep, 40*, 1–12.

Martin, S., Mikutta, C., Leonard, M., Hungate, D., Knight, R., & Pasley, B. (2018). Neural encoding of auditory features during music perception and imagery. *Cerebral Cortex, 28*, 4222–4233. https://doi.org/10.1093/cercor/bhx277.

Marzano, C., Moroni, F., Gorgoni, M., Nobili, L., Ferrara, Michele, & De Gennaro, L. (2013). How we fall asleep: Regional and temporal differences in electroencephalographic synchronization at sleep onset. *Sleep Medicine, 14*, 1112–1122. https://doi.org/10.1016/j.sleep.2013.05.021.

Mason, M., Norton, M., Van Horn, J., Wenger, D., Grafton, S., & Macrae, N. (2007). Wandering minds: The default network and stimulus-independent thought. *Science, 315*, 393–395.

Mason, R., & Just, M. A. (2020). Neural representations of procedural knowledge. *Psychological Science, 31*, 729–740.

Mathes, J., Schredl, M., & Goritz, A. (2014). Frequency of typical dream themes in most recent dreams: An online study. *Dreaming, 24*, 57–66.

Mathur, R., & Douglas, N. (1995). Frequency of EEG arousals from nocturnal sleep in normal subjects. *Sleep, 18*, 330–333.

Matlock, T. (1988). *The metaphorical extension of "see."* Paper presented at the Proceedings of the Western Conference on Linguistics, Berkeley, CA.

Mazandarani, A. A., Aguilar-Vafaie, M. E., & Domhoff, G. W. (2013). Content analysis of Iranian college students' dreams: Comparison with American data. *Dreaming, 23*, 163–174.

Mazzoni, G., & Loftus, E. (1998). Dreaming, believing, and remembering. In J. de Rivera & T. Sarbin (Eds.), *Believed-in imaginings: The narrative construction of reality* (pp. 145–156). Washington, DC: American Psychological Association Press.

Mazzoni, G., Loftus, E., Seitz, A., & Lynn, S. (1999). Changing beliefs and memories through dream interpretation. *Applied Cognitive Psychology, 13*(2), 125–144.

McCarley, R., & Hoffman, E. (1981). REM sleep dreams and the activation-synthesis hypothesis. *American Journal of Psychiatry, 138,* 904–912.

McClone, M. (2001). Concepts as metaphors. In S. Glucksberg (Ed.), *Understanding figurative language: From metaphors to idioms* (pp. 90–107). New York: Oxford University Press.

McDermott, J. H. (2009). What can experiments reveal about the origins of music? *Current Directions in Psychological Science, 18,* 164–168.

McGaugh, J. (2000). Memory—A century of consolidation science. *Science, 287,* 248–251. https://doi.org/10.1126/science.287.5451.248.

McGinty, D., & Szymusiak, R. (2017). Neural control of sleep in mammals. In M. Kryger, T. Roth, & W. Dement (Eds.), *Principles and practices of sleep medicine* (6th ed., pp. 62–77). Philadelphia: Elsevier.

McKendrick, R., Parasuraman, R., & Ayaz, H. (2015). Wearable functional near infrared spectroscopy (fNIRS) and transcranial direct current stimulation (tDCS): Expanding vistas for neurocognitive augmentation. *Frontiers in Systems Neuroscience, 9*(March 9). https://doi.org/10.3389/fnsys.2015.00027.

Meier, B. (1993). Speech and thinking in dreams. In C. Cavallero & D. Foulkes (Eds.), *Dreaming as cognition* (pp. 58–76). New York: Harvester Wheatsheaf.

Meier, C., Ruef, H., Ziegler, A., & Hall, C. (1968). Forgetting of dreams in the laboratory. *Perceptual and Motor Skills, 26,* 551–557.

Menon, V., & Uddin, L. (2010). Saliency, switching, attention and control: A network model of insula function. *Brain Structure & Function, 214,* 655–667.

Merritt, J., Stickgold, R., Pace-Schott, E., Williams, J., & Hobson, J. A. (1994). Emotion profiles in the dreams of men and women. *Consciousness and Cognition, 3,* 46–60.

Meyer, M. L., Hershfield, H. E., Waytz, A. G., Mildner, J. N., & Tamir, D. I. (2019). Creative expertise is associated with transcending the here and now. *Journal of Personality and Social Psychology, 116,* 483–494. https://doi.org/10.1037/pspa0000148.

Meyer, M. L., & Lieberman, M. D. (2018). Why people are always thinking about themselves: Medial prefrontal cortex activity during rest primes self-referential processing. *Journal of Cognitive Neuroscience, 30,* 714–721. https://doi.org/10.1162/jocn_a_01232.

Meyer, S., & Shore, C. (2001). Children's understanding of dreams as mental states. *Dreaming, 11,* 179–194.

Moraczewski, D., Nketia, J., & Redcay, E. (2020). Cortical temporal hierarchy is immature in middle childhood. *Neuroimage, 216,* 116616. https://doi.org/10.1016/j.neuroimage.2020.116616.

Moran, M. (2003). Arguments for rejecting the sequential Bonferroni in ecological studies. *Oikos, 100*, 403–405.

Morikawa, T., Hayashi, M., & Hori, T. (2002). Spatiotemporal variations of alpha and sigma band EEG in the waking-sleeping transition period. *Perceptual and Motor Skills, 95*, 131–154.

Moskowitz, E., & Berger, R. J. (1969, November 8). Rapid eye movements and dream imagery: Are they related? *Nature, 224*(5219), 613–614.

Moss, C. S. (Ed.) (1967). *The hypnotic investigation of dreams*. New York: Wiley & Sons.

Mota-Rolim, S., Pavlou, A., Nascimento, G., Fontenele-Araujo, J., & Ribeiro, S. (2019). Portable devices to induce lucid dreams—Are they reliable? *Frontiers in Neuroscience, 13*, 428. https://doi.org/10.3389/fnins.2019.00428.

Moulton, S. T., & Kosslyn, S. (2011). Imagining predictions: Mental imagery as mental emulation. In M. Bar (Ed.), *Predictions in the brain: Using our past to generate a future* (pp. 95–106). New York: Oxford University Press.

Mysliwiec, V., O'Reilly, B., Polchinski, J., Kwon, H., Germain, A., & Roth, B. J. (2014). Trauma associated sleep disorder: A proposed parasomnia encompassing disruptive nocturnal behaviors, nightmares, and REM without atonia in trauma survivors. *Journal of Clinical Sleep Medicine, 10*, 1143–1148.

Nelson, K. (2004). A welcome turn to meaning in infant development: Commentary on Mandler's *The foundations of mind: Origins of conceptual thought. Developmental Science, 7*, 506–507.

Nelson, K. (2005). Emerging levels of consciousness in early human development. In H. S. Terrace & J. Metcalfe (Eds.), *The missing link in cognition: Origins of self-reflective consciousness* (pp. 116–141). New York: Oxford University Press.

Nelson, K. (2007). *Young minds in social worlds: Experience, meaning, and memory*. Cambridge, MA: Harvard University Press.

Nelson, K. (2011). "Concept" is a useful concept in developmental research. *Journal of Theoretical and Philosophical Psychology, 31*, 96–101.

Nguyen, T., Babawale, O., Kim, T., Jo, H., Liu, H., & Kim, J. (2018). Exploring brain functional connectivity in rest and sleep states: A fNIRS study. *Science Reports, 8*(16144), 1–10. https://doi.org/10.1038/s41598-018-33439-2.

Niedenthal, P., Winkielman, P., Mondillon, L., & Vermeulen, N. (2009). Embodiment of emotion concepts. *Journal of Personality and Social Psychology, 96*, 1120–1136.

Nielsen, T. (2000). A review of mentation in REM and NREM sleep: "Covert" REM sleep as a possible reconciliation of two opposing models. *Behavioral and Brain Sciences, 23*, 851–866.

Nielsen, T., & Levin, R. (2007). Nightmares: A new neurocognitive model. *Sleep Medicine Reviews, 11*, 295–310. https://doi.org/10.1016/j.smrv.2007.03.004.

Nielsen, T., & Powell, R. (1992). The day-residue and dream-lag effect. *Dreaming, 2*, 67–77.

Nielsen, T., Zadra, A., Simard, V., Saucier, S., Stenstrom, P., Smith, C., & Kuiken, D. (2003). The typical dreams of Canadian university students. *Dreaming, 13*, 211–235.

Noegel, S. (2001). Dreams and dream interpreters in Mesopotamia and in the Hebrew Bible [Old Testament]. In K. Bulkeley (Ed.), *Dreams: A reader on religious, cultural, and psychological dimensions of dreaming* (pp. 45–71). New York: Palgrave Macmillan/Springer Nature.

Nofzinger, E., Mintun, M., Wiseman, M., Kupfer, D., & Moore, R. (1997). Forebrain activation in REM sleep: An FDG PET study. *Brain Research, 770*, 192–201.

Noreen, E. (1989). *Computer intensive methods for testing hypotheses: An introduction.* New York: Wiley & Sons.

Norris, C. J. (2021). The negativity bias, revisited: Evidence from neuroscience measures and an individual differences approach. *Social Neuroscience, 16*, 68–82. https://doi.org/10.1080/17470919.2019.1696225.

Nosek, B. A. (2015). Estimating the reproducibility of psychological science. *Science, 349*(6251). https://doi.org/10.1126/science.aac4716.

Oberst, U., Charles, C., & Chamarro, A. (2005). Influence of gender and age in aggressive dream content in Spanish children and adolescents. *Dreaming, 15*, 170–177.

Ochsner, K., & Kosslyn, S. (2014a). Cognitive neuroscience: Where are we going? In K. Ochsner & S. Kosslyn (Eds.), *The Oxford handbook of cognitive neuroscience* (Vol. 2, pp. 477–483). New York: Oxford University Press.

Ochsner, K., & Kosslyn, S. (2014b). Cognitive neuroscience: Where are we now? In K. Ochsner & S. Kosslyn (Eds.), *The Oxford handbook of cognitive neuroscience* (Vol. 2, pp. 1–7). New York: Oxford University Press.

Ogilvie, R., Hunt, H., Tyson, P., Lucescu, M., & Jeakins, D. (1982). Lucid dreaming and alpha activity: A preliminary report. *Perceptual & Motor Skills, 55*, 795–808.

Orne, M. (1962). On the social psychology of the psychological experiment: With particular reference to demand characteristics and their implications. *American Psychologist, 17*, 776–783.

Otgaar, H., Howe, M., Patihis, L., Lilienfeld, S., & Loftus, E. (2019). The return of the repressed: The persistent and problematic claims of long-forgotten trauma. *Perspectives on Psychological Science, 14*, 1072–1095. https://doi.org/10.1177/1745691619862306.

Pagel, J. F. (2003). Non-dreamers. *Sleep Medicine, 4,* 235–241.

Pagel, J. F., Blagrove, M., Levin, R., States, B., Stickgold, R., & White, S. (2001). Definitions of dream: A paradigm for comparing field descriptive specific studies of dream. *Dreaming, 11*(4), 195–202.

Partridge, E. (1937/1984). *A dictionary of slang and unconventional English from the fifteenth century to the present day.* New York: Macmillan.

Patel, A. (2008). *Music, language, and the brain.* New York: Oxford University Press.

Patel, A. (2010). Music, biological evolution, and the brain. In M. Bailar (Ed.), *Emerging disciplines* (pp. 91–144). Houston, TX: Rice University Press.

Patel, A. (2014). The evolutionary biology of musical rhythm: Was Darwin wrong? *PLoS Biology, 12*(5), e1001873.

Payne, J., Kensinger, E., Wamsley, E., Spreng, R. N., & Alger, S. (2015). Napping and the selective consolidation of negative aspects of scenes. *Emotion, 15,* 176–186.

Pearlman, C., & Becker, M. (1974). REM sleep deprivation impairs bar-press acquisition in rats. *Physiology & Behavior, 13,* 813–817. https://doi.org/10.1016/0031-9384 (74)90267-4.

Pearlman, C., & Greenberg, R. (1973). Posttrial REM sleep: A critical period for consolidation of shuttlebox avoidance. *Animal Learning & Behavior, 1,* 49–51. https://doi .org/10.3758/BF03198999.

Perneger, T. (1998). What's wrong with Bonferroni adjustments. *British Medical Journal, 316,* 1236–1238.

Perogamvros, L., Baird, B., Seibold, M., Riedner, B., Boly, M., & Tononi, G. (2017). The phenomenal contents and neural correlates of spontaneous thoughts across wakefulness, NREM sleep, and REM sleep. *Journal of Cognitive Neuroscience, 29,* 1766–1777.

Piazza, E., Hasenfratz, L., Hasson, U., & Lew-Williams, C. (2020). Infant and adult brains are coupled to the dynamics of natural communication. *Psychological Science, 31,* 6–17. https://doi.org/10.1177/0956797619878698.

Picchioni, D., Pixa, M. L., Fukunaga, M., Carr, W., Horovitz, S., Braun, A. R., & Duyn, J. H. (2014). Decreased connectivity between the thalamus and the neocortex during human nonrapid eye movement sleep. *Sleep, 37,* 387–397.

Pitcher, E., & Prelinger, E. (1963). *Children tell stories: An analysis of fantasy.* New York: International Universities Press.

Pivik, R. T. (1986). Sleep: Physiology and psychophysiology. In M. Coles, E. Donchin, & S. Porges (Eds.), *Psychophysiology: Systems, processes and applications* (pp. 378–406). New York: Guilford Press.

Pivik, R. T. (1991). Tonic states and phasic events in relation to sleep mentation. In S. Ellman & J. Antrobus (Eds.), *The mind in sleep: Psychology and psychophysiology* (2nd ed., pp. 214–247). New York: Wiley.

Pivik, R. T., & Foulkes, D. (1968). NREM mentation: Relation to personality, orientation time, and time of night. *Journal of Consulting and Clinical Psychology, 32,* 144–151.

Pohlchen, D., Pawlizki, A., Gais, S., & Schönauer, M. (2020). Evidence against a large effect of sleep in protecting verbal memories from interference. *Journal of Sleep Research, 30*(2). https://doi.org/10.1111/jsr.13042.

Polner-Clark, E., Wager, T., Satpute, A., & Barrett, L. F. (2016). Neural fingerprinting: Meta-analysis, variation, and the search for brain-based essences in the science of emotions. In L. F. Barrett, M. Lewis, & J. Haviland-Jones (Eds.), *Handbook of emotions* (4th ed., pp. 146–165). New York: Guilford Press.

Popp, C., Lubarsky, L., & Crits, C. (1992). The parallel of the CCRT from therapy narratives with the CCRT from dreams. In L. Lubarsky & C. Crits (Eds.), *Understanding transference* (pp. 158–172). New York: Basic Books.

Postman, L. (1963). One-trial learning. In C. Cofer & B. Musgrave (Eds.), *Verbal behavior and learning: Problems and processes* (pp. 295–335). New York: McGraw-Hill.

Power, J., Cohen, A., Nelson, S., & Petersen, S. (2011). Functional organization of the human brain. *Neuron, 72,* 665–678.

Poza, J., & Marti-Masso, J. (2006). Total dream loss secondary to left temporo-occipital brain injury. *Neurologia, 21,* 152–154.

Prasad, B. (1982). Content analysis of dreams of Indian and American college students: A cultural comparison. *Journal of Indian Psychology, 4,* 54–64.

Raichle, M. E., MacLeod, A. M., Snyder, A. Z., Powers, W. J., Gusnard, D. A., & Shulman, G. L. (2001). A default mode of brain function. *Proceedings of the National Academy of Sciences, 98,* 676–682.

Rapp, A., Mutschler, D., & Erb, M. (2012). Where in the brain is nonliteral language? A coordinate-based meta-analysis of functional magnetic resonance imaging studies. *Neuroimage, 63,* 600–610.

Rasch, B., & Born, J. (2015). In search of a role of REM sleep in memory formation. *Neurobiology of Learning and Memory, 122,* 1–3.

Rasch, B., Pommer, J., Diekelmann, S., & Born, J. (2009). Pharmacological REM sleep suppression paradoxically improves rather than impairs skill memory. *Natural Neuroscience, 12,* 396–397.

Raz, G., Touroutoglou, A., Atzil, S., & Barrett, L. F. (2016). Functional connectivity dynamics during film viewing reveal common networks for different emotional experiences. *Cognitive, Affective & Behavioral Neuroscience, 16*, 709–723. https://doi .org/10.3758/s13415-016-0425-4.

Rechtschaffen, A., Hauri, P., & Zeitlin, M. (1966). Auditory awakening thresholds in REM and NREM sleep stages. *Perceptual and Motor Skills, 22*, 927–942.

Rechtschaffen, A., Verdone, P., & Wheaton, J. (1963). Reports of mental activity during sleep. *Canadian Psychiatric Association Journal, 8*, 409–414.

Reese, E. (2013). Culture, narrative, and imagination. In M. Taylor (Ed.), *The Oxford handbook of the development of imagination* (pp. 196–211). New York: Oxford University Press.

Reichers, M., Kramer, M., & Trinder, J. (1970). A replication of the Hall–Van de Castle character scale norms. *Psychophysiology, 7*, 238.

Reinsel, R., Antrobus, J., & Wollman, M. (1985). The phasic-tonic difference and the time-of-night effect. *Sleep Research, 14*, 115.

Reinsel, R., Antrobus, J., & Wollman, M. (1992). Bizarreness in dreams and waking fantasy. In J. S. Antrobus & M. Bertini (Eds.), *The neuropsychology of sleep and dreaming* (pp. 157–184). Hillsdale, NJ: Erlbaum.

Reisberg, D. (1992). *Auditory imagery.* Hillsdale, NJ: Erlbaum.

Resnick, J., Stickgold, R., Rittenhouse, C., & Hobson, J. A. (1994). Self-representation and bizarreness in children's dream reports collected in the home setting. *Consciousness and Cognition, 3*, 30–45.

Revonsuo, A. (2000a). Did ancestral humans dream for their lives? *Behavioral & Brain Sciences, 23*, 1063–1082.

Revonsuo, A. (2000b). The reinterpretation of dreams: An evolutionary hypothesis of the function of dreaming. *Behavioral and Brain Sciences, 23*(6), 877–901.

Revonsuo, A., & Salmivalli, C. (1995). A content analysis of bizarre elements in dreams. *Dreaming, 5*, 169–187.

Revonsuo, A., Tuominen, J., & Valli, K. (2015). The avatars in the machine: Dreaming as a simulation of social reality. In T. Metzinger & J. Windt (Eds.), *Open MIND: 32(T)* (pp. 1–28). Frankfurt am Main: MIND Group.

Revonsuo, A., & Valli, K. (2000). Dreaming and consciousness: Testing the threat simulation theory of the function of dreaming. *Psyche, 6*, 1–31.

Revonsuo, A., & Valli, K. (2008). How to test the threat-simulation theory. *Consciousness and Cognition, 17*, 1292–1296.

Reynolds, H. (1984). *Analysis of nominal data.* Newbury Park, CA: Sage.

Robert, G., & Zadra, A. (2014). Thematic and content analysis of idiopathic nightmares and bad dreams. *Sleep, 37*, 409–417.

Rosch, E. (1973). Natural categories. *Cognitive Psychology, 4*, 328–350. https://doi.org /10.1016/0010-0285(73)90017-0.

Rosch, E., & Mervis, C. (1975). Family resemblances: Studies in the internal structure of categories. *Cognitive Psychology, 7*, 573–605. https://doi.org/10.1016/0010-0285 (75)90024-9.

Rosch, E., Mervis, C., Gray, W., Johnson, D., & Boyes-Braem, P. (1976). Basic objects in natural categories. *Cognitive Psychology, 8*, 382–439. https://doi.org/10.1016/0010 -0285(76)90013-X.

Roseman, I., & Evdokas, A. (2004). Appraisals cause experienced emotions: Experimental evidence. *Cognition and Emotion, 18*, 1–28. https://doi.org/10.1080/02699930244000390.

Roseman, I., & Smith, C. (2001). Appraisal theory: Overview, assumptions, varieties, controversies. In K. Scherer, A. Schorr, & T. Johnstone (Eds.), *Appraisal processes in emotion: Theory, methods, research* (pp. 3–19). New York: Oxford University Press.

Roseman, M. (1991). *Healing sounds from the Malaysian rainforest: Temiar music and medicine*. Berkeley: University of California Press.

Rosenthal, R. (1976). *Experimenter effects in behavioral research*. New York: Irvington.

Rosenthal, R., & Ambady, N. (1995). Experimenter effects. In A. Manstead & M. Hewstone (Eds.), *Encyclopedia of social psychology* (pp. 230–235). Oxford: Blackwell.

Rosenthal, R., & Rubin, D. B. (1982). A simple general purpose display of magnitude of experimental effect. *Journal of Educational Psychology, 74*, 166–169.

Rosnow, R. L., & Rosenthal, R. (1997). *People studying people: Artifacts and ethics in behavioral research*. New York: W. H. Freeman.

Roussy, F. (1998). *Testing the notion of continuity between waking experience and REM dream content*. Doctoral dissertation, University of Ottawa.

Roussy, F., Brunette, M., Mercier, P., Gonthier, I., Grenier, J., Sirois-Berliss, M., & De Koninck, J. (2000). Daily events and dream content: Unsuccessful matching attempts. *Dreaming, 10*, 77–83.

Roussy, F., Camirand, C., Foulkes, D., De Koninck, J., Loftis, M., & Kerr, N. (1996). Does early-night REM dream content reliably reflect presleep state of mind? *Dreaming, 6*, 121–130.

Roussy, F., Raymond, I., & De Koninck, J. (2000). Affect in REM dreams: Exploration of a time-of-night effect. *Sleep, 23*, A174–A175.

Röver, S. A., & Schredl, M. (2017). Measuring emotions in dreams: Effects of dream length and personality. *International Journal of Dream Research, 10*, 65–68.

Rozin, P., & Royzman, E. B. (2001). Negativity bias, negativity dominance, and contagion. *Personality and Social Psychology Review, 5*, 296–320. https://doi.org/10.1207 /S15327957PSPR0504_2.

Rudofsky, S., & Wotiz, J. (1988). Psychologists and the dream accounts of August Kekulé. *Journal of Human Behavior & Learning, 5*, 1–11.

Ryals, A., & Voss, J. (2015). The outer limits of implicit memory. In D. Addis & M. Barense (Eds.), *The Wiley handbook on the cognitive neuroscience of memory* (pp. 44–59). New York: Wiley-Blackwell.

Sacks, O. (2013). *Hallucinations*. New York: Knopf.

Sakai, K., Petitjean, F., & Jouvet, M. (1976). Effects of ponto-mesencephalic lesions and electrical stimulation upon PGO waves and EMPs in unanesthetized cats. *Electroencephalography and Clinical Neurophysiology, 41*, 49–63.

Sala, G., & Gobet, F. (2017). Does far transfer exist? Negative evidence from chess, music, and working memory training. *Current Directions in Psychological Science, 26*, 515–520.

Sala, G., Tatlidil, K. S., & Gobet, F. (2017). Video game training does not enhance cognitive ability: A comprehensive meta-analytic investigation. *Psychological Bulletin*, advance online publication. https://doi.org/10.1037/bul0000139.

Saline, S. (1999). The most recent dreams of children ages 8–11. *Dreaming, 9*, 173–181.

Sämann, P., Wehrle, R., Hoehn, D., Spoormaker, V., Peters, H., Tully, C., & Czisch, M. (2011). Development of the brain's default mode network from wakefulness to slow wave sleep. *Cerebral Cortex, 21*, 2082–2093.

Sanday, P. (1981). *Female power and male dominance: On the origins of sexual inequality*. New York: Cambridge University Press.

Sandor, P., Szakadát, S., Kertész, K., & Bódizs, R. (2015). Content analysis of 4 to 8 year-old children's dream reports. *Frontiers in Psychology, 6*(534). https://doi.org/10.3389/fpsyg.2015.00534.

Sarbin, T. (1993). Whither hypnosis? A rhetorical analysis. *Contemporary Hypnosis, 10*, 1–9.

Sastre, J.-P., Sakai, K., & Jouvet, M. (1981). Are the gigantocellular tegmental field neurons responsible for paraxodical sleep? *Brain Research, 229*(1), 147–161.

Sato, J. R., Salum, G. A., Gadelha, A., Picon, F. A., Pan, P. M., Vieira, G., Zugmanch, A., Queiroz Hoexter, M., Anés, M., Mourach, L. M., Del'Aquilla, M. A. G., Amaro, E., Jr., McGuire, P., Crossley, N., Lacerda, A., Rohde, L. A., Miguel, E. C., Bressan, R. A., & Jackowskich, A. P. (2014). Age effects on the default mode and control networks in typically developing children. *Journal of Psychiatric Research, 58*, 89–95.

Scarpelli, S., Alfonsi, V., Mangiaruga, A., Musetti, A., Quattropani, M., Lenzo, V., & Franceschini, C. (2021). Pandemic nightmares: Effects on dream activity of the COVID-19 lockdown in Italy. *Journal of Sleep Research, 30*(5). https://doi.org/10.1111/jsr.13300.

Scarpelli, S., D'Atri, A., Mangiaruga, A., Marzano, C., Gorgoni, M., Schiappa, C., & De Gennaro, L. (2017). Predicting dream recall: EEG activation during NREM sleep or shared mechanisms with wakefulness? *Brain Topography, 30,* 629–638. https://doi .org/10.1007/s10548-017-0563-1.

Schacter, D., Addis, D., & Buckner, R. (2008). Episodic simulation of future events: Concepts, data, and applications. *Annals of the New York Academy of Sciences, 1124,* 39–60. https://doi.org/10.1196/annals.1440.001.

Schachter, S., & Singer, J. L. (1962). Cognitive social and physiological determinants of emotional state. *Psychological Review, 69,* 379–399. https://doi.org/10.1037 /h0046234.

Schaefer, A., Kong, R., Eickhoff, S., & Yeo, B. (2018). Local-global parcellation of the human cerebral cortex from intrinsic functional connectivity MRI. *Cerebral Cortex, 28,* 3095–3114. https://doi.org/10.1093/cercor/bhx179.

Schmidt, F. (1996). Statistical significance testing and cumulative knowledge in psychology: Implications for training of researchers. *Psychological Methods, 1*(2), 115–129.

Schmidt, M. (2005). Neural mechanisms of sleep-related penile erections. In M. Kryger, T. Roth, & W. Dement (Eds.), *Principles and practices of sleep medicine* (4th ed., pp. 305–317). Philadelphia: W. B. Saunders.

Schmidt, S. (2009). Shall we really do it again? The powerful concept of replication is neglected in the social sciences. *Review of General Psychology, 13*(2), 90–100.

Schneider, A. (1999). Regular expressions. http://www.dreambank.net/regex.html.

Schneider, D. (1969). The dream life of the Yir Yiront. In D. Schneider & L. Sharp (Eds.), *The dream life of a primitive people: The dreams of the Yir Yoront of Australia* (pp. 12–56). Washington, DC: American Anthropological Association.

Schneider, E. (1953). *Coleridge, opium, and Kubla Khan.* Chicago: University of Chicago Press.

Schredl, M. (2010). Characteristics and contents of dreams. *International Review of Neurobiology, 92,* 135–154.

Schredl, M., & Bulkeley, K. (2020). Dreaming and the COVID-19 pandemic: A survey in a U.S. sample. *Dreaming, 30,* 189–198. https://doi.org/10.1037%2Fdrm0000146.

Schredl, M., & Doll, E. (1998). Emotions in diary dreams. *Consciousness & Cognition, 7*(4), 634–646.

Schredl, M., & Göritz, A. (2017). Dream recall frequency, attitude toward dreams, and the Big Five personality factors. *Dreaming, 27,* 49–58.

Schredl, M., Petra, C., Bishop, A., Golitz, E., & Buschtons, D. (2003). Content analysis of German students' dreams: Comparison to American findings. *Dreaming, 13,* 237–243.

Schweickert, R. (2007a). Properties of the organization of memory for people: Evidence from dream reports. *Psychonomic Bulletin & Review, 14,* 270–276.

Schweickert, R. (2007b). Social networks of characters in dreams. In D. Barrett & P. McNamara (Eds.), *The new science of dreaming: Cultural and theoretical perspectives* (Vol. 3, pp. 277–297). Westport, CT: Praeger.

Schweickert, R. (2007c). The structure of semantic and phonological networks and the structure of a social network in dreams. In J. S. Nairne (Ed.), *The foundations of remembering: Essays in honor of Henry L. Roediger, III* (pp. 281–296). New York: Psychology Press.

Schweickert, R., Zhuangzhuang, X., Viau-Quesnel, C., & Zheng, X. (2020). Power law distribution of frequencies of characters in dreams explained by random walk on semantic network. *International Journal of Dream Research, 13,* 192–201.

Schweitzer, P., & Randgazzo, A. (2017). Drugs that disturb sleep and wakefulness. In M. Kryger, T. Roth, & W. Dement (Eds.), *Principles and practices of sleep medicine* (6th ed., pp. 480–498). Philadelphia: Elsevier.

Seeley, W., Menon, V., Schatzberg, A., Keller, J., Glover, G., Kenna, H., & Grecius, M. (2007). Dissociable intrinsic connectivity networks for salience processing and executive control. *Journal of Neuroscience, 27,* 2349–2356.

Shanks, D. R. (2003). Attention and awareness in "implicit" sequence learning. In L. Jimenez (Ed.), *Attention and implicit learning* (pp. 11–42). Philadelphia: John Benjamins.

Shapiro, D., & Notowitz, D. (1988/2016). Night wars: The nightmares of Vietnam veterans. YouTube. https://www.youtube.com/watch?v=R_aURpo3Als.

Sherman, L., Rudie, J., Pfeifer, J., Masten, C., McNealy, K., & Dapretto, M. (2014). Development of the default mode and central executive networks across early adolescence: A longitudinal study. *Developmental Cognitive Neuroscience, 10,* 148–159.

Shevrin, H. (1986). Subliminal perception and dreaming. *Journal of Mind & Behavior, 7*(2–3), 379–395.

Shevrin, H. (1996). *Conscious and unconscious processes: Psychodynamic, cognitive, and neurophysiological convergences.* New York: Guilford Press.

Shevrin, H. (2003). *Subliminal explorations of perception, dreams, and fantasies: The pioneering contributions of Charles Fisher.* Madison, CT: International Universities Press.

Shevrin, H., & Eiser, A. (2000). Continued vitality of the Freudian theory of dreaming. *Behavioral and Brain Sciences, 23*(6), 1004–1006.

Siclari, F., Baird, B., Perogamvros, L., Bernardi, G., Boly, M., & Tononi, G. (2017). The neural correlates of dreaming. *Nature Neuroscience 20,* 872–878.

Siegel, E., Sands, M. K., Chang, Y., & Barrett, L. F. (2018). Emotion fingerprints or emotion populations? A meta-analytic investigation of autonomic features of emotion categories. *Psychological Bulletin, 144*, 343–393. https://doi.org/10.1037/bul 0000128.

Siegel, J. (2001). The REM sleep-memory consolidation hypothesis. *Science, 294*, 1058–1063.

Siegel, J. (2017a). Rapid eye movement sleep. In M. Kryger, T. Roth, & W. Dement (Eds.), *Principles and practices of sleep medicine* (6th ed., pp. 78–95). Philadelpha: Elsevier.

Siegel, J. (2017b). Sleep in animals: A state of adaptive inactivity. In M. Kryger, T. Roth, & W. Dement (Eds.), *Principles and practices of sleep medicine* (6th ed., pp. 103–114). Philadelphia: Elsevier.

Siegel, J. (2021). Memory consolidation is similar in waking and sleep. *Current Sleep Medicine Reports, 7*, 15–18.

Siegel, J., & McGinty, D. (1977). Pontine reticular formation neurons: Relationship of discharge to motor activity. *Science, 196*, 678–680.

Siegel, J., & McGinty, D. (1986). Location of the systems generating REM sleep: Lateral versus medial pons. *Behavioral and Brain Sciences, 9*, 420–421.

Siegel, J., McGinty, D., & Breedlove, S. M. (1977). Sleep and waking activity of pontine gigantocellular field neurons. *Experimental Neurology, 56*, 553–573.

Sikka, P. (2020). *Dream affect: Conceptual and methodological issues in the study of emotions and moods experienced in dreams*. Turku, Finland: University of Turku Press. http://urn.fi/URN:ISBN:978-951-29-7939-4.

Sikka, P., Feilhauer, D., Valli, K., & Revonsuo, A. (2017). How you measure is what you get: Differences in self- and external ratings of emotional experiences in home dreams. *American Journal of Psychology 130*, 367–384.

Sikka, P., Revonsuo, A., Sandman, N., Tuominen, J., & Valli, K. (2018). Dream emotions: A comparison of home dream reports with laboratory early and late REM dream reports. *Journal of Sleep Research, 27*, 206–214. https://doi.org/10.1111/jsr .12555.

Sikka, P., Valli, K., Virta, T., & Revonsuo, A. (2014). I know how you felt last night, or do I? Self- and external ratings of emotions in REM sleep dreams. *Consciousness and Cognition, 25*, 51–66.

Simor, P., Horváth, K., Ujma, P. P., Gombos, F., & Bodizs, R. (2013). Fluctuations between sleep and wakefulness: Wake-like features indicated by increased EEG alpha power during different sleep stages in nightmare disorder. *Biological Psychology, 94*, 592–600.

Singer, J. L. (1975). *The inner world of daydreaming*. New York: Harper and Row.

Singh, M. (2018). The cultural evolution of shamanism. *Behavioral and Brain Sciences, 41*, 1–61. https://doi.org/10.1017/S0140525X17001893.

Slipp, S. (2000). Subliminal stimulation research and its implications for psychoanalytic theory and treatment. *Journal of the American Academy of Psychoanalysis, 28*(2), 305–320.

Smith, Carlyle. (1985). Sleep states and learning: A review of the animal literature. *Neuroscience & Biobehavioral Reviews, 9*(2), 157–168.

Smith, Charles. (2000). Content analysis and narrative analysis. In H. Reis & C. Judd (Eds.), *Handbook of research methods in social and personality psychology* (pp. 313–335). New York: Cambridge University Press.

Smith, M., & Hall, C. S. (1964). An investigation of regression in a long dream series. *Journal of Gerontology, 19*, 66–71.

Snyder, F. (1970). The phenomenology of dreaming. In L. Madow & L. Snow (Eds.), *The psychodynamic implications of the physiological studies on dreams* (pp. 124–151). Springfield, IL: Thomas.

Solms, M. (1997). *The neuropsychology of dreams: A clinico-anatomical study*. Hillsdale, NJ: Erlbaum.

Solms, M. (2000). Dreaming and REM sleep are controlled by different brain mechanisms. *Behavioral and Brain Sciences, 23*, 843–850.

Solms, M. (2002). Dreaming: Cholinergic and dopaminergic hypotheses. In E. Perry, H. Ashton, & A. Young (Eds.), *Neurochemistry of consciousness* (pp. 123–131). Philadelphia: Benjamins.

Solms, M., & Turnbull, O. (2002). *The brain and the inner world*. New York: Other Press.

Sporns, O. (2011). *Networks of the brain*. Cambridge, MA: MIT Press.

Spreng, R. N., Madore, K., & Schacter, D. (2018). Better imagined: Neural correlates of the episodic simulation boost to prospective memory performance *Neuropsychologia, 113*, 22–28. https://doi.org/10.1016/j.neuropsychologia.2018.03.025.

Steen, G., Dorst, A., Herrmann, J. B., Kaal, A., Krennmayr, T., & Pasma, T. (2010). *A method for linguistic metaphor identification*. Philadephia: John Benjamins.

Stevner, A., Vidaurre, D., Cabral, J., Rapuano, K., Nielsen, S., Tagliazucchi, E., & Kringelbach, M. (2019). Discovery of key whole-brain transitions and dynamics during human wakefulness and non-REM sleep. *Nature Communications, 10*(1035), 1–14. https://doi.org/10.1038/s41467-019-08934-3.

Stickgold, R. (1998). Sleep: off-line memory reprocessing. *Trends in Cognitive Sciences, 2*, 484–492. https://doi.org/10.1016/S1364-6613(98)01258-3.

Stickgold, R., James, L., & Hobson, J. A. (2000). Visual discrimination learning requires sleep after training. *Natural Neuroscience, 3*, 1237–1238.

Stickgold, R., Scott, L., Rittenhouse, C., & Hobson, J. A. (1999). Sleep-induced changes in associative memory. *Journal of Cognitive Neuroscience, 11*, 182–193.

Stickgold, R., & Wamsley, E. (2017). Why we dream. In M. Kryger, T. Roth, & W. Dement (Eds.), *Principles and practices of sleep medicine* (6th ed., pp. 509–514). Philadelphia: Elsevier.

Strauch, I. (2003). Träume im Alter (Dreams in old age). In B. Boothe & B. Ugolini (Eds.), *Lebenshorizont Alter* (Life horizons of the elderly) (pp. 171–187). Zurich, Switzerland: Hochschulverlag AG an der ETH Zürich.

Strauch, I. (2004). *Träume im Übergang von der Kindheit ins Jugendalter: Ergebnisse einer Langzeitstudie* (Dreams in the transition from childhood to adolescence: Results from a longitudinal study). Bern, Switzerland: Huber.

Strauch, I. (2005). REM dreaming in the transition from late childhood to adolescence: A longitudinal study. *Dreaming, 15*, 155–169.

Strauch, I. (2014). *The most recent dreams of elderly men and women.* Zurich: Department of Psychology, University of Zurich.

Strauch, I., & Lederbogen, S. (1999). The home dreams and waking fantasies of boys and girls ages 9–15. *Dreaming, 9*, 153–161.

Strauch, I., & Meier, B. (1996). *In search of dreams: Results of experimental dream research.* Albany: State University of New York Press.

Sullivan, K., & Sweetser, E. (2010). Is "generic is specific" a metaphor? In F. Parrill, V. Tobin, & M. Turner (Eds.), *Meaning, form and body* (pp. 309–328). Stanford, CA: CSLI Publications.

Supekar, K., Uddin, L. Q., Prater, K., Amin, H., Greicius, M. D., & Menon, V. (2010). Development of functional and structural connectivity within the default mode network in young children. *Neuroimage, 52*, 290–301.

Tagliazucchi, E., von Wegner, F., Morzelewski, A., Brodbeck, V., Jahnke, K., & Laufs, H. (2013). Breakdown of long-range temporal dependence in default mode and attention networks during deep sleep. *Proceedings of the National Academy of Sciences, 110*, 15419–15424.

Tamaki, M., Nittono, H., Hayashi, M., & Hori, T. (2005). Examination of the first-night effect during the sleep-onset period. *Sleep, 28*, 195–202.

Tarun, A., Wainstein-Andriano, D., Perogamvros, L., Solms, M., Schwartz, S., & Van De Ville, D. (2021). NREM sleep stages specifically alter dynamical integration of large-scale brain network. *iScience, 24*, 101923. https://doi.org/10.1016/j.isci.2020.101923.

Taschereau-Dumouchel, V., Kawato, M., & Lau, H. (2020). Multivoxel pattern analysis reveals dissociations between subjective fear and its physiological correlates. *Molecular Psychiatry, 25,* 2342–2354. https://doi.org/10.1038/s41380-019-0520-3.

Taylor, M. (2013). Imagination. In P. D. Zelazo (Ed.), *The Oxford handbook of developmental psychology, Vol. 1: Body and mind* (pp. 793–831). New York: Oxford University Press.

Tonay, V. (1990/1991). California women and their dreams: A historical and subcultural comparison of dream content. *Imagination, Cognition, and Personality, 10,* 83–97.

Tonay, V. (1993). Personality correlates of dream recall: Who remembers? *Dreaming, 3,* 1–8.

Touroutoglou, A., Hollenbeck, M., Dickerson, B., & Barrett, L. F. (2012). Dissociable large-scale networks anchored in the right anterior insula subserve affective experience and attention. *Neuroimage, 60,* 1947–1958. https://doi.org/10.1016/j.neuroimage.2012.02.012.

Touroutoglou, A., Lindquist, K., Dickerson, B. C., & Barrett, L. F. (2015). Intrinsic connectivity in the human brain does not reveal networks for "basic" emotions. *Social Cognitive and Affective Neuroscience, 10,* 1257–1265.

Tremblay, R., Hartup, W., & Archer, J. (2007). *Developmental origins of aggression.* New York: Guilford Press.

Tulving, E. (2005). Episodic memory and autonoesis. In H. S. Terrace & J. Metcalf (Eds.), *The missing link in cognition: Origins of self-reflective consciousness* (pp. 3–56). New York: Oxford University Press.

Tururen, M. (2016). *Conceptual metaphors in slang: A study on the words shit, piss, and blood.* Master's thesis, University of Tampere, Finland.

Tylor, E. B. (1871/1958). *Primitive culture* (Vol. 2). New York: Harper & Row.

Tyson, P., Ogilvie, R., & Hunt, H. T. (1984). Lucid, prelucid, and nonlucid dreams related to the amount of EEG alpha activity during REM sleep. *Psychophysiology, 21*(4), 442–451.

Uddin, L. (2015). Salience processing and insular cortical function and dysfunction. *Nature Reviews Neuroscience, 16,* 55–61.

Uddin, L., Yeo, B., & Spreng, R. (2019). Towards a universal taxonomy of macroscale functional human brain networks. *Brain Topography, 32,* 926–942. https://doi.org/10.1007/s10548-019-00744-6.

Uebersax, J. (1987). Diversity of decision-making models and the measurement of interrater agreement. *Psychological Bulletin, 101,* 140–146.

Uebersax, J. (2014). Statistical methods for diagnostic agreement: Kappa coefficients, a critical appraisal. http://www.john-uebersax.com/stat/kappa.htm.

Uitermarkt, B. D., Bruss, J., Hwang, K., & Boes, A. D. (2020). Rapid eye movement sleep patterns of brain activation and deactivation occur within unique functional networks. *Human Brain Mapping* (June), 1–9. https://doi.org/10.1002/hbm.25102.

Ullman, M. (1959). The adaptive significance of the dream. *Journal of Nervous and Mental Disease, 129*, 144–149.

Ünal, G., & Hohenberger, A. (2017). The cognitive bases of the development of past and future episodic cognition in preschoolers. *Journal of Experimental Child Psychology, 162*, 242–258.

Vallat, R., Eichenlaub, J. B., Nicolas, A., & Ruby, P. (2018). Dream recall frequency is associated with medial prefrontal cortex white-matter density. *Frontiers in Psychology, 9*(1856), 1–5. https://doi.org/10.3389/fpsyg.2018.01856.

Valli, K., & Revonsuo, A. (2009). The threat simulation theory in light of recent empirical evidence: A review. *American Journal of Psychology, 122*, 17–38.

Van de Castle, R. (1969). Problems in applying methodology of content analysis. In M. Kramer (Ed.), *Dream psychology and the new biology of dreaming* (pp. 185–197). Springfield, IL: Charles C. Thomas.

Van de Castle, R. (1977). Sleep and dreams. In B. Wolman (Ed.), *Handbook of parapsychology* (pp. 473–499). New York: Van Nostrand Reinhold.

Van de Castle, R. (1983). Animal figures in fantasies and dreams. In A. Katcher & A. Beck (Eds.), *New perspectives on our lives with companion animals* (pp. 148–173). Philadelphia: University of Pennsylvania Press.

Van Gennep, A. (1909/1960). *The rites of passage.* Chicago: University of Chicago Press.

Vartanian, O. (2012). Dissociable neural systems for analogy and metaphor: Implications for the neuroscience of creativity. *British Journal of Psychology, 103*, 302–316.

Vaughn, B., & D'Cruz, O. N. (2017). Cardinal manifestations of sleep disorders. In M. Kryger, B. J. Roth, & W. Dement (Eds.), *Principles and practices of sleep medicine* (Vol. 6, pp. 576–586). Philadelphia: Elsevier.

Vecchio, F., Miraglia, F., Gorgoni, M., Ferrara, M., & De Gennaro, L. (2017). Cortical connectivity modulation during sleep onset: A study via graph theory on EEG data. *Human Brain Mapping, 38*, 5456–5464. https://doi.org/10.1002/hbm.23736.

Verghese, A., Garner, K. G., Mattingley, J., & Dux, P. E. (2016). Prefrontal cortex structure predicts training-induced improvements in multitasking performance. *Journal of Neuroscience, 36*, 2638–2645.

Vertes, R. (1977). Selective firing of gigantocellular neurons during movement and REM sleep. *Brain Research, 128*, 146–152.

Vertes, R. (1979). Brain stem gigantocellular neurons: Patterns of activity during behavior and sleep in the freely moving rat. *Journal of Neurophysiology, 42*, 214–228.

Vertes, R. (1995). Memory consolidation in REM sleep: Dream on. *Bulletin of the Sleep Research Society, 1*, 27–32.

Vertes, R., & Eastman, K. (2000). The case against memory consolidation in REM sleep. *Behavioral and Brain Sciences, 23*, 867–876.

Voss, U., Holzmann, R., Tuin, I., & Hobson, J. A. (2009). Lucid dreaming: A state of consciousness with features of both waking and non-lucid dreaming. *Sleep, 32*, 1191–1200.

Wald, A., & Wolfowitz, J. (1940). On a test whether two samples are from the same population. *Annals of Mathematical Statistics, 11*, 147–162.

Walsh, K. (1994). *Neuropsychology: A clinical approach.* New York: Churchill Livingstone.

Wamsley, E., Hirota, Y., Tucker, M., Smith, M., Doan, T., & Antrobus, J. (2007). Circadian and ultradian influences on dreaming: A dual rhythm model. *Brain Research Bulletin, 71*, 347–354.

Ware, J. C., Hirshkowitz, M., Salis, P., & Karacan, I. (1997). Sleep-related erections: Absence of change following presleep sexual arousal. *Journal of Psychosomatic Research, 42*(6), 547–553.

Waters, F., Dirk Blom, J., Dang-Vu, T., Cheyne, A., Alderson-Day, B., Woodruff, P., & Collerton, D. (2016). What is the link between hallucinations, dreams, and hypnagogic–hypnopompic experiences? *Schizophrenia Bulletin, 42*, 1098–1109. https://doi.org/10.1093/schbul/sbw076.

Watts, D., & Strogatz, S. (1998). Collective dynamics of small world networks. *Nature, 393*, 440–442.

Webb, E., Campbell, D., Schwartz, R., Sechrest, L., & Grove, J. (1981). *Nonreactive measures in the social sciences.* Boston: Houghton Mifflin.

Weiss, H. (1944). Oneirocritica Americana. *Bulletin of the New York Public Library, 48*, 519–541.

Weiss, H. (1959). Oneicritica Americana (abridged version). In M. DeMartino (Ed.), *Dreams and personality dynamics* (pp. 29–41). Springfield, IL: Charles C. Thomas.

Weisz, R., & Foulkes, D. (1970). Home and laboratory dreams collected under uniform sampling conditions. *Psychophysiology, 6*, 588–596.

Welsh, M., Pennington, B., & Groisser, D. (1993). A normative-developmental study of executive function: A window on prefrontal function in children. *Developmental Neuropsychology, 7*, 131–149.

Whitley, B. E., & Kite, M. E. (2013). *Principles of research in behavioral science* (3rd ed.). New York: Routledge/Taylor & Francis Group.

Williams, J., Merritt, J., Rittenhouse, C., & Hobson, J. A. (1992). Bizarreness in dreams and fantasies: Implications for the activation-synthesis hypothesis. *Consciousness and Cognition, 1*, 172–185.

Wilson, Margaret. (2002). Six views of embodied cognition. *Psychonomic Bulletin & Review, 9*, 625–636.

Wilson, Matthew, & McNaughton, B. (1994). Reactivation of hippocampal ensemble memories during sleep. *Science, 265*, 676–679.

Windt, J. (2015). *Dreaming: A conceptual framework for philosophy of mind and empirical research.* Cambridge, MA: MIT Press.

Winget, C., & Kramer, M. (1979). *Dimensions of dreams.* Gainesville: University Presses of Florida.

Witkin, H. (1969). Presleep experiences and dreams. In J. Fisher & L. Breger (Eds.), *The meaning of dreams* (pp. 1–37). California Mental Health Research Symposium, No. 3. Sacramento: Department of Mental Hygiene.

Witkin, H., & Lewis, H. (1967). Presleep experiences and dreams. In H. Witkin & H. Lewis (Eds.), *Experimental studies in dreaming* (pp. 148–201). New York: Random House.

Wong, W., Noreik, V., Moro, L., Revonsuo, A., Windt, J., Valli, K., & Tsuchiya, N. (2020). The Dream Catcher experiment: Blinded analyses failed to detect markers of dreaming consciousness in EEG spectral power. *Neuroscience of Consciousness, 6*(1), 1–19. https://doi.org/10.1093/nc/niaa006.

Woolley, J. D. (1995). Young children's understanding of fictional versus epistemic mental representations: Imagination and belief. *Child Development, 66*(4), 1011–1021.

Woolley, J. D., & Boerger, E. (2002). Development of beliefs about the origins and controllability of dreams *Developmental Psychology, 38*, 24–41.

Woolley, J. D., & Nissel, J. (2020). Development of the fantasy-reality distinction. In A. Abraham (Ed.), *The Cambridge handbook of the imagination* (pp. 479–499). New York: Cambridge University Press.

Wyatt, R., Fram, D., Kupfer, D., & Snyder, F. (1971). Total prolonged drug-induced REM sleep suppression in anxious-depressed patients. *Archives of General Psychiatry, 24*, 145–155.

Yamanaka, T., Morita, Y., & Matsumoto, J. (1982). Analysis of the dream contents in college students by REM-awakening technique. *Folia Psychiatrica et Neurologica Japonica, 36,* 33–52.

Yeo, B., Krienen, F., Sepulcre, J., Sabuncu, M., Lashkari, D., Hollinshead, M., & Buckner, R. (2011). The organization of the human cerebral cortex estimated by intrinsic functional connectivity. *Journal of Neurophysiology, 106,* 1125–1165.

Yu, C. (2008). Typical dreams experienced by Chinese people. *Dreaming, 18,* 1–10. https://doi.org/10.1037/1053-0797.18.1.1.

Zabelina, D., & Andrews-Hanna, J. R. (2016). Dynamic network interactions supporting internally-oriented cognition. *Current Opinion in Neurobiology, 40,* 86–93.

Zadra, A., Desjardins, S., & Marcotte, Â. (2006). Evolutionary function of dreams: A test of the threat simulation theory in recurrent dreams. *Consciousness and Cognition, 15*(2), 450–463.

Zadra, A., & Donderi, D. (2000a). Nightmares and bad dreams: Their prevalence and relationship to well-being. *Journal of Abnormal Psychology, 109,* 273–281.

Zadra, A., & Donderi, D. (2000b). Threat perceptions and avoidance in recurrent dreams. *Behavioral and Brain Sciences, 23,* 1017–1018.

Zadra, A., Nielsen, T., & Donderi, D. (1998). Prevalence of auditory, olfactory and gustatory experiences in home dreams. *Perceptual & Motor Skills, 87*(3, Pt. 1), 819–826.

Zadra, A., Pilon, M., & Donderi, D. (2006). Variety and intensity of emotions in nightmares and bad dreams. *Journal of Nervous and Mental Disease, 194*(4), 249–254.

Zepelin, H. (1972). *Comparison of dreams recalled in the laboratory and at home.* Paper presented at the Association for the Psychophysiological Study of Sleep, Chicago.

Zepelin, H. (1980). Age differences in dreams: I. Men's dreams and thematic apperceptive fantasy. *International Journal of Aging and Human Development, 12,* 171–186.

Zepelin, H. (1981). Age differences in dreams: II. Distortion and other variables. *International Journal of Aging and Human Development, 13,* 37–41.

Zhang, R., & Volkow, N. (2019). Brain default-mode network dysfunction in addiction. *NeuroImage, 200,* 313–331. https://doi.org/10.1016/j.neuroimage.2019.06.036.

Zhang, S., & Li, C. S. (2012). Functional connectivity mapping of the human precuneus by resting state fMRI. *Neuroimage, 59,* 3548–3562.

Zheng, X., & Schweickert, R. (2021). Comparing Hall Van de Castle coding and Linguistic Inquiry and Word Count using canonical correlation analysis. *Dreaming, 31,* 207–224. https://doi:10.1037/drm0000173.

Zhong, J., Rifkin-Graboi, A., Ta, A. T., Yap, K. L., & Chuang, K.-H. (2014). Functional networks in parallel with cortical development associate with executive functions in children. *Cerebral Cortex, 24*, 1937–1947.

Zimmerman, W. B. (1970). Sleep mentation and auditory awakening thresholds. *Psychophysiology, 6*, 540–549.

Zmigrod, L., Garrison, J. R., Carr, J., & Simons, J. S. (2016). The neural mechanisms of hallucinations: A quantitative meta-analysis of neuroimaging studies. *Neuroscience Biobehavioral Review, 69*, 113–123.

Index

Note: Page numbers followed by *f* and *t* indicate figures and tables.